Mechanisms of Vasodilatation

Satellite Symposium to the 27th International Congress of Physiological Sciences
Wilrijk, July 10–12, 1977

Mechanisms of Vasodilatation

Editors
Paul M. Vanhoutte, Wilrijk, and *I. Leusen,* Gent

94 figures and 9 tables, 1978

S. Karger · Basel · München · Paris · London · New York · Sydney

Cataloging in Publication
Mechanisms of vasodilatation
Editors: Paul M. Vanhoutte and I. Leusen. – Basel; New York: Karger, 1978
Proceedings of a symposium held in Wilrijk, July 10–12, 1977 as a satellite symposium to the 27th International Congress of Physiological Sciences
1. Vasodilation – congresses I. Vanhoutte, Paul M., ed. II. Leusen, I., ed.
WG 106 M486 1977
ISBN 3–8055–2841–8

Printed in Switzerland by Thür AG Offsetdruck, Pratteln
ISBN 3–8055–2841–8

Contents

Therapeutic Agents Causing Vasodilatation

Local Factors Causing Vasodilatation

Neurogenic Vasodilatation

Foreword

This monograph contains a series of articles by the participants of the symposium 'Mechanisms of Vasodilatation', which was held in Wilrijk, Belgium, from July 10–12, 1977, as an official Satellite Symposium of the 27th International Congress of Physiological Sciences. Each author has been asked to produce a brief overview of the topic he discusses, rather than focus only on his own data. The editors want to thank all the authors for the excellence of their papers, and for the rapidity of their responses to the editorial comments.

The symposium was made possible by official grants of the Nationaal Fonds voor Wetenschappelijk Onderzoek, the Ministerie van Nationale Opvoeding en Nederlandse Cultuur and the Universitaire Instelling Antwerpen, and by the generous support of Boehringer Ingelheim, Brocades, Continental Pharma, Imperial Chemical Industries, Mead-Johnson, Merck Institute for Therapeutic Research, Merck, Sharp & Dohme, Pfizer, The Upjohn Company and the Wellcome Research Foundation. Several sponsors wish to remain anonymous. In addition, it should be noted that several nonpharmaceutical companies contributed financially to the symposium, as did the attendants.

The symposium would not have been realized without the dedication and expertise of Ms. *Van den Eynde,* who carried all of the administrative load, and without the enthusiastic help of *T. Verbeuren, W. Janssens, J. De Mey, L. Zonnekeyn, F. Jordaens, A. Zaman and G. Bauwens* who were responsible for the technical aspects. The participants will not easily forget the dedication of Mrs. *Collis,* Mr. *De Boeck* (Sabena, Antwerpen) and Mr. *De Smet* (American Express, Antwerpen). The organizing committee wishes to gratefully acknowledge their invaluable assistance. As editors of the monograph, they want to thank both those who helped them to review the manuscripts, and the S. Karger Company for moving the completed text quickly to publication.

P.M. Vanhoutte
I. Leusen

Mechanisms of Vasodilatation. Satellite Symp., 27th Int. Congr. Physiol. Sci., Wilrijk 1977, pp. 1–11 (Karger, Basel 1978)

Prolegomenon

John T. Shepherd
Mayo Clinic and Mayo Foundation, Rochester, Minn.

Thanks to the scientific knowledge and organizing ability of Profs. *Paul Vanhoutte* and *Isidore Leusen,* and the hospitality of the University of Antwerp, skilled investigators from around the world have met here to enunciate current knowledge of the mechanisms by which blood vessels dilate.

When I started my career in cardiovascular research, the application of venous occlusion plethysmography and calorimetry permitted the investigation of the circulation in human limbs. My first teacher, *Henry Barcroft* (2), had demonstrated the presence of sympathetic fibers to human muscle blood vessels and my colleagues and I were interested in examining the various reflexes which utilized these nerves in the control of skeletal muscle blood flow. From these studies, we learned that cardiovascular reflexes in humans caused changes in the caliber of the resistance vessels by alteration in the activity of efferent sympathetic adrenergic nerve fibers. There are also cholinergic fibers to specific regions which when activated cause vasodilatation, for example, those to the sweat glands which are stimulated by thermal stress. During severe emotion, there is a marked increase in forearm blood flow which has been attributed in part to activation of cholinergic fibers (18). In this symposium, we will learn of other mechanisms of neurogenic vasodilatation including β-adrenergic, histaminergic, dopaminergic, purinergic and peptidergic. Whether any of these operate in the human remains to be determined.

Following the work of *Lewis* (14) and *Grant* (9), we were intrigued also with the local regulation of blood flow, and particularly with the phenomena of *cold vasodilatation* and of exercise and reactive hyperemia, as we observed them in man. If a subject's finger (or toe) is placed in ice-cold water, the blood flow in the immersed part first decreases nearly to zero; this initial vasoconstriction is followed by a pronounced vasodilatation, which in the typical case alternates with episodes of intense vasoconstriction, the so-called hunting reaction (fig. 1). We still await a satisfactory explanation for the phenomenon.

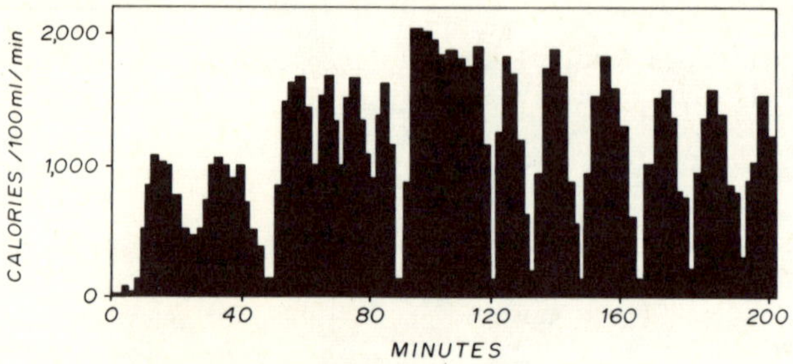

Fig. 1. The heat loss in cal/100 ml/min from the right index finger to water at 0–6 °C. From *Greenfield et al.* (20), by permission.

An even more intriguing question is what causes the *exercise hyperemia* of skeletal muscles. In man, the resting blood flow in the latter is around 3 ml/100 g of tissue per minute. Immediately after *sympathetic nerve* block, the resting blood flow increases about threefold, whereas immediately after cessation of severe exercise, the blood flow is more than 10–15 times the resting value. It has been calculated that during severe exercise the total blood flow in the skeletal muscles increases from a resting value of about 1 to about 20 l/min. The increase in flow for a given level of work is similar in the normal and the sympathectomized limb (fig. 2). Thus, as was already evident from the animal studies of *Gaskell* (7), the vasodilatation is mediated locally and results from some product(s) of muscle activity which decreases the resistance of the blood vessels as oxygen demand increases.

Since that time, many mediators have been proposed, and many rejected. Before concluding that any vasoactive substance is the mediator, it should meet the following criteria when infused into the brachial artery: (1) the vasodilatation in the forearm muscles should be equal to the maximum obtained immediately after strong exercise; (2) the increase in flow should be sustained during a prolonged infusion; (3) the subject should not be aware of any local or general sensations during the infusion; (4) the substance should not appear in the systemic circulation in concentrations sufficient to cause general hemodynamic effects; and (5) the concentration infused to cause maximal dilatation should correspond to concentrations believed to occur in the exercising muscles.

Oxygen lack is often referred to as a major factor. This implies that with muscle contraction, the oxygen content of the tissues falls to a level insufficient to maintain the contractile process in the smooth muscle cells of the arterioles. However, with a single contraction of the forearm muscles of about 0.3 sec duration, an increase in flow can be detected within a second following the end of the contraction, due to dilatation of the muscle vessels (fig. 3). This response is unchanged when the subject breathes oxygen instead of air (fig. 2). Also,

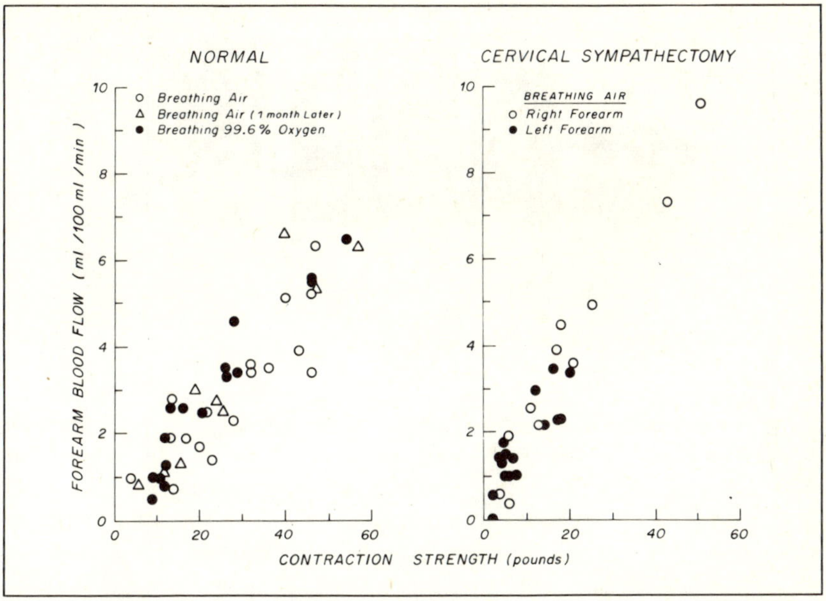

Fig. 2. Increase in flow in forearm immediately after single brief contraction of forearm muscles in a normal subject and in a subject who had had cervical sympathectomy 3 years previously. The increase in forearm flow was obtained by subtracting the resting flow before each contraction from the flow measured just after contraction. Oxygen capacity of radial artery blood was 19.4 vol%; O_2 content was 19.0 vol% while breathing air and 20.8 vol% while breathing O_2. From *Corcondilas et al.* (21), by permission.

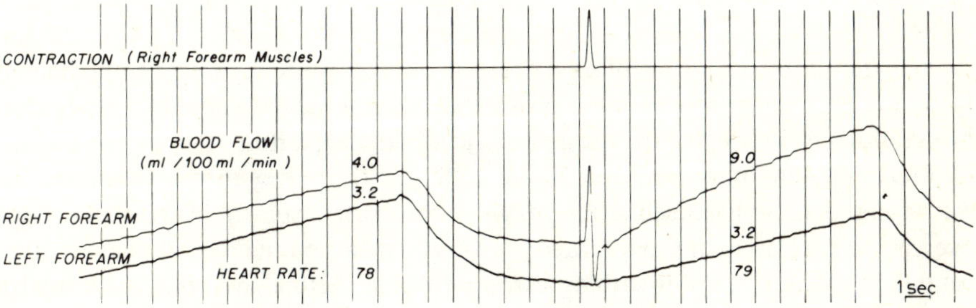

Fig. 3. Plethysmographic records of blood flow through both forearms before and after brief contraction of muscles of the right forearm. At time of contraction, there is an artifact on the plethysmographic tracing due to muscle movement. Numbers above each plethysmogram represent the forearm blood flow (in ml/100 ml/min). From *Corcondilas et al.* (21), by permission.

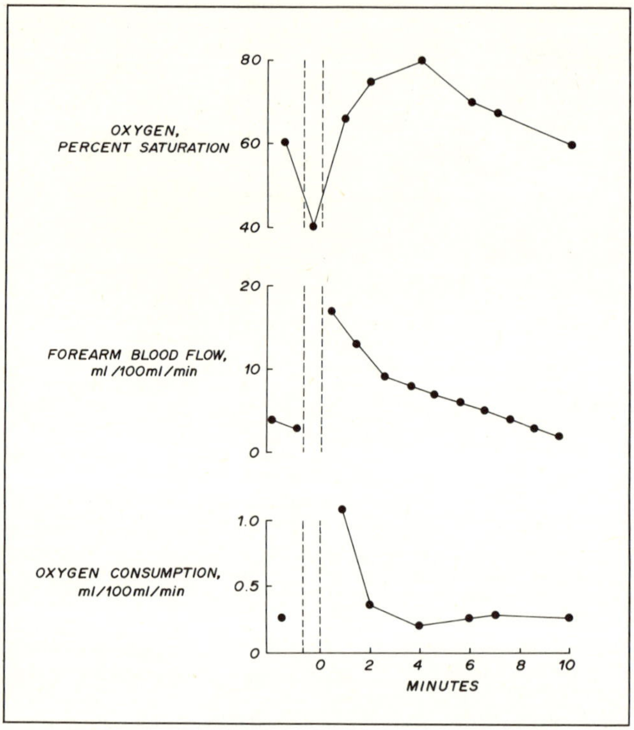

Fig. 4. The effects on rhythmic contractions for 1 min on the percentage saturation with oxygen of deep venous blood, the forearm blood flow rate, and the forearm oxygen consumption. Exercise was performed during the period represented between the vertical lines. Blood flows are plotted as minute averages. From *Love* (22), by permission.

immediately after cessation of exercise, the oxygen content in the venous blood from the forearm increases well above the control values, at a time when the blood flow is in excess of the resting level (fig. 4). Thus, it is unlikely that hypoxia plays any important role in the increased flow after contraction.

A comparison has been made of the vasodilator response of the forearm to exercise and to ischemia with that produced by a comparable increase in the venous P_{CO_2} induced by breathing carbon dioxide. Although *hypercapnia* might account for two thirds of the responses to ischemia, its contribution to exercise hyperemia appears to be much less (13). The hyperemia of exercise is not altered in human subjects with an abnormally low phosphorylase activity in whom muscle glycogen cannot be broken down during activity, so that *lactic acid* is not liberated and there is no change in the *pH* of the venous effluent (15). Thus, the role of pH, P_{CO_2} and lactic acid in exercise hyperemia probably is a minor one.

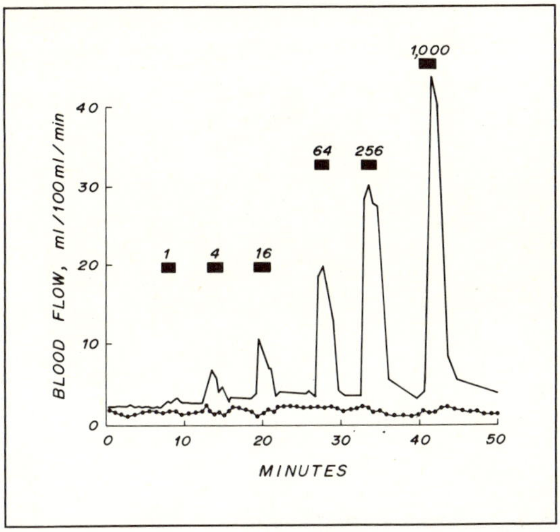

Fig. 5. The effect on forearm blood flow of various doses of magnesium adenosine triphosphate injected over 2-min periods into the brachial artery of a normal subject. Continuous line = injected side; dotted line = control side. Doses are in μg/min. From *Duff et al.* (23), by permission.

It has been suggested that *potassium* ions might be responsible for exercise hyperemia (12). This ion appears in the venous blood in the early phases of exercise, but as the latter continues and the blood flow remains increased, the potassium content falls toward normal. With the intra-arterial infusion of the ion in doses that can be tolerated without severe pain, there is only about a twofold increase in forearm blood flow (8).

The hyperemia of exercise might be mediated in part by the liberation of *inorganic phosphate* since, during vigorous sustained or rhythmic exercise of the forearm muscles, plasma phosphate in the venous effluent from the active muscles increases about 20% (10). However, the infusion of sodium phosphate into the brachial artery which raises plasma phosphate in the venous effluent by 400% has no effect upon forearm blood flow (1).

Following intra-arterial infusion of prostaglandins of the E series at a rate of 1 ng/kg body weight per minute, there is a tenfold increase in forearm blood flow (3). After the administration of indomethacin, an agent which inhibits prostaglandin synthesis, the forearm blood flow after sustained or rhythmic contractions is diminished, suggesting that prostaglandins of the E series may play some, but probably not a major, role (11).

When magnesium *adenosine triphosphate* (ATP) is infused, the dilatation achieved equals that found after severe exercise (fig. 5). This occurs without discomfort to the subjects, and contrasts with the tingling and bursting sensa-

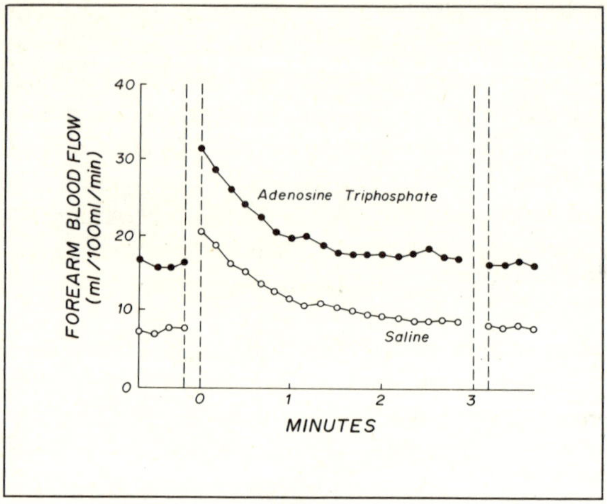

Fig. 6. Forearm blood flow before and after rhythmic exercise of the forearm muscles carried out during the interval indicated by the first pair of dotted lines. The second pair of dotted lines indicate a 2-min interval. ○ = Interval during the infusion of isotonic saline into the brachial artery; • = interval during the infusion of adenosine triphosphate. From *Patterson and Shepherd* (24), by permission.

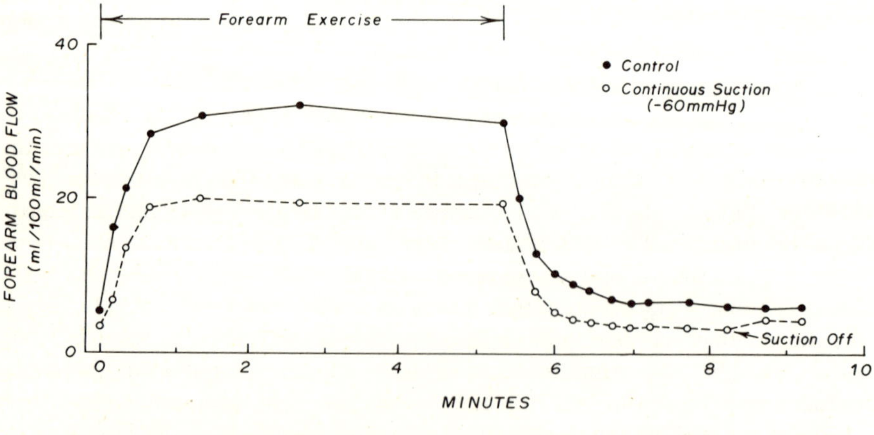

Fig. 7. Effect of increased sympathetic outflow to forearm blood vessels on blood flow through the forearm during and after rhythmic exercise of the forearm muscles. Increased sympathetic outflow induced by application of subatmospheric pressure to lower body (continuous suction). From *Strandell and Shepherd* (25), by permission.

tions observed with concentrations of histamine and acetylcholine, respectively, which cause similar dilatations. Adenosine triphosphate has been identified in the venous effluent from the exercising forearm in the presence of arterial and venous occlusion, but not in the venous effluent from the occluded forearm without exercise (6). The amounts calculated to be released could satisfy the vasodilator requirements of active skeletal muscle. Recently, adenosine diphosphate (ADP) and adenosine monophosphate (AMP) have also been identified in the venous blood from exercising muscle; the total rise in the venous concentration of the adenine nucleotides bears a quantitative relationship to the severity of the exercise (16).

Other possible mediators of exercise hyperemia which have been examined in animals, but not yet in the human, and which will be discussed in this symposium include *hyperosmolarity* and *adenosine*. It has further been demonstrated that combinations such as potassium, hyperosmolarity, and low oxygen are more effective than either agent separately, which opens up the possibility that the vasodilatation may result from a multiplicity of local changes occurring simultaneously or in sequence (19).

Another approach to the mechanism of the vasodilatation with exercise has been the analysis of the recovery of the blood flow immediately after the cessation of muscle contractions. Several intriguing observations have been obtained. For example, lowering the effective local arterial pressure to approximately 50% does not lead to any prolongation of either post-exercise or post-ischemic hyperemia, although all flows are reduced by this procedure (4). Also, dilatation of the muscle vessels with a variety of agents does not reduce the hyperemia of exercise. The infusion of either ATP, histamine, or acetylcholine into the brachial artery before, during, and after rhythmic exercise of the forearm muscles, at a rate sufficient to more than double the resting flow through the forearm, has no effect on the amount and rate of blood-debt repayment (fig. 6). The same is true when the muscle blood flow is increased reflexly or by an intravenous infusion of epinephrine. Thus, it seems that the additional blood supplied to the muscles is not used for metabolic purposes during the exercise. This is substantiated by the finding that the oxygen content of the blood from the exercising leg muscles is much higher during an intravenous infusion of epinephrine than when no infusion is given (18). In order to account for the fact that the excess flow after exercise is independent of the resting level, it is necessary to assume that a given amount of metabolites produce a very similar increase in blood flow regardless of the initial level.

If the activity of the sympathetic vasoconstrictor nerves to the forearm is reflexly augmented by the application of subatmospheric pressures to the lower part of the body, the hyperemia produced by mild exercise of the forearm muscles is reduced, as is the recovery blood flow in the post-exercise period (fig. 7). The lessened increase in blood flow caused by the reflex increase in

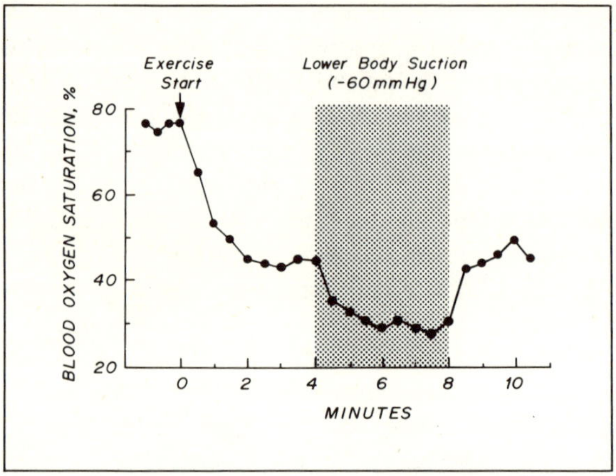

Fig. 8. Effect of increased sympathetic outflow to forearm blood vessels on oxygen saturation in a deep forearm vein during rhythmic exercise of forearm muscle. Increased sympathetic outflow induced by application of subatmospheric pressure to lower body (lower body suction). From *Strandell and Shepherd* (25), by permission.

sympathetic activity during and after the muscle contractions is compensated for by greater oxygen extraction (fig. 8), so that no further debt occurs and the metabolic recovery is not prolonged. This implies that during mild exercise the increase in flow is in excess of the metabolic demands.

An equally intriguing question is the mechanism(s) mediating *reactive hyperemia.* Is this the same as or different from that causing exercise hyperemia? While the mechanisms remain unknown, I suspect the latter is true. After a period of ischemia, there is a pronounced but relatively brief vasodilatation, whereas following isometric and dynamic forearm muscle activity sufficient to cause a similar initial vasodilatation, the blood flow remains above the resting level for a much longer period (fig. 9). In man, the myogenic response to the decrease in *vascular transmural pressure* with circulatory arrest plays a role in the subsequent hyperemia. When extra blood is brought into the forearm by suction and trapped there while the circulation is arrested, the pressure in the brachial artery distal to the occluding cuff is 45 mm Hg compared to about 10 mm Hg in the nonpacked arm. The subsequent hyperemia is much less in the packed arm, even when allowance is made for the extra blood trapped in the forearm (17). However, although the flow in excess of the control level after a period of occlusion was reduced by maintaining a higher intravascular pressure, it was not abolished; in fact, the peak flow was often little changed.

Antihistamines can reduce the increase in flow after 10–25 min but not after a 5 min arrest. Thus, release of *histamine* is not the cause of the hyperemia that follows a short period of circulatory arrest, but after prolonged arrest,

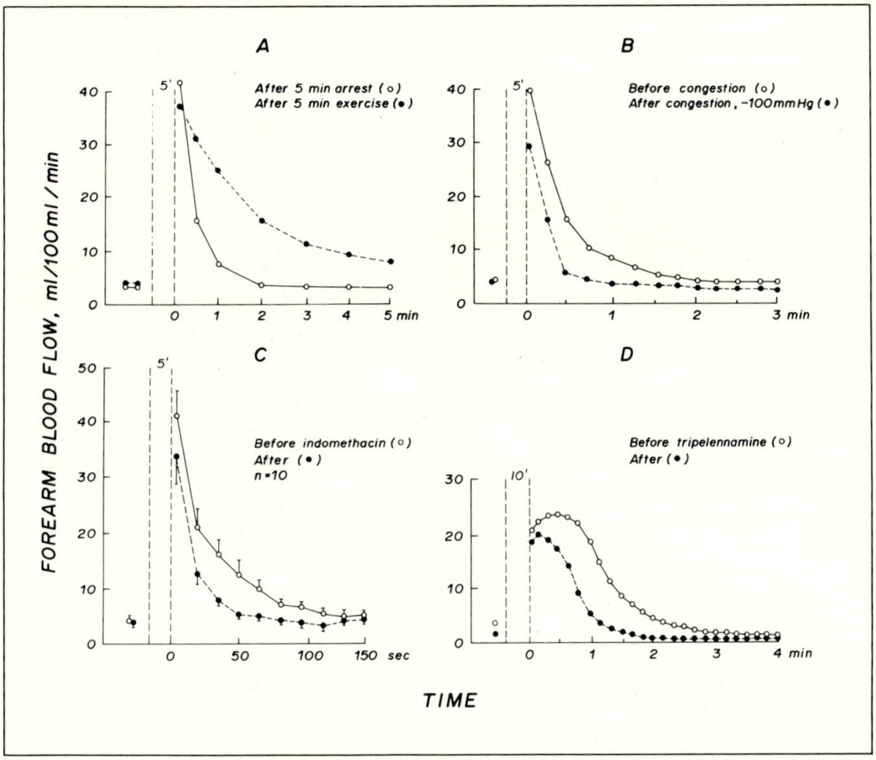

Fig. 9. Studies on reactive hyperemia in the human forearm. The vertical dotted lines indicate the period of circulatory arrest by a pneumatic cuff around the arm. *A* Following circulatory arrest there is a more rapid return of forearm blood flow to resting values as compared to the more prolonged increase in flow after 5 min of rhythmic exercise of the forearm muscles. *B* Myogenic response. When the transmural pressure in the forearm vessels distal to the site of circulatory arrest is maintained at a higher level by trapping blood in the forearm during the period of arrest, the subsequent hyperemia is less. For details, see text. *C* Inhibition of prostaglandin synthesis by indomethacin reduces the reactive hyperemia. *D* Antihistamines diminish the reactive hyperemia following 10 min or more of circulatory arrest. Data in *B* from ***Patterson*** (17); data in *C* from ***Kilbom and Wennmalm*** (11); data in *D* from ***Duff et al.*** (5), by permission.

approximately 50% of the hyperemia may be due to local release of histamine (5). More recently, the contribution of endogenously formed *prostaglandins* of the E series (PGE) has been investigated. The increase in blood flow through the forearm after 5 min of arterial occlusion was accompanied by a marked elevation of PGE-like substances (PLS) in the venous effluent. After the administration of indomethacin, an inhibitor of prostaglandin synthesis, the release of PLS was suppressed and the peak level of flow as well as the duration of hyperemia were decreased (11). Thus, the present evidence suggests that both changes in transmural pressure and vasodilator substance(s) play a role in reactive hyperemia (fig. 9).

In this introduction, because of time and because I do not wish to encroach on the many aspects to be discussed, I have limited my remarks to observations on humans. In so doing, I recognize the many studies in animals and in isolated tissues which bear on these problems. It is my hope that as we leave this symposium, enriched by the current understanding of the intrinsic factors involved in the relaxation of vascular smooth muscle and of the local humoral and neural events leading to vasodilatation, we will return to our laboratories with new approaches to unravel these most intriguing questions in circulatory physiology. This knowledge, together with the data to be presented on the mechanisms by which therapeutic agents cause vasodilatation, will allow us to find better treatments for those clinical situations where the impairment of the blood supply to the tissues is a major problem.

References

1 *Barcroft, H.; Foley, T.H., and McSwiney, R.R.:* Experiments on the liberation of phosphate from the muscles of the human forearm during vigorous exercise and on the action of sodium phosphate on forearm muscle vessels. J. Physiol., Lond. *213:* 411–420 (1971).

2 *Barcroft, H. and Swan, H.J.C.:* Sympathetic control of human blood vessels (Arnold, London 1953).

3 *Bevegård, S. and Orö, L.:* Effect of prostaglandin E_1 on forearm blood flow. Scand. J. clin. Lab. Invest. *23:* 347–353 (1967).

4 *Dornhorst, A.C. and Whelan, R.F.:* The blood flow in muscle following exercise and circulatory arrest: the influence of reduction in effective local blood pressure, of arterial hypoxia and of adrenaline. Clin. Sci. *12:* 33–40 (1953).

5 *Duff, F.; Patterson, G.C., and Whelan, R.F.:* Effect of intra-arterial antihistamines on the hyperemia following temporary arrest of the circulation in the human forearm. Clin. Sci. *14:* 267–273 (1955).

6 *Forrester, T.:* An estimate of the adenosine triphosphate release into the venous effluent from the exercising human forearm muscle. J. Physiol., Lond. *224:* 611–628 (1972).

7 *Gaskell, W.H.:* The involuntary nervous system. (Longmans, Green, London 1920).

8 *Glover, W.E.; Roddie, I.C., and Shanks, R.G.:* Effect of intra-arterial potassium chloride infusions on vascular reactivity in the human forearm. J. Physiol., Lond. *163:* 22P (1962).

9 *Grant, R.T.:* Observations on the blood circulation in voluntary muscle in man. Clin. Sci. *3:* 157–173 (1938).

10 *Hilton, S.M. and Vrbovà, G.:* Magnesium phosphate – a new candidate for mediator of functional dilatation in skeletal muscle. J. Physiol., Lond. *206:* 29–30P (1970).

11 *Kilbom, Å. and Wennmalm, Å.:* Endogenous prostaglandins as local regulators of blood flow in man: effect of indomethacin on reactive and functional hyperemia. J. Physiol., Lond. *257:* 109–121 (1976).
12 *Kjellmer, I.:* The potassium ion as a vasodilator during muscular exercise. Acta physiol. scand. *63:* 460–468 (1965).
13 *Kontos, H.A.:* Role of hypercapnia acidosis in the local regulation of blood flow in skeletal muscle. Circulation Res. *28:* suppl. 1, pp. 98–105 (1971).
14 *Lewis, T.:* Observations upon the reactions of the vessels of the human skin to cold. Heart *15:* 177–208 (1930).
15 *McArdle, B.:* Myopathy due to a defect in muscle glycogen breakdown. Clin. Sci. *10:* 13–35 (1951).
16 *Parkinson, P.I.:* The effect of graded exercise on the concentration of adenine nucleotides in plasma. J. Physiol., Lond. *234:* 72P–74P (1973).
17 *Patterson, G.C.:* Role of extra vascular pressure in the causation of reactive hyperemia in the human forearm. Clin. Sci. *15:* 17–25 (1956).
18 *Shepherd, J.T.:* Physiology of the circulation in human limbs in health and disease (Saunders, Philadelphia 1963).
19 *Skinner, N.S., jr. and Costin, J.C.:* Interactions between oxygen, potassium, and osmolality in regulation of skeletal muscle blood flow. Circulation Res. *28:* suppl. 1, pp. 73–85 (1971).
20 *Greenfield, A.D.M.; Shepherd, J.T., and Whelan, R.F.:* Cold vasoconstriction and vasodilatation. Irish J. med. Sci. *309:* 415–419 (1951).
21 *Corcondilas, A.; Koroxenidis, G.T., and Shepherd, J.T.:* Effect of a brief contraction of forearm muscles on forearm blood flow. J. appl. Physiol. *19:* 142–146 (1964).
22 *Love, A.H.G.:* The rate of blood flow and the oxygen saturation of the effluent blood following contraction of the muscles of the human forearm. Clin. Sci. *14:* 275–283 (1955).
23 *Duff, F.; Patterson, G.C., and Shepherd, J.T.:* A quantitative study of the response to adenosine triphosphate of the blood vessels of the human hand and forearm. J. Physiol., Lond. *125:* 581–589 (1954).
24 *Patterson, G.C. and Shepherd, J.T.:* The effects of continuous infusions into the brachial artery of adenosine triphosphate, histamine and acetylcholine on the amount and rate of blood debt repayment following rhythmic exercise of the forearm muscles. Clin. Sci. *13:* 85–91 (1954).
25 *Strandell, T. and Shepherd, J.T.:* The effect in humans of increased sympathetic activity on the blood flow to active muscles. Acta med. scand. *472:* suppl., pp. 146–167 (1967).

J.T. Shepherd, MD, DSc, Mayo Clinic and Mayo Foundation, *Rochester, MN 55901* (USA)

Intrinsic Factors Concerned in the Relaxation of Vascular Smooth Muscle

Mechanisms of Vasodilatation. Satellite Symp., 27th Int. Congr. Physiol. Sci., Wilrijk 1977, pp. 12–20 (Karger, Basel 1978)

The Contractile Apparatus of Vascular Smooth Muscle[1]

Avril V. Somlyo

Pennsylvania Muscle Institute, Presbyterian-University of Pennsylvania Medical Center, Philadelphia, Pa.

The filamentous organization of contractile proteins and their linkage through attachment sites are the ultrastructural bases of the sliding filament mechanism of contraction. Historically, the structural organization of the contractile proteins in smooth muscle was characterized following the biochemical demonstration of actin and myosin (12, 19, 21, 33, 39) after prolonged, and often less than amiable, controversy (41, 43). In the following, I shall review the known features of the thick (myosin) and the thin (actin) filaments, the dense bodies and the 'intermediate' (10 nm diameter) filaments.

1. Myosin Filaments

The myosin filaments of mammalian smooth muscle are approximately 15–18 nm in diameter and in rabbit portal-anterior mesenteric vein, the only system where accurate measurements have been made, their length is approximately 2.2 μm (1). Preliminary studies on rabbit main pulmonary artery smooth muscle (*Ashton et al.*, unpublished observations) show a similar thick filament length, but the possibility of other species or organ variations cannot be excluded. Reliable length measurements require stereo electron micrograph pairs of longitudinal sections containing entire thick filaments and preferably correlated with serial transverse sections. Such studies should be based on preparatory techniques that do not involve enzymatic digestion and/or tissue fractionation, since the latter are prone to lead to fragmentation and/or aggregation of thick filaments (11, 27). The propensity of short bipolar segments of artificial myosin

[1] Supported by HL-15835 to the Pennsylvania Muscle Institute and by GM 00092.

filaments to make very long aggregates has been demonstrated (47), and the artefactual distortion that may be caused by enzymes is also well known (11, 27; *Somlyo and Somlyo,* unpublished observations).

The myosin filaments of smooth, as of striated, muscle have tapered ends, but a central bare zone has not been unequivocally demonstrated (1). The presence of a central bare zone would imply that the filaments are bipolar structures similar to those found in striated muscle. One would not expect a central bare zone if the *crossbridges* along the entire filament length were oriented in opposite directions on the two faces of the filament. Crossbridges have been observed on the thick filaments of smooth muscle (7, 38, 41, 43), but their precise arrangements have yet to be established. The suggestion of a helical arrangement of the crossbridges on thick filaments in fragmented single cells (43) cannot be accepted as *prima facie* evidence of the normal structure, in view of the above-noted susceptibility of thick filaments to artefact formation during cell isolation and fractionation. Furthermore, while artificial myosin filaments assembled *in vitro* often show considerable fine structure (6, 36, 47), the relationship of these structures to the native filament remains to be established. The presence of a 14.3 nm meridional reflection in X-ray diagrams of vertebrate smooth muscle (18, 32), under some conditions, is highly suggestive of crossbridge arrangements with this repeat.

The *distribution* of thick filaments in at least some vertebrate smooth muscles, while not as regular as in striated muscle, is not totally lacking in order. Optical diffraction patterns of micrographs of rabbit portal-mesenteric vein smooth muscle show that thick filaments are organized into an approximately 60- to 80-nm lattice (29), and examination of serial semi-thick (0.25–0.5 μm) sections shows that thick filaments are grouped into 3–5 neighboring filaments that may form the part of a 'mini-sarcomere'. The possibility that such filamentous groupings form the basis of 'fibrils' in birefringence light micrographs (9, 34) has been considered, but more recently it has been suggested that the birefringent 'fibrils' are due to invaginations of the surface membrane (10).

The notorious propensity of the thick filaments of smooth muscle to *artefact* formation has hindered initial investigations in this field. The first source of artefact is the lability of thick filaments in swollen smooth muscle fibers during osmium fixation (17, 37, 40). The absence of thick filaments in osmium-*fixed* preparations has been noted in smooth muscles swollen by intentional (KCl substitution for NaCl-, Ca-, and Mg-free solutions) or by unintentional (dissection) procedures (5, 17, 37). It must be emphasized that the thick filaments in swollen muscles are destroyed during fixation, but are still present in the swollen smooth muscles *in vivo* as evidenced by their ability to develop tension (17). The second source of artefact is due to the aggregation of the thick filaments in severely stretched muscles and particularly under the influence of hypertonic solutions. The artefactual nature of these aggregates of thick fila-

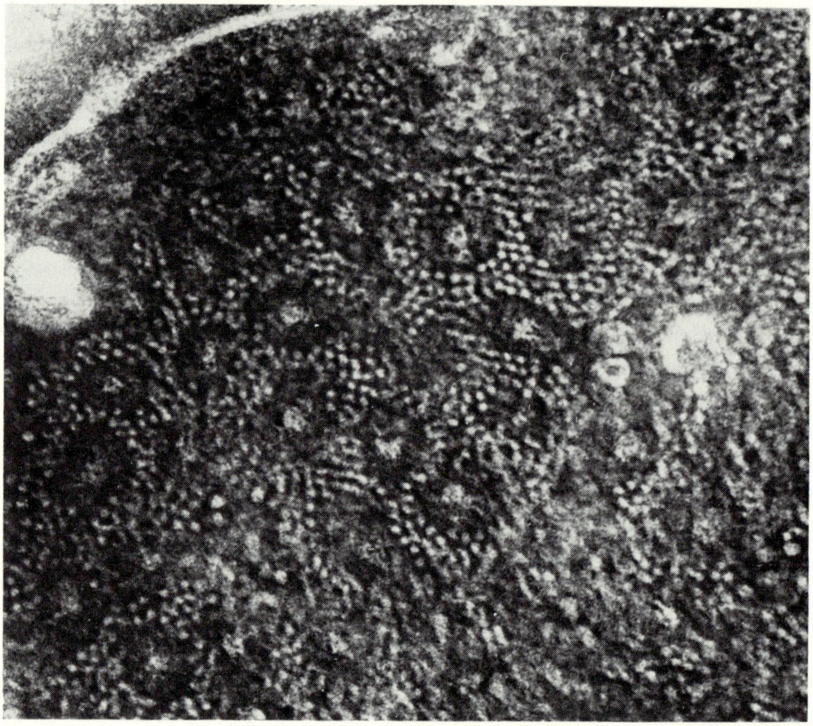

Fig. 1. Transverse section of rabbit portal-anterior mesenteric vein, smooth muscle. The muscle was fixed in 2% gluteraldehyde for 30 min, frozen in supercooled Freon 22, sectioned at –110 °C, picked up from the dry knife onto a drop of frozen sucrose solution, allowed to melt and negatively stained with 3% ammonium molybdate. This technique excludes the effects of osmium and dehydration. Rosettes of thick filaments surrounded by thin filaments are present as well as a few 10-nm, intermediate filaments. From *Somlyo and Ashton* (unpublished). × 20,000.

ments has been demonstrated some time ago (17, 40, 42), and the large ribbons are not considered to be the normal structure of organized myosin in vertebrate smooth muscle (1, 45).

In spite of considerable progress in characterizing the myosin filament in mammalian smooth muscles, conventional methods of fixation and/or cell fractionation may not yield the ideal preservation of *in vivo* fine structure. Of the chemical fixatives, osmium is the most prone to destroy fine structure and recent efforts in our laboratory have been towards the characterization of smooth muscle ultrastructure without osmimum fixation. Preliminary results with glutaraldehyde-stabilized cryo-sections – tissues rapidly frozen and sectioned at $-130°$; for method, see *Somlyo et al.* (44) – show the thick filaments (and other structures) in material that has not been exposed to osmium (fig. 1); crossbridges are present in such cryo-sections.

2. Thin Filaments

Thin (5- to 8-nm diameter, actin) filaments frequently surround the myosin filaments forming a rosette (fig. 2). The average ratio of actin:myosin filaments is approximately 15:1. The electron microscopically obtained actin:myosin ratios are comparable to those obtained by biochemical methods (12, 19, 20, 35, 46). These filaments are identified as actin through their globular substructure, through decorating them with HMM (15, 16) or with the S_1 subfragments of myosin (43), and through a spacing comparable to that of the equatorial X-ray pattern of actin (8). X-ray studies show that in vertebrate smooth muscle, as in striated muscle, contraction is associated with a movement of the tropomyosin in the groove of the actin filament (13). The length of the actin filaments has not been directly measured.

3. Dense Bodies

Dense bodies are relatively amorphous structures either attached to the plasma membrane or 'free-floating' within the myofiber (1, 24, 25, 30, 40, 41). It is generally agreed that the plasma membrane-bound dense bodies are attachment sites for the thin (actin) filaments. The role of 'free-floating' dense bodies as attachment sites for thin filaments (analogous to the Z lines of striated muscle) has been well debated (4), but there are several lines of evidence supporting this view. The insertion of the actin filaments into free-floating dense bodies has been demonstrated in stereo pairs of semi-thick longitudinal sections of smooth muscle (1, fig. 12) as well as in serial transverse sections (1, fig. 14). Actin filaments can be seen entering isolated dense bodies (22, 23, fig. 6; 28, fig. 14). Finally, antibody studies visualizing antialpha-actinin with peroxidase enabled *Schollmeyer et al.* (30) to demonstrate the presence of α-actinin within the membrane-bound and free-floating dense bodies of vertebrate smooth muscle. α-Actinin is a protein of the Z line of striated muscle, hence most of the ultrastructural and immunological evidence supports the role of dense bodies as attachment sites of actin filaments analogous to the Z lines. Nevertheless, in the absence of antibody staining, the dense body is a sufficiently amorphous and non-specifically stained structure so that 'dense bodies' identified by conventional ultrastructural criteria may represent more than one biochemical and functional entity (41).

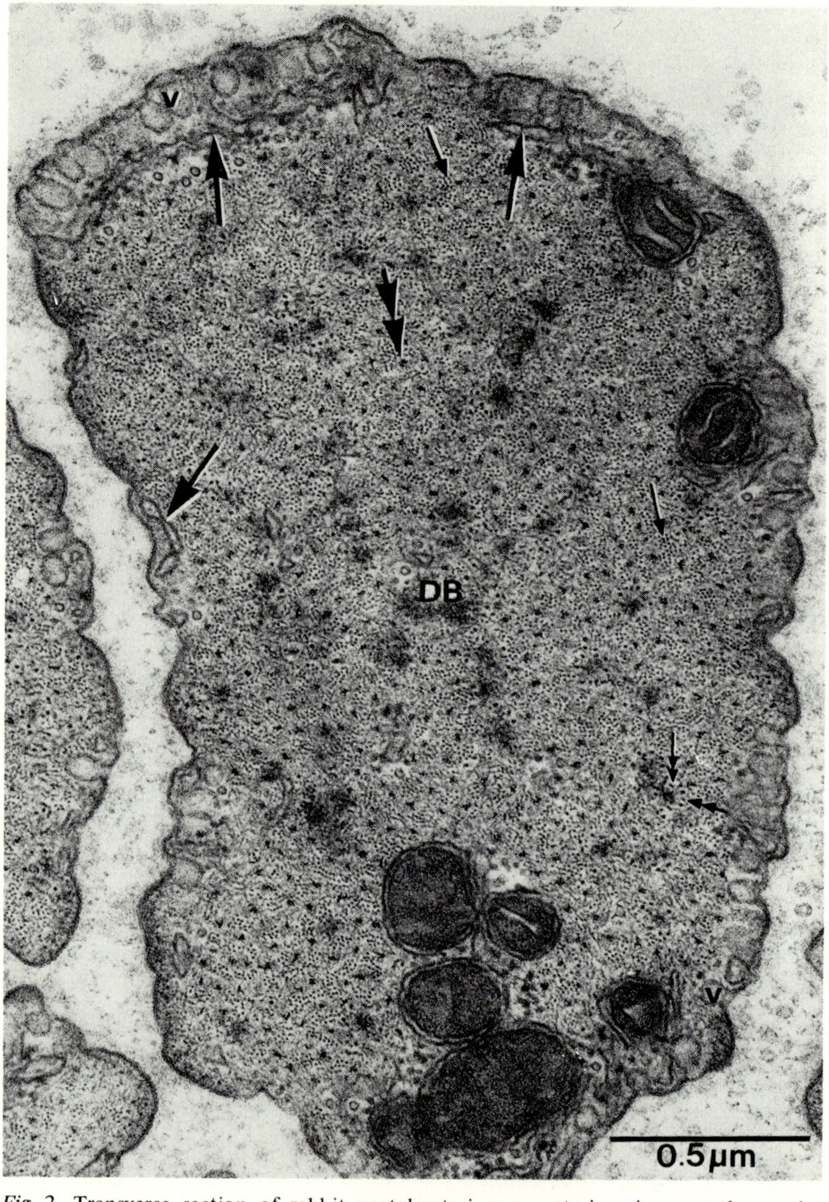

Fig. 2. Transverse section of rabbit portal-anterior mesenteric vein, smooth muscle. Arrays of thick, myosin (large double arrows) are surrounded by thin, actin filaments (small arrows). Dense bodies (DB) are associated with 10-nm filaments. The sparse sarcoplasmic reticulum (SR) (large single arrows) is characteristic of this muscle and taenia coli. Surface vesicles (v) are open to the extracellular space. From *Somlyo et al.* (48). × 55,000.

4. Intermediate Filaments

The intermediate filaments are the third type of filamentous protein found in smooth muscle. They are generally associated with dense bodies, often surrounding them (1, 4, 28, 38). While this association makes the recognition of the actin filaments inserted on cytoplasmic dense bodies technically more difficult, it does not completely obscure it (1). Intermediate filaments are approximately 10 nm in diameter, may run in bundles over several micra and are rather resistant to procedures, e.g. fixation, that may destroy the thick filament lattice. Similar filaments have been observed in a great variety of muscle and non-muscle cells, e.g. neurofilaments (31), fibroblasts (16), endothelial cells (2), and their numbers can be markedly increased by combined treatment with colchicine and cytochalasin (14). The intermediate filaments are continuous with neither actin nor myosin filaments and do not appear to play a direct role in the contractile process (1). It is difficult to draw a conclusion regarding their function in smooth muscle (or in other cells), since their numbers and location are variable: in abnormal smooth muscle fibers, large masses of intermediate filaments may 'crowd out' the thin and thick filament lattice (38, 40). We have pointed out some time ago (42) that the intermediate filaments resemble the cytoskeleton of obliquely striated invertebrate muscle (26). This possibility has been actively pursued (4, 5), but the evidence supporting it (or of the role of a cytoskeleton) is not conclusive. The composition of the intermediate filaments in smooth muscle was first studied by *Rice and Brady* (28) and shows a major protein of approximately 55,000 daltons (4). The cross-reactivity between neurofilaments and the intermediate filaments of other cells suggests that the 2 structures are related or at least share a common antigenic group (2). In contrast to their relative resistance to swelling and fixation, the intermediate filaments are solubilized by 6–8 *M* urea or acetic acid (4) and there is a possibility that fragments of these filaments may be redeposited at other sites during isolation of cells with proteolytic enzymes or during prolonged incubation of smooth muscle. In this context, it is interesting to note that the diameter of conventionally fixed thick filaments appears to increase after prolonged incubation (3, 11).

5. Conclusion

The presence of approximately 2.2 μm long, tapered, thick (15- to 18-nm diameter, myosin) filaments associated with thin (5–8 nm, actin) filaments, with the latter inserting on dense bodies, is compatible with a sliding filament mechanism of contraction in smooth muscle. The presence of crossbridges on the myosin filaments also points to a fundamentally similar contractile mechanism to that described in striated muscle, although details of the molecular assembly of the *in vivo* thick filament remain to be established. The intermediate (10-nm diameter) filaments are an ubiquitous feature of a variety of contractile and non-contractile cells and may occur in increased numbers under abnormal conditions: their function is not known.

References

1 *Ashton, F.T.; Somlyo, A.V., and Somlyo, A.P.:* The contractile apparatus of vascular smooth muscle: intermediate high voltage stereo electron microscopy. J. molec. Biol. *98:* 17–29 (1975).

2 *Blose, S.H.; Shelanski, M.L., and Chacko, S.:* Localization of bovine brain filament antibody on intermediate (100 Å) filaments in guinea pig vascular endothelial cells and chick cardiac muscle cells. Proc. natn. Acad. Sci. USA *74:* 662–665 (1977).

3 *Campbell, G.R. and Chamley, J.H.:* Thick filaments in vertebrate smooth muscle. Cell Tissue Res. *156:* 201–216 (1975).

4 *Cooke, P.:* A filamentous cytoskeleton in vertebrate smooth muscle fibers. J. Cell Biol. *68:* 539–556 (1976).

5 *Cooke, P.H. and Fay, F.:* Correlation between fiber length, ultrastructure, and the length-tension relationship of mammalian smooth muscle. J. Cell Biol. *52:* 105–116 (1972).

6 *Craig, R. and Megerman, J.:* Assembly of smooth muscle myosin into side-polar filaments. J. Cell Biol. *75:* 990–996 (1977).

7 *Devine, C.E. and Somlyo, A.P.:* Thick filaments in vascular smooth muscle. J. Cell Biol. *49:* 636–649 (1971).

8 *Elliot, G.F. and Lowy, J.:* Organization of actin in a mammalian smooth muscle. Nature, Lond. *219:* 156–157 (1968).

9 *Fisher, B.A. and Bagby, R.M.:* Reorientation of myofilaments during contraction of a vertebrate smooth muscle. Am. J. Physiol. *232:* C5–C14 (1977).

10 *Gabella, G.:* Structural changes in smooth muscle cells during isotonic contraction. Cell Tissue Res. *170:* 187–201 (1976).

11 *Gabella, G. and Raeymaekers, L.:* Effect of collagenase on mechanical activity and fine structure of an intestinal smooth muscle. Cell Tissue Res. *173:* 29–44 (1976).

12 *Hartshorne, D.J. and Aksoy, M.:* Introduction. Biochemistry of the contractile proteins in smooth muscle: a survey of current knowledge; in *Stephens* The biochemistry of smooth muscle, pp. 363–378 (University Park Press, Baltimore 1977).

13 *Haselgrove, J.C.:* Structural changes in smooth and striated muscle during contraction; in *Bolis, Maddrell and Schmidt-Nielsen* Comparative physiology – functional aspects of structural materials, pp. 127–138 (North-Holland, Amsterdam 1975).

14 *Holtzer, H.; Croop, J.; Dienstman, S.; Ishikawa, H., and Somlyo, A.P.:* The effects of cytochalasin-B and colcimide on myogenic cultures. Proc. natn. Acad. Sci. USA *72:* 513–517 (1975).

15 *Ishikawa, H.:* Arrowhead complexes in a variety of cell types. Int. Congr. Ser. No. 333, pp. 37–50 (Excerpta Medica, Amsterdam 1973).

16 *Ishikawa, H.; Bischoff, R., and Holtzer, H.:* Mitosis and intermediate-sized filaments in developing skeletal muscle. J. Cell Biol. *38:* 538 (1968).

17 *Jones, A.W.; Somlyo, A.P., and Somlyo, A.V.:* Potassium accumulation in smooth

muscle and associated ultrastructural changes. J. Physiol., Lond. *232:* 247–273 (1973).

18 *Lowy, J.; Polsen, F.R., and Vibert, P.J.:* Myosin filaments in vertebrate smooth muscle. Nature, Lond. *225:* 1053 (1970).

19 *Murphy, R.A.:* Contractile system function in mammalian smooth muscle. Blood Vessels *13:* 1–23 (1976).

20 *Murphy, R.A.; Herlihy, J.T., and Megerman, J.:* Force generating capacity and contractile protein content of arterial smooth muscle. J. gen. Physiol. *64:* 691–705 (1974).

21 *Needham, D.M. and Shoenberg, C.F.:* Proteins of the contractile mechanism in vertebrate smooth muscle; in *Code* Handbook of physiology; 4th ed., sect. 6, pp. 1793–1810 (Am. Physiological Society, Washington 1968).

22 *Nonomura, Y. and Ebashi, S.:* Isolation and identification in smooth muscle contractile proteins; in *Daniel and Paton* Methods in pharmacology; 3rd ed., pp. 141–162 (Plenum Publishing, New York 1975).

23 *Panner, B.J. and Honig, C.R.:* Filament ultrastructure and organization in vertebrate smooth muscle: contraction hypothesis based on localization of actin and myosin. J. Cell Biol. *35:* 303–321 (1967).

24 *Pease, D.C. and Molinari, S.:* Electron microscopy of muscular arteries; pial vessels of the cat and monkey. J. Ultrastruct. Res. *3:* 447–468 (1960).

25 *Prosser, C.L.; Burnstock, G., and Kahn, J.:* Conduction in smooth muscle: comparative structural properties. Am. J. Physiol. *199:* 545–582 (1960).

26 *Rosenbluth, J.:* Obliquely striated muscle, III. Contraction mechanism of Ascaris body muscle. J. Cell Biol. *34:* 15–33 (1967).

27 *Rosenbluth, J.:* Myosin-like aggregates in trypsin-treated smooth muscle cells. J. Cell Biol. *48:* 174–188 (1971).

28 *Rice, R.V. and Brady, A.C.:* Biochemical and ultrastructural studies on vertebrate smooth muscle; in The mechanism of muscle contraction, Cold Spring Harb. Symp. Quant. Biol.; 37th ed., pp. 429–438 (Cold Spring Harbor Laboratory, New York 1973).

29 *Rice, R.V.; McManus, G.M.; Devine, C.E., and Somlyo, A.P.:* A regular organization of thick filaments in mammalian smooth muscle. Nature new Biol. *231:* 242–243 (1971).

30 *Schollmeyer, J.E.; Furcht, L.T.; Goll, D.E.; Robson, R.M., and Stromer, M.H.:* Localization of contractile proteins in smooth muscle cells and in normal and transformed fibroblasts; in *Goldman, Pollard and Rosenbaum* Cell motility (A), Cold Spring Harb. Conf. Cell Proliferation; 3rd ed., pp. 361–388 (Cold Spring Harbor Laboratory, New York 1976).

31 *Shelanski, M.L.; Yen, S.H., and Lee, V.M.:* Neurofilaments and glial filaments; in *Goldman, Pollard and Rosenbaum* Cell motility, Cold Spring Harb. Conf. Cell Proliferation; 3rd ed., pp. 1007–1020 (Cold Spring Harbor Laboratory, New York 1976).

32 *Shoenberg, C.F. and Haselgrove, J.C.:* Filaments and ribbons in vertebrate smooth muscle. Nature, Lond. *249:* 152–154 (1974).

33 *Shoenberg, C.F. and Needham, D.M.:* A study of the mechanism of contraction in vertebrate smooth muscle. Biol. Rev. *51:* 53–104 (1976).

34 *Small, J.V.:* Contractile units in vertebrate smooth muscle cells. Nature, Lond. *249:* 325–327 (1974).

35 *Small, J.V. and Sobieszek, A.:* Studies on the function and composition of the 10 nm (100 Å) filaments of vertebrate smooth muscle. J. Cell Sci. *23:* 243–268 (1977).

36 *Sobieszek, A. and Small, J.V.:* The assembly of ribbon-shaped structures in low ionic strength extracts obtained from vertebrate smooth muscle. Phil. Trans. R. Soc. B *265:* 203–212 (1973).

37 *Somlyo, A.P.; Devine, C.E., and Somlyo, A.V.:* Thick filaments in unstretched mammalian smooth muscle. Nature new Biol. *233:* 218–219 (1971).

38 *Somlyo, A.P.; Devine, C.E.; Somlyo, A.V., and Rice, R.V.:* Filament organization in vertebrate smooth muscle. Phil. Trans. R. Soc. B *265:* 223–229 (1973).

39 *Somlyo, A.P. and Somlyo, A.V.:* Vascular smooth muscle. I. Normal structure, pathology, biochemistry and biophysics. Pharmacol. exp. Ther. *159:* 129–145 (1968).

40 *Somlyo, A.P. and Somlyo, A.V.:* Ultrastructure of smooth muscle; in *Daniel and Paton* Methods in pharmacology; 3rd ed., pp. 3–43 (Plenum Publishing, New York 1975).

41 *Somlyo, A.P.; Somlyo, A.V.; Ashton, F.T., and Vallieres, J.:* Vertebrate smooth muscle: ultrastructure and function; in *Goldman, Pollard and Rosenbaum* Cell motility. Cold Spring Harb. Conf. Cell Proliferation; 3rd ed., pp. 165–183 (Cold Spring Harbor Laboratory, New York 1976).

42 *Somlyo, A.P.; Somlyo, A.V.; Devine, C.E., and Rice, R.V.:* Aggregation of thick filaments into ribbons in mammalian smooth muscle. Nature new Biol. *231:* 243–246 (1971).

43 *Somlyo, A.V.; Ashton, F.T.; Lemanski, L.; Vallieres, J., and Somlyo, A.P.:* Filament organization and dense bodies in vertebrate smooth muscle; in *Stephens* The biochemistry of smooth muscle, pp. 445–471 (University Park Press, Baltimore 1977).

44 *Somlyo, A.V.; Shuman, H., and Somlyo, A.P.:* Elemental distribution in striated muscle and effects of hypertonicity: electron probe analysis of cryo sections. J. Cell Biol. *74:* 828–857 (1977).

45 *Squire, J.M.:* Muscle filament structure and muscle contraction. Annu. Rev. Biophys. Bioengr. *4:* 137–163 (1975).

46 *Tregear, R.T. and Squire, J.M.:* Myosin content and filament structure in smooth and striated muscle. J. molec. Biol. *77:* 279–290 (1973).

47 *Wachsberger, P.R. and Pepe, F.A.:* Purification of uterine myosin and synthetic filament formation J. molec. Biol. *88:* 385–391 (1974).

48 *Somlyo, A.P.; Devine, C.E., and Somlyo, A.V.:* In *Betz* Symp. Vascular Smooth Muscle, pp. 119–121 (Springer, Berlin 1972).

A.V. Somlyo, PhD, Pennsylvania Muscle Institute, Presbyterian-University of Pennsylvania Medical Center, *Philadelphia, PA 19104* (USA)

Mechanisms of Vasodilatation. Satellite Symp., 27th Int. Congr. Physiol. Sci., Wilrijk 1977, pp. 21–29 (Karger, Basel 1978)

The Role of Organelles in Regulating Cytoplasmic Calcium in Vascular Smooth Muscle[1]

Andrew P. Somlyo

Pennsylvania Muscle Institute, Presbyterian-University of Pennsylvania Medical Center, Philadelphia, Pa.

The modulation of contractile activity in smooth, as in striated, muscles is mediated primarily through alterations in the level of cytoplasmic free calcium concentration ([Ca]) (for review see 5, 19, 24, 28). While the influx of extracellular Ca and the active extrusion of cytoplasmic Ca may also contribute to activation or relaxation (6, 16, 39, and for review see 2, 4, 5, 24, 28), considerable evidence also supports the role of intracellular organelles in regulating the level of cytoplasmic free Ca (for review see 19, 29, 33). We shall review here the properties of the sarcoplasmic reticulum and of the mitochondria in vascular smooth muscle and the evidence supporting their role as Ca storage sites.

1. Sarcoplasmic Reticulum (SR)

The SR is a system of tubules approximately 30 nm in diameter (fig. 1) and expanding into flat cisternae in some species – e.g. turtle (25) – and constituting from 2 to 7.5% of the cell volume in glutaraldehyde-fixed preparations (9, 12, 23). Extracellular markers (e.g. colloidal lanthanum, horseradish peroxidase or ferritin) enter the surface vesicles but *not* the SR (9, 10, 25, 27); nor is the SR directly continuous with the lumen of the Golgi apparatus (15, 26). The relative volume of the SR in different smooth muscles correlates with their ability to contract in the absence of extracellular Ca (9), but this correlation may be fortuitous. The ability of smooth muscle to be activated by intracellularly stored (presumably by the SR) Ca is maintained longer at relatively low temperatures (25).

[1] Supported by HL-15835 to the Pennsylvania Muscle Institute and by GM00092.

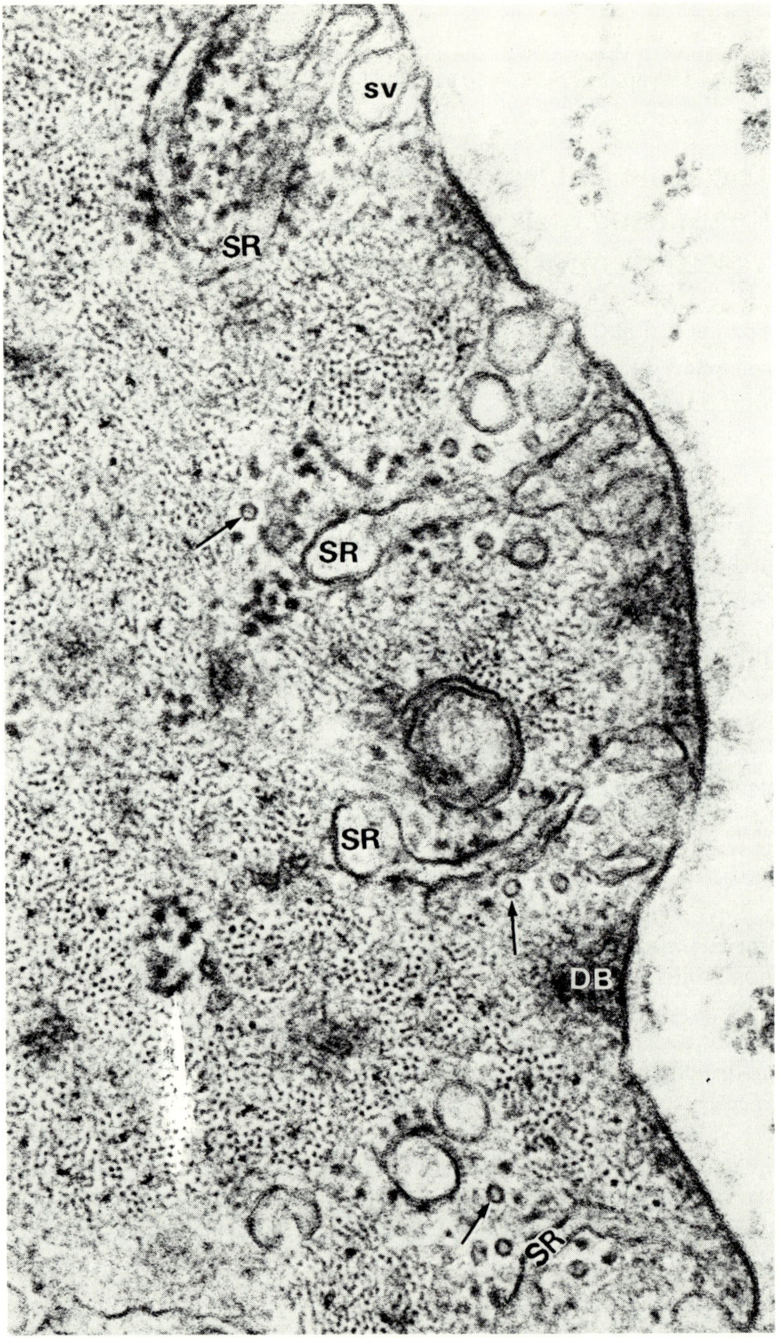
sv
SR
SR
SR
DB
SR

The SR forms fenestrations around surface vesicles (see below), and it has been speculated that these SR and the mitochondrial-surface vesicle contacts may be sites of Ca extrusion (30, 45). There are no periodic electron opaque processes comparable to those found at the anatomical couplings (see below) between the surface vesicle and the adjacent SR membrane. The central and peripheral, and the rough (covered with ribosomes) and the smooth SR are in direct continuity (9, 27), although the question whether the Ca storage capacities of the rough and the smooth SR are identical has not been answered. In the one published electron micrograph showing strontium deposits in a region of continuous rough and smooth SR (27, fig. 20), the deposit is present in the smooth, but not in the rough portion.

Couplings are structural specializations where elements of the peripheral SR come to within 10–12 nm of the surface membrane: the gap between the two membrane systems is traversed by quasi-periodic, electron-opaque processes (9, 25, 27). The junctional SR of the couplings is the most probable site where Ca is released by electrical signals (e.g. action potentials or graded depolarization) spreading along the surface membrane of smooth muscle (37, 38).

The fact that the SR of vascular smooth muscle can actively accumulate divalent cations has been demonstrated through electron microscopy and electron probe analysis (22, 30, 32, 38). Thus, strontium, a relatively electron opaque 'Ca substitute', is concentrated in central and peripheral elements of the SR of vascular smooth muscle incubated in Sr-containing solutions prior to fixation (30, 38). Sequestered Ca has been demonstrated in osmium-fixed smooth muscles preloaded with Ca in the presence of oxalate (22), and in unfixed cryo-sections (fig. 2, 3) (29, 32, 34). The latter approach avoids the possibility of elemental translocations that could occur during chemical fixation, and provides unequivocal evidence that the SR (with the continuous perinuclear space) is a major Ca storage site in vascular smooth muscle. Furthermore, considering that maximal activation of smooth muscle requires from 0.1 to 0.23 m*M* Ca/kg wet fiber weight (11, 40, and for review see 19), the release of the Ca stored in an SR of only 1% of cell volume and having a [Ca] of 66 mmole/kg dry weight would appear to be sufficient for the activation of the contractile apparatus. The latter [Ca] have been measured in the *in situ* SR of vertebrate *striated* muscle (35, 36), although comparable studies on smooth muscle have not been completed. Nevertheless, only the uptake of Ca^{++} into the

Fig. 1. Transverse section of part of a smooth muscle cell from rabbit main pulmonary artery. Four portions of sarcoplasmic reticulum (SR) approach the surface membrane near surface vesicles (sv). Surface-dense bodies (DB) alternate with the peripheral SR and SV. Some microtubules (arrows) are also present. The basement membrane is prominent especially in the regions of the surface dense bodies. × 104,500. From *Somlyo and Somlyo* (27).

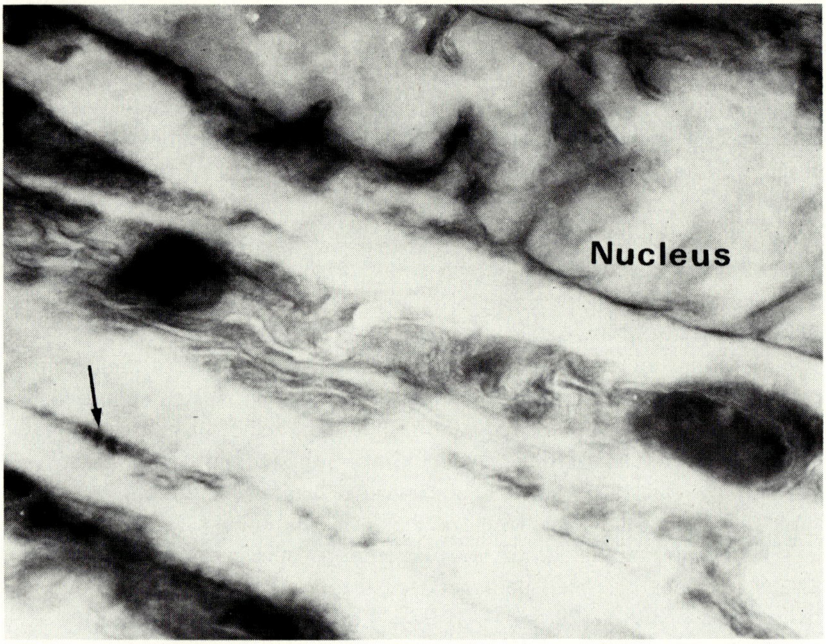

Fig. 2. Frozen-dried thin section from an unfixed K-depolarized, rabbit portal-anterior mesenteric vein, sectioned at -110 °C. The dried section was stained with osmium vapor *in vacuo* (10^{-5} Torr). The sarcoplasmic reticulum (arrow) containing dense material was analyzed as shown in figure 3. × 36,000. From *Somlyo et al.* (34).

SR of vascular smooth muscle has been unequivocally demonstrated with direct electron optical techniques, but not its release. Recent studies showing that the intracellular ^{45}Ca released, respectively, by norepinephrine and by caffeine in aortic smooth muscle originates from the same kinetic pool (8), the known actions of caffeine (3, 43) and the presence of the high [Ca] in the SR are certainly compatible with a mechanism in which the SR is a major source and sink of activator Ca in vascular smooth muscle. A number of studies have also been reported showing active (ATP supported) uptake of Ca by microsomal fractions isolated from a variety of smooth muscles (1, 7, 17, 18, 21). However, there are considerable technical problems associated with preparing these fractions from tissues containing large amounts of collagen and elastin, but usually only small amounts of SR. Therefore, the identity (whether SR, surface membrane, Golgi, etc.) and condition of these microsomal preparations and their often conflicting (in various laboratories) responses to drugs and cyclic nucleotides are difficult to evaluate.

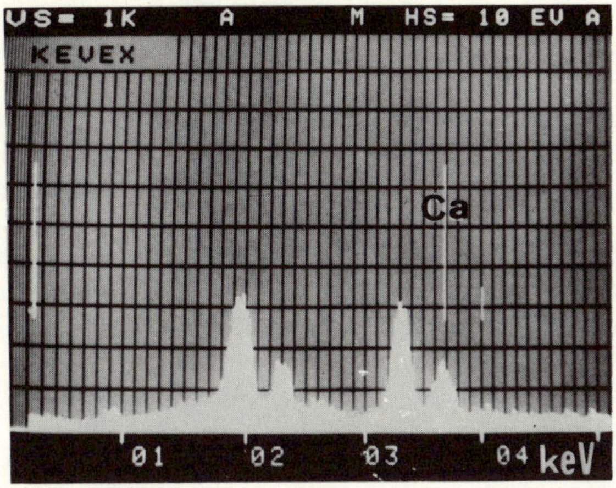

Fig. 3. X-Ray spectrum from the SR shown in figure 2. The counts on the ordinate are shown for the energies in keV on the abscissa. There is a significant Ca peak at 3.69 keV. From *Somlyo et al.* (34).

2. Mitochondria

Mitochondria, in addition to the SR, have been considered as potential Ca storage sites in vascular smooth muscle. Mitochondria in vascular smooth muscle incubated with Sr or Ba accumulate granules containing these cations in association with P, and this process can be blocked by inhibitors of active mitochondrial transport (30, 38). Mitochondria isolated from bovine main pulmonary artery and mesenteric vein (in the presence of EDTA) contain high concentrations (90–140 nmole/mg protein) of Ca and can accumulate it at rates up to 12.5 nmol/sec mg protein (42). In well-coupled mitochondria, these uptake rates can be supported by oxidation of respiratory substrates and do not require the presence of ATP. Similar results have been observed recently with mitochondria isolated from uterine smooth muscle (44). However, the apparent K_m for mitochondrial Ca uptake (42) is relatively high: 17 μM. Therefore, these results did not support a major role of mitochondria in regulating [Ca] during twitch contractions, when such occur in smooth muscles. Furthermore, electron probe analysis of cryo-sections of resting rabbit portal-anterior mesenteric vein smooth muscle suggested that the mitochondrial [Ca] *in situ* is low (less than 3.5 mmole/kg dry mitochondrial weight) (31). Comparison of mitochondria isolated from rabbit vascular smooth muscle with bovine mitochondria shows major differences in endogenous mitochondrial Ca content (unpublished observations with *Sloane and Scarpa*). Furthermore, both in resting muscles and in smooth muscles depolarized with high K solutions, mitochondria containing high [Ca] in the form of granules are found only in fibers that have abnormally low cytoplasmic [K] and high [Na], compatible with cell damage (31). Electron-opaque deposits of Ca have been observed in the SR of depolarized fibers in which mitochondrial Ca granules were absent. It is possible that mitochondria serve in vascular smooth muscle, and in numerous other cell systems, as reserve Ca pools during periods of excessive influx. In contrast to the above observations on adult vascular smooth muscle, mitochondrial Ca granules are frequent (in up to 75% of cells) in 'normal' cultured guinea pig aortic cells, suggesting that much of the cell Ca in these cells may be located in the mitochondria (13, 14). In skeletal (20) and in cardiac (41) muscles, the Ca transport system of the SR is relatively inactive during the earlier stages of morphogenesis. Whether the observations on cultured vascular smooth muscle are representative of a normal, early phase in morphogenesis or due to an abnormality imposed by cell culture conditions has not been established. Furthermore, since guinea pig aortic cells cultured under the conditions of the cited study do not contract, no implication should be drawn regarding mitochondria as sources of activator Ca even in immature smooth muscle. There remains the important possibility that similar mitochondrial Ca loading may occur under pathological conditions (e.g. atherosclerosis) in dedifferentiated vascular smooth muscle.

3. Conclusion

The SR is probably the major Ca storage site mediating relaxation in vascular smooth muscle, and its role as a source of activator Ca is also supported by reasonable, although less direct evidence. Mitochondria (in rabbit portal-anterior mesenteric vein) do not appear to control free cytoplasmic [Ca] during the physiological contraction-relaxation cycle, although they may serve as calcium storage sites during periods of excessive Ca loading. The effects of species and age differences and of the type of smooth muscle on the relative roles played by cell organelles and by transmembrane movements in the regulation of cytoplasmic Ca^{++} remain to be determined.

References

1 *Batra, S.C. and Daniel, E.E.:* Effect of multivalent cations and drugs on Ca uptake by the rat myometrial microsomes. Comp. Biochem. Physiol. A *38:* 285 (1971).

2 *Biamino, G. and Johansson, B.:* Effects of calcium and sodium on contracture tension

in the smooth muscle of the rat portal vein. Pflügers Arch. ges. Physiol. *321:* 143–158 (1970).

3 *Bianchi, C.P.:* Cell calcium; 1st ed., pp. 1–131 (Butterworth, London 1968).

4 *Blaustein, M.P.:* Sodium ions, calcium ions, blood pressure regulation, and hypertension: a reassessment and a hypothesis. Am. J. Physiol. *232:* C165–C173 (1977).

5 *Bohr, D.F.:* Vascular smooth muscle; in *Johnson* The peripheral circulation (Wiley & Sons, Chichester, in press).

6 *Breemen, C. van; Farinas, B.R.; Casteels, R.; Gerba, P.; Wuytack, F., and Deth, R.:* Factors controlling cytoplasmic Ca^{2+} concentration. Phil. Trans. R. Soc. *265:* 57–71 (1973).

7 *Carsten, M.E.:* Role of calcium binding by sarcoplasmic reticulum in the contraction and relaxation of uterine smooth muscle. J. gen. Physiol. *53:* 414 (1969).

8 *Deth, R. and Casteels, R.:* A study of releasable Ca fractions in smooth muscle cells of the rabbit aorta. J. gen. Physiol. *69:* 401–416 (1977).

9 *Devine, C.E.; Somlyo, A.V., and Somlyo, A.P.:* Sarcoplasmic reticulum and excitation-contraction coupling in mammalian smooth muscle. J. Cell Biol. *52:* 690–718 (1972).

10 *Devine, C.E.; Somlyo, A.V., and Somlyo, A.P.:* Sarcoplasmic reticulum and mitochondria as cation-accumulating sites in smooth muscle. Phil. Trans. R. Soc. B *265:* 17–23 (1973).

11 *Ford, G.D.; Litten, R.Z., and Solaro, R.J.:* Calcium binding to actomyosin from bovine aorta. Biophys. J. *17:* 266a (1977).

12 *Gabella, G.:* Quantitative morphological study of smooth muscle cells of the guinea-pig taenia coli. Cell Tissue Res. *170:* 161–186 (1976).

13 *Garfield, R.E. and Somlyo, A.P.:* Electron probe analysis and ultrastructure of cultured, freeze-dried vascular smooth muscle; in *Bailey and Arceneaux* Proc. Elec. Micros. Soc. Am. 33rd Ann. Meet., pp. 558–559 (Claitor's Publishing Division, Baton Rouge 1974).

14 *Garfield, R.E. and Somlyo, A.P.:* Ultrastructure, electron probe and spectrophotometric microanalysis of cultured vascular smooth muscle; in Histochemistry and cytochemistry. Proc. 5th Int. Congr. Histochem. and Cytochem., Bucharest 1976, p. 124.

15 *Garfield, R.E. and Somlyo, A.P.:* Golgi apparatus and lectin-binding sites: effects of lasalocid (X537A). Expl. Cell Res. *109:* 163–179 (1977).

16 *Godfraind, T.:* Calcium exchange in vascular smooth muscle; action of noradrenaline and lanthanum. J. Physiol., Lond. *260:* 21–35 (1976).

17 *Hess, M.L. and Ford, G.D.:* Calcium accumulation by subcellular fractions from vascular smooth muscle. J. mol. cell. Cardiol. *6:* 275 (1974).

18 *Hurwitz, L.; Fitzpatrick, D.F.; Debbas, G., and Landon, E.S.:* Localization of calcium pump activity in smooth muscle. Science *179:* 334 (1973).

19 *Johansson, B. and Somlyo, A.P.:* Electrophysiology and excitation-contraction coupling; in *Bohr, Sparks and Somlyo* The handbook of physiology. Vascular smooth muscle (Williams & Wilkins, Baltimore, in press).

20 *Martonosi, A.:* Development of sarcoplasmic reticulum in cultured chicken muscle. J. biol. Chem. *252:* 318–322 (1977).

21 *Nilsson, K.B.:* Ca^{2+}-binding and Ca^{2+}-releasing microsomal fractions from intestinal smooth muscle; influence of contracting and relaxing agents and cyclic nucleotides; med. diss. No. 49, pp. 5–411 Linköping (1977).

22 *Popescu, L.M. and Diculescu, I.:* Calcium in smooth muscle SR *in situ:* conventional and X-ray analytical electron microscopy. J. Cell Biol. *67:* 911–918 (1975).

23 *Popescu, L.M.; Diculescu, I.; Zelck, U., and Ionescu, I.:* Ultrastructural distribution of

calcium in smooth muscle cells of guinea-pig taenia coli. Cell Tissue Res. *154:* 357 (1974).

24 *Somlyo, A.P.:* Vascular smooth muscle; in *Narahashi* Cellular pharmacology of excitable tissues, pp. 360–407 (Thomas, Springfield 1975).

25 *Somlyo, A.P.; Devine, C.E.; Somlyo, A.V., and North, S.R.:* Sarcoplasmic reticulum and the temperature-dependent contraction of smooth muscle in calcium-free solutions. J. Cell Biol. *51:* 722–741 (1971).

26 *Somlyo, A.P.; Garfield, R.E.; Chacko, S., and Somlyo, A.V.:* Golgi organelle response to the antibiotic X537A. J. Cell Biol. *66:* 425–443 (1975).

27 *Somlyo, A.P. and Somlyo, A.V.:* Ultrastructure of smooth muscle; in *Daniel and Paton* Methods in pharmacology; 3rd ed., pp. 3–45 (Plenum Publishing, New York 1975).

28 *Somlyo, A.P. and Somlyo, A.V.:* Calcium and magnesium in vascular smooth muscle function; in *Genest, Kolw and Juckel* Int. textbook on hypertension, pp. 440–452 (McGraw-Hill, New York 1977).

29 *Somlyo, A.P.; Somlyo, A.V.; Ashton, F.T., and Vallières, J.:* Vertebrate smooth muscle: ultrastructure and function; in *Goldman, Pollard and Rosenbaum* Cell motility. Cold Spring Harb. Conf. Cell Proliferation; 3rd ed., pp. 165–183 (Cold Spring Harbor Laboratory, New York 1976).

30 *Somlyo, A.P.; Somlyo, A.V.; Devine, C.E.; Peters, P.D., and Hall, T.A.:* Electron microscopy and electron probe analysis of mitochondrial cation accumulation in smooth muscle. J. Cell Biol. *61:* 723–742 (1974).

31 *Somlyo, A.P.; Somlyo, A.V., and Shuman, H.:* Ca and Cl distribution in vascular smooth muscle: electron probe analysis. Blood Vessels (in press).

32 *Somlyo, A.P.; Somlyo, A.V.; Shuman, H., and Garfield, R.E.:* Calcium compartments in vascular smooth muscle: electron probe analysis; in *Betz* Ionic actions on vascular smooth muscle with special regard to brain vessels, pp. 17–20 (Springer, Berlin 1976).

33 *Somlyo, A.P.; Vallières, J.; Garfield, R.E.; Shuman, H.; Scarpa, A., and Somlyo, A.V.:* Calcium compartmentalization in vascular smooth muscle: electron probe analysis and studies in isolated mitochondria; in *Stephens* Biochemistry of smooth muscle, pp. 563–583 (University Park Press, Baltimore 1977).

34 *Somlyo, A.V.; Shuman, H., and Somlyo, A.P.:* Electron probe analysis of vertebrate smooth muscle: distribution of Ca and Cl; in *Bailey* Proc. Elec. Micros. Soc. Am. 34th Annu. Meet., pp. 334–335 (Claitor's Publishing Division, Baton Rouge 1976).

35 *Somlyo, A.V.; Shuman, H., and Somlyo, A.P.:* Elemental distributions in striated muscle and effects of hypertonicity: electron probe analysis of cryo sections. J. Cell Biol. *74:* 828–857 (1977).

36 *Somlyo, A.V.; Shuman, H., and Somlyo, A.P.:* The composition of the sarcoplasmic reticulum *in situ:* electron probe X-ray microanalysis of cryo sections. Nature, Lond. *268:* 556–558 (1977).

37 *Somlyo, A.V. and Somlyo, A.P.:* Electromechanical and pharmacomechanical coupling in vascular smooth muscle. J. Pharmac. exp. Ther. *159:* 129–145 (1968).

38 *Somlyo, A.V. and Somlyo, A.P.:* Strontium accumulation by sarcoplasmic reticulum and mitochondria in vascular smooth muscle. Science *174:* 955–958 (1971).

39 *Somlyo, A.P. and Somlyo, A.V.:* Electrophysiological correlates of the inequality of maximal vascular smooth muscle contraction elicited by drugs; in *Bevan, Furchgott, Maxwell and Somlyo* Vascular neuroeffector systems, pp. 216–226 (Karger, Basel 1971).

40 *Sparrow, M.P. and Simmonds, W.J.:* The relationship of the calcium content of smooth muscle to its contractility in response to different modes of stimulation. Biochim. biophys. Acta *109:* 503–511 (1965).

41 *Sulakhe, P.V. and Dhalla, N.S.:* Excitation-contraction coupling in heart. IV. Energy-dependent calcium transport in the myocardium of developing rat. Life Sci. *9:* 1367–1370 (1970).

42 *Vallières, J.; Scarpa, A., and Somlyo, A.P.:* Subcellular fractions of smooth muscle I. Isolation, substrate utilization and Ca^{++} transport by main pulmonary artery and mesenteric vein mitochondria. Archs Biochem. Biophys. *170:* 659–669 (1975).

43 *Weber, A. and Herz, R.:* The relationship between caffeine contracture of intact muscle and the effect of caffeine on reticulum. J. gen. Physiol. *52:* 750 (1968).

44 *Wikström, M.; Ahonen, P., and Luukkainen, T.:* The role of mitochondria in uterine contractions. FEBS Lett. *56:* 120–123 (1975).

45 *Wootton, G.S. and Goodford, P.J.:* An association between mitochondria and vesicles in smooth muscle. Cell Tissue Res. *161:* 119–132 (1975).

A.P. Somlyo, MD, Pennsylvania Muscle Institute, Presbyterian-University of Pennsylvania Medical Center, *Philadelphia, PA 19104* (USA)

Mechanisms of Vasodilatation. Satellite Symp., 27th Int. Congr. Physiol. Sci., Wilrijk 1977, pp. 30–36 (Karger, Basel 1978)

Control of the Actin-Myosin Interaction by Ca^{++} in Vascular Smooth Muscle

R.A. Murphy

Department of Physiology, University of Virginia School of Medicine, Charlottesville, Va.

Cellular regulation of vasoactivity consists of two processes. One involves the membrane-mediated events influencing smooth muscle cytoplasmic Ca^{++} levels (fig. 1, panel I). The other is concerned with the mechanisms conferring a Ca^{++} dependence on the cyclic interactions between myosin and actin. This review is concerned with the latter events. Evidence for the regulatory action of Ca^{++} being mediated through actions on (1) a thin filament protein, (2) myosin, and (3) indirectly through a kinase-catalyzed phosphorylation of myosin has been reported for vertebrate smooth muscles (fig. 1, panels, II–IV). The current status of research on these putative regulatory mechanisms is assessed in the following sections.

1. Regulation via a Thin Filament Ca^{++}-Binding Protein

The basic evidence for thin filament regulation comes from studies of protein fractions isolated from several smooth muscles which restore Ca^{++} dependence to actomyosin systems lacking regulation (5, 7–10, 19, 26, 34). This action resembles that of skeletal muscle tropomyosin and troponin which can restore Ca^{++} dependence to desensitized smooth muscle actomyosin (19, 30, 34), although this effect may be small (9). Smooth muscle protein fractions reportedly can sensitize skeletal actomyosin to Ca^{++} (5, 8), but this role was strongly challenged (9, 19, 32). Consequently, many investigators suggested that the regulatory system in smooth muscle consists of a thin filament Ca^{++}-binding protein whose function is analogous to the troponin system in vertebrate striated muscles (fig. 1, panel II) (5, 9, 10, 18, 19, 34).

However, smooth muscles do not contain stoichiometric amounts (compared to tropomyosin) of a protein having the same subunit mobilities on

SDS-acrylamide gels as skeletal troponin (2–4, 7, 17, 27, 32, 34), although fairly small amounts of protein having some characteristics of troponin C and I can be obtained (18). The Ca^{++}-sensitizing 'troponin' described by *Ebashi* and his colleagues (9, 23) has an estimated molecular weight of 80,000, which differs from skeletal muscle troponin subunits, and appears to bind to actin. Even very small amounts of this smooth muscle protein stimulate actin-myosin interactions in the presence of Ca^{++} (9, 10, 19, 23). In contrast, troponin in skeletal muscle acts by inhibiting actin-myosin interactions in the absence of Ca^{++} when stoichiometrically bound to tropomyosin.

The current status of the hypothesis of thin filament regulation in smooth muscle is equivocal. A troponin-type of control is possible, as skeletal troponin can bind to smooth muscle thin filaments to produce a regulated actomyosin,

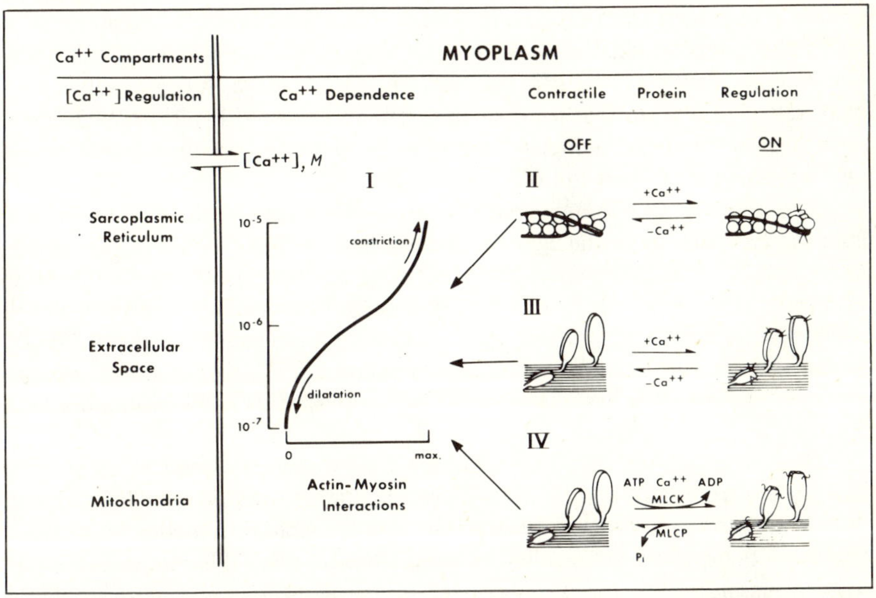

Fig. 1. Events implicated in the regulation of contraction in vascular smooth muscle. Panel I shows the general dependence of contractile response on the myoplasmic Ca^{++} concentration (26), controlled by membrane-mediated processes. Panel II illustrates thin filament control where binding of Ca^{++} (>) to a Ca^{++}-binding protein (elipsoid) alters the conformation of the thin filament. The case illustrated is that in vertebrate striated muscles where Ca^{++} binding to troponin causes a shift in tropomyosin (heavy line) to expose sites on the actin (circles) which can interact with myosin. Panel III indicates how Ca^{++} binding (>) to a site on myosin in the thick filaments can alter its conformation and permit the interaction with actin. Panel IV depicts a similar effect mediated by phosphorylation (~) in the presence of Ca^{++} by a myosin light chain kinase (MLCK), and reversed by the action of a myosin light chain phosphatase (MLCP).

and proteins similar to troponin C and I can be detected (18). However, the presence of a troponin-like protein in smooth muscle in adequate amounts to explain Ca^{++} regulation has not been confirmed.

2. Regulation via a Myosin Ca^{++}-Binding Site

The discovery of a Ca^{++}-binding site on myosin regulating the actin-myosin interaction in molluscan muscles (fig. 1, panel III) (20) and the development of indirect methods to test for thin filament or myosin types of control (22) stimulated new studies in vertebrate smooth muscle. Competition assays revealed that rabbit skeletal muscle myosin added to smooth muscle actomyosin (2, 3, 13, 25, 33) or thin filaments (4) in the absence of Ca^{++} showed an unrestrained ATPase activity. This would be expected if the smooth muscle thin filaments had no endogenous Ca^{++} regulating system which functioned in the presence of the added myosin. The implication that regulation was via myosin fits the observation that Ca^{++}-dependence was retained in the ATPase activity of mixtures of purified smooth muscle myosin and actin (4, 17, 25, 33). Unfortunately, this critical test is highly dependent on the purity of the smooth muscle myosin, which is difficult to isolate.

The specific Ca^{++}-binding site involved with regulation of actin-myosin interactions is on one of the light chains associated with each head of the myosin molecule (21). Thus, there are two regulatory light chains having Ca^{++}-binding sites associated with each myosin molecule in molluscs exhibiting myosin-linked regulation. A chemically similar light chain with a molecular weight of 20,000 obtained from vertebrate smooth muscles restored Ca^{++}-dependence to molluscan myosin which had been desensitized by removal of its native regulatory light chains (21).

In summary, there is strong evidence that vertebrate smooth and striated muscles contain myosin with a Ca^{++}-binding site on one class of light chains. This light chain can regulate the actin-myosin interaction in molluscan muscles. However, there is no comparable evidence that it serves this role in vertebrate smooth muscles. This caveat is suggested by the absence of a regulatory role for the similar skeletal muscle light chain (21).

3. Regulation via a Ca^{++}-Dependent Myosin Light Chain Kinase and Phosphorylation

Three myofibrillar proteins can be phosphorylated and dephosphorylated by enzymes present in muscle: troponin I, troponin T, and myosin (14, 28). Phosphorylation of the troponin subunits occurs at fairly low rates by phosphorylase kinase and a 3′,5′-cyclic AMP-dependent protein kinase (28). These reactions may play some role in modulating the actin-myosin interaction in muscles having a thin filament regulatory system, particularly for fairly slow inotropic changes. In addition, the regulatory light chain of myosin for striated and smooth muscles is specifically and rapidly phosphorylated by myosin light chain kinase, an enzyme associated with myosin prepared by standard methods (6, 11, 12, 15, 28, 29, 31). A light chain phosphatase is also associated with myosin, giving the potential for a phosphorylation-mediated regulation (fig. 1, panel IV) (6, 11, 24, 28).

Two additional factors implicate this system in the control of vertebrate smooth muscles. The first is that myosin light chain kinase activity exhibits a Ca^{++}-dependence comparable to that seen for the contractile process (1, 6, 12, 15, 29, 31). The second point is that myosin from vertebrate smooth muscles is activated by actin only when phosphorylated (6, 15, 31). This contrasts with troponin-regulated vertebrate striated muscles which show no important dependence of their actomyosin ATPase activity on the phosphorylation of the myosin light chain (24, 28).

The very recent studies reviewed in this section provide biochemical evidence for the hypothesis that contraction might be initiated by a Ca^{++}-dependent phosphorylation of the regulatory light chain of myosin, and relaxation would follow dephosphorylation by a myosin light chain phosphatase. This hypothesis would be greatly strengthened by physiological evidence for a correlation between light chain phosphorylation and contraction. If the light chains are always phosphorylated *in vivo,* this mechanism would be unlikely to play a primary regulatory role. A preliminary report suggests there is an increase in phosphorylated light chains preceding contraction in uterine muscle (16).

4. Assessment

None of the hypotheses for regulation has met all of the following criteria for acceptance: (1) the relevant sites where Ca^{++} acts must be identified; (2) the affinities of these sites must be appropriate for Ca^{++} binding and removal over the physiological range of concentrations; (3) the actin-myosin interactions must be dependent on Ca^{++} binding or some intermediate Ca^{++}-dependent process, and (4) it must be shown that a particular regulatory mechanism operates in the living cell.

The evidence for a thin filament regulatory protein meets few of these criteria, and the possibility that thin filament regulatory protein preparations contain myosin light chain kinase (1, 15) or another cofactor – as in some non-muscle contractile systems (35) – must be considered, as the molecular weights are similar. However, thin filament and myosin-linked regulation can coexist (22), and a mechanism cannot be excluded on the basis that alternative Ca^{++} regulatory systems are present.

Biochemical evidence for the remaining two mechanisms is accumulating. However, if the findings reviewed above (sections 2 and 3) are confirmed, it implies that Ca^{++} serves two roles, to: (1) activate myosin light chain kinase with phosphorylation of the regulatory light chain, and (2) bind to a regulatory site on the same light chain. Evidence for such a dual role is seen in the Ca^{++} requirement for the actin-myosin interaction of phosphorylated myosin apparently free of kinase and phosphatase (6). However, this was not confirmed with proteins differing in source and preparative procedures (*Hartshorne and Górecka,* personal commun.).

The significance of different types of Ca^{++} regulation in various muscles is unknown, but obviously may be related to functional properties. It should be noted that biochemical studies have concentrated on chicken gizzard which may not be representative of all smooth muscles. The only certainty in this active field is that many answers will be forthcoming, and will probably be accompanied by the discovery of currently unrecognized aspects of the process.

Acknowledgments

The author appreciates the many constructive comments on drafts of this paper by Dr. *D.J. Hartshorne,* Carnegie-Mellon University, and Dr. *S.P. Driska,* University of Virginia; and acknowledges the research support provided by NIH Grants HL 14547 and HL 19242.

References

1 *Aksoy, M.O.; Williams, D.; Sharkey, E.M., and Hartshorne, D.J.:* A relationship between Ca^{2+} sensitivity and phosphorylation of gizzard actomyosin. Biochem. biophys. Res. Commun. *69:* 35–41 (1976).

2 *Borejdo, J. and Oplatka, A.:* Evidence for myosin-linked regulation in guinea pig taenia coli muscle. Pflügers Arch. *366:* 177–184 (1976).

3 *Bremel, R.D.:* Myosin linked calcium regulation in vertebrate smooth muscle. Nature, Lond. *252:* 404–407 (1974).

4 *Bremel, R.D.; Sobieszek, A., and Small, J.V.:* Regulation of actin-myosin interaction in vertebrate smooth muscle; in *Stephens* The biochemistry of smooth muscle, pp. 533–549 (University Park Press, Baltimore 1977).

5 *Carsten, M.E.:* Uterine smooth muscle: troponin. Archs Biochem. Biophys. *147:* 353–357 (1971).

6 *Chacko, S.; Conti, M.A., and Adelstein, R.S.:* The effect of phosphorylation of smooth muscle myosin on actin activation and Ca^{2+} regulation. Proc. natn. Acad. Sci. USA *74:* 129–133 (1977).

7 *Driska, S. and Hartshorne, D.J.:* The contractile proteins of smooth muscle. Properties and components of a Ca^{2+}-sensitive actomyosin from chicken gizzard. Archs Biochem. Biophys. *167:* 203–212 (1975).

8 *Ebashi, S.; Iwakura, H.; Nakajima, H.; Nakamura, R., and Ooi, Y.:* New structural proteins from dog heart and chicken gizzard. Biochem. Z. *345:* 201–211 (1966).

9 *Ebashi, S.; Toyo-Oka, T., and Nonomura, Y.:* Gizzard troponin. J. Biochem. *78:* 859–861 (1975).

10 *Ebashi, S.; Toyo-Oka, T., and Nonomura, Y.:* Gizzard troponin; in *Stephens* The biochemistry of smooth muscle, pp. 551–556 (University Park Press, Baltimore 1977).

11 *Frearson, N.; Focant, B.W.W., and Perry, S.V.:* Phosphorylation of a light chain component of myosin from smooth muscle. FEBS Lett. *63:* 27–32 (1976).

12 *Frearson, N.; Solaro, R.J., and Perry, S.V.:* Changes in the phosphorylation of P light chain of myosin in perfused rabbit heart. Nature, Lond. *264:* 801–802 (1976).

13 *Frederiksen, D.W.:* Myosin-mediated Ca^{++}-regulation of actomyosin-adenosinetriphosphatase from porcine aorta. Proc. natn. Acad. Sci. USA *73:* 2706–2710 (1976).

14 *Gergely, J.:* Troponin-tropomyosin-dependent regulation of muscle contraction by calcium; in *Goldman, Pollard and Rosenbaum* Cell motility, book A, pp. 137–149 (Cold Spring Harbor Laboratory, New York 1976).

15 *Górecka, A.; Aksoy, M.O., and Hartshorne, D.J.:* The effect of phosphorylation of gizzard myosin on actin activation. Biochem. biophys. Res. Commun. *71:* 325–331 (1976).

16 *Gualtieri, R.T. and Janis, R.A.:* Protein phosphorylation of spontaneously contracting rat myometrium. Fed. Proc. Fed. Am. Socs exp. Biol. *36:* 603 (1977).

17 *Hartshorne, D.J.; Abrams, L.; Aksoy, M.; Dabrowska, R.; Driska, S., and Sharkey, E.:* Molecular basis for the regulation of smooth muscle actomyosin; in *Stephens* The biochemistry of smooth muscle, pp. 513–532 (University Park Press, Baltimore 1977).

18 *Head, J.F.; Weeks, R.A., and Perry, S.V.:* Affinity-chromatographic isolation and some properties of troponin C from different muscle types. Biochem. J. *161:* 465–471 (1977).

19 *Ito, N.; Takagi, T., and Hotta, K.:* Regulatory protein of vascular smooth muscle. J. Biochem. *80:* 899–901 (1976).

20 *Kendrick-Jones, J.; Lehman, W., and Szent-Györgyi, A.G.:* Regulation in molluscan muscles. J. molec. Biol. *54:* 313–326 (1970).

21 *Kendrick-Jones, J.; Szentkiralyi, E.M., and Szent-Györgyi, A.G.:* Regulatory light chains in myosins. J. molec. Biol. *104:* 747–775 (1976).

22 *Lehman, W.; Kendrick-Jones, J., and Szent-Györgyi, A.G.:* Myosin-linked regulatory systems: comparative studies. Cold Spring Harb. Symp. quant. Biol. *37:* 319–330 (1973).

23 *Mikawa, T.; Toyo-Oka, T.; Nonomura, Y., and Ebashi, S.:* Essential factor of gizzard 'troponin' fraction. J. Biochem. *81:* 273–275 (1977).

24 *Morgan, M.; Perry, S.V., and Ottaway, J.:* Myosin light-chain phosphatase. Biochem. J. *157:* 687–697 (1976).

25 *Mrwa, U. and Rüegg, J.C.:* Myosin-linked calcium regulation in vascular smooth muscle. FEBS Lett. *60:* 81–84 (1975).

26 *Murphy, R.A.:* Structural proteins in the myofilaments and regulation of contraction in vertebrate smooth muscle. Fed. Proc. Fed. Am. Socs exp. Biol. *35:* 1302–1306 (1976).

27 *Murphy, R.A.; Herlihy, J.T., and Megerman, J.:* Force generating capacity and contractile protein content of arterial smooth muscle. J. gen. Physiol. *64:* 691–705 (1974).

28 *Perry, S.V.; Cole, H.A.; Morgan, M.; Moir, A.J.G., and Pires, E.:* Phosphorylation of the proteins of the myofibril; in *Biro* Proteins of contractile systems, pp. 163–176 (North-Holland/American Elsevier, New York 1975).

29 *Pires, E.; Perry, S.V., and Thomas, M.A.W.:* Myosin light-chain kinase, a new enzyme from striated muscle. FEBS Lett. *41:* 292–296 (1974).

30 *Russell, W.E.:* Insolubilization and activation of arterial actomyosin by bivalent cations. Eur. J. Biochem. *33:* 459–466 (1973).

31 *Sobieszek, A.:* Ca-linked phosphorylation of a light chain of vertebrate smooth-muscle myosin. Eur. J. Biochem. *73:* 477–483 (1977).

32 *Sobieszek, A. and Bremel, R.D.:* Preparation and properties of vertebrate smooth-muscle myofibrils and actomyosin. Eur. J. Biochem. *55:* 49–60 (1975).

33 *Sobieszek, A. and Small, J.V.:* Myosin-linked calcium regulation in vertebrate smooth muscle. J. molec. Biol. *102:* 75–92 (1976).

34 *Sparrow, M.P. and Bockxmeer, F.M. van:* Arterial tropomyosin and a relaxing protein fraction from vascular smooth muscle. J. Biochem. *72:* 1075–1080 (1972).

35 *Stossel, T.P. and Hartwig, J.H.:* Phagocytosis and the contractile proteins of pulmonary macrophages; in *Goldman, Pollard and Rosenbaum* Cell motility, book B, pp. 529–544 (Cold Spring Harbor Laboratory, New York 1976).

R.A. Murphy, PhD, Department of Physiology, University of Virginia School of Medicine, *Charlottesville, VA 22901* (USA)

Mechanisms of Vasodilatation. Satellite Symp., 27th Int. Congr. Physiol. Sci., Wilrijk 1977, pp. 37–47 (Karger, Basel 1978)

Lowering of Activator Calcium Concentration in the Relaxation of Vascular Smooth Muscle[1]

D.F. Bohr and R.C. Webb[2]

Department of Physiology; University of Michigan Medical School, Ann Arbor, Mich.

Physiological and pharmacological regulation of contraction or relaxation of vascular smooth muscle is determined by the concentration of activator calcium in the sarcoplasm. Contraction is initiated when this calcium concentration exceeds 10^{-7} *M* and reaches its maximum when the concentration exceeds 10^{-5} *M* (8). Under normal conditions *in situ,* vascular smooth muscle is partially contracted, a fact that can be interpreted as indicating that the concentration of activator calcium in these smooth muscle cells falls in a range between 10^{-7} and 10^{-5} *M.* Regulatory interventions that cause contraction do so by increasing the concentration of activator calcium whereas those that cause relaxation decrease the concentration of activator calcium. It is the purpose of this chapter to review evidence bearing on the mechanism by which the concentration of activator calcium is altered, with special emphasis on the mechanisms which cause a decrease in calcium concentration and hence relaxation.

Figure 1 depicts a model of the systems that are involved in this regulation. Two systems increase calcium concentration: *(1a) Passive influx:* Since the concentration of calcium outside of the cell is greater than 10^{-3} *M,* whereas that in the partially contracted smooth muscle cell is approximately 10^{-6} *M,* there is a 1,000-fold concentration gradient across the plasma membrane. It should also be noted that the inside of the cell is electronegative. Therefore, both the chemical and electrical gradients force calcium through special channels in this semipermeable plasma membrane to increase the concentration of activator calcium. *(1b) Release from internal sources:* Calcium is known to be sequestered in large amounts in both the sarcoplasmic reticulum and the mitochondria, and possibly also in the plasma membrane. Under the influence of appropriate

[1] Supported by a grant from the National Heart, Lung, and Blood Institute, HL-18575.

[2] Dr. *Webb* is a Post-Doctoral Research Fellow of the Michigan Heart Association.

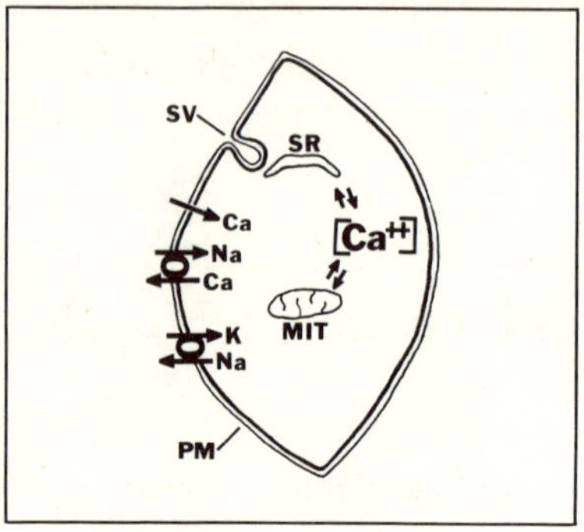

Fig. 1. Calcium movements that determine the concentration of activator calcium [Ca^{++}] in the vascular smooth muscle cell. It is increased by passive influx down its electrochemical gradient (1a in text) and by release from sequestered sites (1b). It is decreased by sequestration into intracellular sinks (2a) and by extrusion from the cell (2b). The external coat of the plasma membrane may be a significant regulator of calcium movement.

stimuli, this calcium is released into the sarcoplasm. Attention has recently been given to calcium bound to the plasma membrane. This structure consists of two layers, the outer glycoprotein and the inner phospholipid. The outer layer (perhaps its sialic acid component) may play an important role as a source or sink for activator calcium (9, 17).

Because of the enormous electrochemical gradient against which calcium must be moved in order to reduce the concentration of activator calcium and produce relaxation, this process requires energy. Two important systems are involved in removing calcium from the sarcoplasm: *(2a) sequestration into intracellular structures,* including both mitochondria and sarcoplasmic reticulum, and *(2b) efflux across the plasma membrane.*

This brief review deals primarily with the changes in rates of calcium movement effected by each of these four systems (1a, 1b, 2a, and 2b identified above) involved in vascular smooth muscle relaxation.

A word of caution is worth introducing. In much of the evidence bearing on the regulation of the concentration of activator calcium the degree of contraction of the vascular smooth muscle is used as an index of the concentration of activator calcium. This, of course, implies that this concentration is the primary, if not the only determinant of the degree of contraction. This is probably a safe assumption, but in each case the investigator must be concerned that such variables as magnesium concentration, pH, cooperative protein interaction, and phosphorylation of myosin or troponin might be interfering factors.

Straightforward observations may be made on isolated vascular smooth muscle to demonstrate that activator calcium may arise from either extracellular

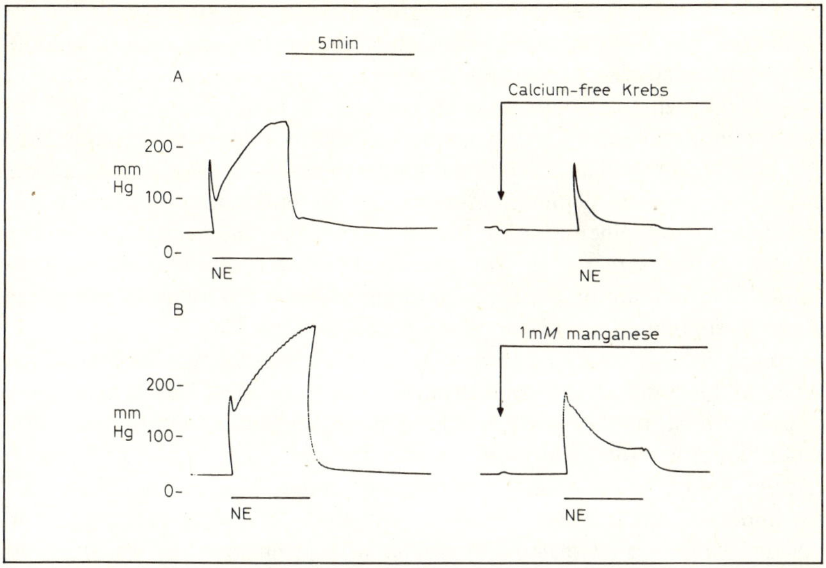

Fig. 2. Effect of a calcium-free environment and of manganese on the constrictor response of rabbit ear artery to norepinephrine (NE – 3 × 10^{-8} g/ml). Under control conditions the response to norepinephrine consists of an initial rapid response, a relaxation, and a continued slow constriction. In a calcium-free environment or in the presence of manganese the slow continued response is virtually eliminated but the initial rapid response is not significantly altered. From *Steinsland et al.* (24).

or intracellular sources. If vascular smooth muscle is placed in a calcium-free salt solution and treated with EGTA in order to remove all calcium available for activation, then treated by a depolarizing solution of potassium chloride, it will not contract. If calcium is now added to the muscle bath, the muscle will contract in a concentration-dependent fashion. Activator calcium initiating this response must have had its origin from the extracellular pool. By means of a second simple experiment, the intracellular origin of activator calcium can be demonstrated. Vascular smooth muscle placed in a calcium-free bath but not treated with EGTA will contract in a concentration-dependent fashion to the addition of norepinephrine or other constrictor agents to the bath. Since there was no calcium in the bath, this activation must have been caused by calcium of cellular origin.

In the isolated muscle bath, in contrast to its *in situ* environment, most vascular smooth muscle remains completely relaxed in a normal physiological salt solution. As in the two examples just cited, an increase in activator calcium concentration to a threshold level for contraction must be induced by some change in the normal bath composition. Some interventions such as a depolarizing solution of potassium chloride act by opening channels in the plasma membrane for an influx of extracellular calcium. Others, such as angiotensin (14), seem to employ primarily calcium of intracellular origin. Activation of the α-adrenergic receptors results in an increase in activator calcium both by an increase in plasma membrane permeability and by a delivery of calcium from its cellularly sequestered sites. This dual source of activator calcium is demonstrated by the experiment depicted in figure 2. An initial fast component of this response to norepinephrine is brought about by the release of sequestered calcium to the sarcoplasm. The subsequent slow component of the response results from calcium of extracellular origin as evidenced by its absence when the muscle is stimulated in a calcium-free 'Krebs' solution (24).

Even in this review, which is focused on mechanisms of relaxation and hence of lowering of the concentration of activator calcium, it is necessary to consider these two systems which increase calcium concentration since the actual concentration at any one time is the net result of rates of delivery of calcium into the sarcoplasm and rates of removal of calcium from it. The remainder of the review, therefore, will deal first with examples in which calcium delivery is decreased and second with mechanisms by which removal of activator calcium is accelerated.

1. Decrease in Delivery of Activator Calcium

(a) Passive influx. When vascular smooth muscle is partially contracted as it is *in situ,* any intervention which decreases the rate of delivery of calcium to the sarcoplasm should result in a decrease in the concentration of activator calcium and hence relaxation. The lower two tracings in figure 2 demonstrate that manganese eliminates the slow component of the norepinephrine response. Since this component of the response is recognized to be due to an influx of extracellular calcium, it may be inferred that manganese blocks the calcium channels in the plasma membrane. In figure 3 are presented the results of an experiment which confirms these inferences. A 15-min exposure of rabbit aortic strips to a concentration of 10^{-6} g/ml of norepinephrine caused a 52% increase in ^{45}Ca content of the strips. This increase was prevented if the solution contained, in addition to the norepinephrine, manganese in a concentration of 2.4 m*M.* Manganese alone resulted in a 21% decrease in ^{45}Ca content of the strip (19). Extensive studies over the past 8 years indicate that there is an important group of inorganic ions and organic compounds which are effective in blocking calcium channels (9, 12) and hence causing relaxation.

Recent studies have characterized the opening of calcium channels in the plasma membrane effected by norepinephrine. *Godfraind* (11) demonstrated that this agent increases, in a similar manner, both influx and efflux of calcium.

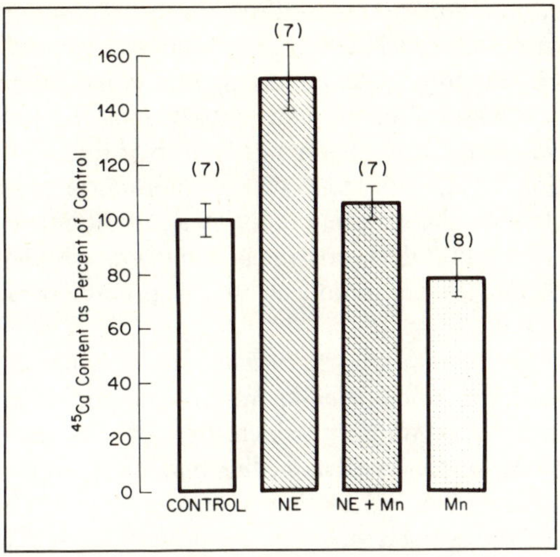

Fig. 3. ^{45}Ca content of strips of rabbit aorta exposed to this isotope for 15 min in physiological salt solutions containing norepinephrine (NE), NE plus Mn, and Mn. Contents are shown as mean ± SEM for the number of strips shown in parentheses. From *Keene et al.* (19).

This observation suggests that a channel across the plasma membrane is opened up by norepinephrine permitting an increase in passive diffusion of calcium in both directions. Since the concentrations of norepinephrine used are known to produce a graded depolarization of some types of vascular smooth muscle (13, 15), it is reasonable to question whether the increase in passive permeability of the membrane to calcium is secondary to this membrane depolarization. However, it seems equally sound to postulate that the change in membrane potential is secondary to changes in permeability to other ions and not required for the increase in concentration of activator calcium. The latter postulate is substantiated by the observation that constrictor and dilator agents are effective in causing contraction and relaxation of vascular smooth muscle when the cell is completely depolarized and when there is no change in membrane potential in response to these agonists (18).

Some biochemical characterization of the calcium channel has been reported by *Toda* (26). He observed the contractile response of rabbit aorta in a calcium-free potassium chloride-depolarizing solution when either calcium or barium was added. Cadmium in a concentration of 0.02–0.1 m*M* suppressed contractions caused by either of the other two bivalent cations. He concluded that cadmium suppressed these responses and those of other stimulating agents by interfering with the influx of bivalent cations. He also observed that the inhibitory effect of cadmium on aortic contractility could be reversed by cysteine and other thiol agents and concluded that 'membrane constituents containing sulfhydryl groups appear to play an important role in regulating the transmembrane flux of bivalent cations in vascular smooth muscle'. Related observations were made by *Needleman and Johnson* (27) in their finding that a tolerance to the relaxant action of glycerol trinitrate on rabbit aorta could be reversed by treating the muscle with the disulfide reducing agent dithiothreitol. This observation suggests that the sulfhydryl group is in some way essential for the reduction in activator calcium concentration by which glycerol trinitrate produces relaxation of vascular smooth muscle.

(b) Release of calcium from sarcoplasmic reticulum. Strong evidence has been presented for the conclusion that calcium leaves the sarcoplasmic reticulum by passive diffusion (6). Under conditions in which this internal calcium-sequestering system is stimulated to release calcium, this may occur by one of two mechanisms (6, 7): (1) calcium-induced calcium release in which a small increment in calcium outside of the sarcoplasmic reticulum causes the reticulum to release its internal calcium in a regenerative fashion, and (2) 'depolarization'-induced release of calcium. These two calcium release systems can be differentiated experimentally since 10 m*M* procaine inhibits the calcium-induced calcium release without affecting that produced by depolarization. On the other hand, a 40 m*M* concentration of sucrose or glucose inhibits 'depolarization'-induced release of calcium while having little effect on the calcium-induced release. It is probable that the physiological importance of these two mechanisms differs in different types of muscles. Inferences can be made from the results of two studies (5, 22) that calcium release from sequestered stores in smooth muscle may be 'calcium-induced'. It is possible that high concentrations of procaine may produce a relaxing effect on vascular smooth muscle by preventing release of calcium from sequestered stores. Ryanodine (24) is another agent that may act at this site.

2. Removal of Activator Calcium

(a) Calcium sequestration. It is probable that calcium is sequestered in at least four intracellular sites: sarcoplasmic reticulum, mitochondria, surface vesicles, and non-specific proteins that bind calcium. Two quite different approaches are currently being used to characterize the involvement of the various components of the vascular smooth muscle cell in the sequestration of calcium: (1) the study of subcellular fractions in vascular smooth muscle homogenates, and (2) the localization of calcium in ultrastructural studies.

In the past 5 years, extensive studies have been carried out on vascular smooth muscle indicating that large amounts of calcium may be sequestered in mitochondria (28), sarcoplasmic reticulum (10), and surface vesicles (30). The uptake in mitochondria appears to be somewhat greater than that in the microsomal fraction (sarcoplasmic reticulum and surface vesicles). From studies of a microsomal fraction, *Ford and Hess* (10) indicated that the microsomes held sufficient calcium 'to have the capability to be both sink and source for the activator calcium in excitation-contraction coupling'.

Insight into the molecular mechanism for sequestration of calcium has evolved from studies of sarcoplasmic reticulum from other tissues (21). Calcium appears to activate a 100,000-dalton ATPase; when the ATP is hydrolyzed the enzyme is phosphorylated and this leads to the delivery of calcium to the inside of the vesicle. Two molecules of calcium are transported per one ATP molecule hydrolyzed. The process of phosphorylation by a cyclic AMP-dependent protein kinase is thought to accelerate this calcium sequestration and hence facilitate relaxation (25).

Somlyo et al. (23) have presented extensive evidence demonstrating that barium and strontium are taken up in the mitochondria and sarcoplasmic reticulum *in situ.* These elements were studied initially because they are more electron opaque than calcium and were visualized as electron opaque deposits in these intracellular organelles. By use of electron probe X-ray microanalysis, these authors have confirmed that these deposits were indeed strontium and barium and that the deposit of barium in the mitochondria *in situ* is energy-dependent. Their X-ray microanalysis also provides the first direct evidence of calcium storage in intracellular organelles in vascular smooth muscle *in situ.*

(b) Calcium extrusion. Since, under physiological conditions, the extrusion of calcium from the cell into the extracellular space must be effected against a high electrochemical gradient, this process must be energy-requiring. Based on extensive studies by *Blaustein* (1) and others, a strong case has been established that an important source of this energy is the counterflow of sodium into the cell along its electrochemical gradient. In a tightly linked transport system, this downhill flow of sodium into the cell supplies energy that extrudes calcium from the cell. One characteristic of this counterflow exchange system seems to be that calcium and sodium compete on both sides of the membrane for the carrier. Three interventions that suppress the calcium extruding function of this system illustrate the operation of this exchange: (1) In a *sodium-free bath,* there is no longer an inwardly directed electrochemical gradient for sodium, hence this process cannot energize calcium extrusion. The concentration of activator calcium remains high and relaxation is impaired (3). (2) In the *Bowditch staircase phenomenon,* intracellular sodium accumulates with each membrane activation. The elevation in intracellular sodium decreases its transmembrane gradient and hence decreases the driving force for calcium extrusion. This leads to an accumulation of activator calcium and a facilitation of vascular smooth muscle contraction (20). (3) Cardiac glycoside *inhibition of* Na^{+}*–*K^{+} *ATPase*

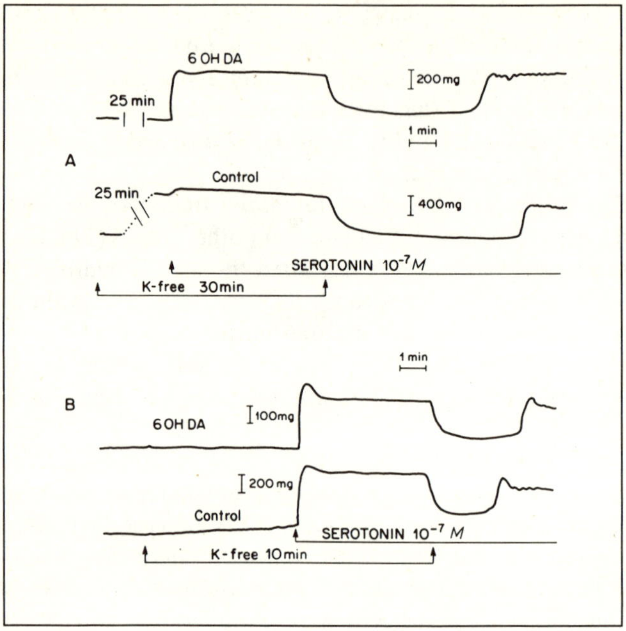

Fig. 4. The effect of duration of exposure to a potassium-free environment on the magnitude of potassium relaxation. The figure compares two studies in which potassium was added to the muscle bath at a time when the muscle had contracted maximally in response to serotonin in a potassium-free solution. In each study the innervated (control) and denervated (6-OHDA-pretreated) strips were mounted in the same bath. The magnitude of the potassium relaxation was greater in *A* in which the muscle had been deprived of potassium for 30 min than it was in *B* in which it had been deprived for only 10 min. Potassium relaxation was similar in the innervated and denervated strips. From *Bonaccorsi et al.* (4).

also leads to an intracellular accumulation of sodium. This intervention again results in both delayed relaxation and potentiation of constrictor responses (3).

The sodium-potassium ATPase is also inactivated in a potassium-free solution. This intervention again leads to sodium accumulation with similar effects on the contractile system. The magnitude of the effects will depend on the length of time that the sodium-potassium pump has been inactive and hence on the extent of accumulation of intracellular sodium. The Na^+–K^+ ATPase can be reactivated by returning potassium to the external environment. When this is done, the pump becomes hyperactive due to the high internal sodium concentration. Since this is an electrogenic pump, the plasma membrane is now hyperpolarized. The hyperpolarization results in relaxation of the contracted muscle (4, 29). Figure 4 indicates that the degree of relaxation depends on the duration

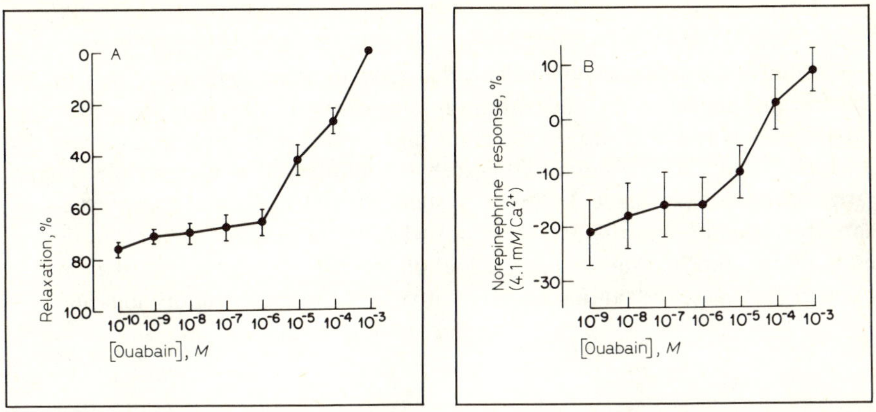

Fig. 5. Effects of ouabain on potassium-(A) and on calcium-(B) relaxation. Strips of rat tail artery were stimulated with norepinephrine 10^{-7} g/ml. 10 min prior to the addition of potassium (15 m*M*) or calcium (20.1 m*M*) ouabain was added to the bath. The shapes of the concentration-response curves for the inhibitory effect of ouabain on these two relaxation processes were similar.

of the inactivation of the sodium-potassium ATPase and hence on the extent of the accumulation of intracellular sodium. A relevant 'aside' of this observation deals with the mechanism by which this hyperpolarization causes the decrease in activator calcium responsible for relaxation. Either or both of the following mechanisms may be involved: (1) hyperpolarization stabilizes the membrane thereby decreasing the channels for free calcium diffusion into the cell; (2) hyperpolarization increases the transmembrane electrical gradient for sodium thereby increasing sodium influx which is linked to the extrusion of activator calcium.

Recent and as yet unreported studies in our laboratory have given insight into another intervention which may produce relaxation by altering the activity of Na^+–K^+ ATPase. It has been known that high concentrations of calcium in physiological salt solution will diminish the magnitude of contraction of vascular smooth muscle (2) and cause relaxation of already contracted vascular smooth muscle (16). We have recently observed that this relaxation is inhibited by ouabain (fig. 5) in concentrations that parallel the inhibition by ouabain of potassium-relaxation. This dependence of the relaxing effect of high concentrations of calcium on Na^+–K^+ ATPase suggests two possible mechanisms for calcium relaxation: (1) Calcium stimulates Na^+–K^+ ATPase to produce hyperpolarization. (2) Calcium decreases membrane conductance thereby augmenting the hyperpolarizing effect of a normal Na^+–K^+ ATPase electrogenic pump. We do not have evidence to differentiate between these two mechanisms.

3. Conclusion

From this brief review of systems that reduce the concentration of activator calcium and hence lead to relaxation, it is evident that the number of possible mechanisms is not very great. Questions that remain to be answered include the intricate molecular mechanisms of each calcium-handling system, the quantitative contribution made by each system in altering the concentration of activator calcium, and the definition of variability in the relative contribution of each of the mechanisms to the reduction of the concentration of activator calcium that causes relaxation of different types of vascular smooth muscle.

References

1 *Blaustein, M.P.:* Sodium-calcium exchange and the regulation of cell calcium in muscle fibers. Physiologist *19:* 525–540 (1976).

2 *Bohr, D.F.:* Vascular smooth muscle: dual effect of calcium. Science *139:* 597–599 (1963).

3 *Bohr, D.F.; Seidel, C., and Sobieski, J.:* Possible role of sodium-calcium pumps in tension development of vascular smooth muscle. Microvasc. Res. *1:* 335–343 (1969).

4 *Bonaccorsi, A.; Hermsmeyer, K.; Aprigliano, O.; Smith, C.B., and Bohr, D.F.:* Mechanism of potassium relaxation of arterial muscle. Blood Vessels *14:* 261–276 (1977).

5 *Cheng, J.T.:* 'Calcium-induced release of calcium' in rectal smooth muscle of mice. Jap. J. Pharmacol. *26:* 73–78 (1976).

6 *Endo, M. and Thorens, S.:* Mechanism of release of calcium from the sarcoplasmic reticulum; in Calcium transport in contraction and secretion, pp. 359–366 (North-Holland, Amsterdam 1975).

7 *Fabiato, A. and Fabiato, F.:* Calcium release from the sarcoplasmic reticulum. Circulation Res. *40:* 119–129 (1977).

8 *Filo, R.S.; Bohr, D.F., and Ruegg, J.C.:* Glycerinated skeletal and smooth muscle: calcium and magnesium dependence. Science *147:* 1581–1583 (1965).

9 *Fleckenstein, A.:* Specific pharmacology of calcium in myocardium, cardiac pacemakers, and vascular smooth muscle. Annu. Rev. Pharmacol. Toxicol. *17:* 149–166 (1977).

10 *Ford, G.D. and Hess, M.L.:* Calcium-accumulating properties of subcellular fractions of bovine vascular smooth muscle. Circulation Res. *37:* 580–587 (1975).

11 *Godfraind, T.:* Calcium exchange in vascular smooth muscle, action of noradrenaline and lanthanum. J. Physiol., Lond. *260:* 21–35 (1976).

12 *Godfraind, T. and Kaba, A.:* Blockade or reversal of the contraction induced by calcium and adrenaline in depolarized arterial smooth muscle. Br. J. Pharmacol. *36:* 549–560 (1969).

13 *Häusler, G.:* The relationship between norepinephrine-induced depolarization and contraction in vascular smooth muscle. Blood Vessels (in press).

14 *Healy, J.K. and Bohr, D.F.:* Angiotensin II responses of rabbit aortic strips compared with fast and slow epinephrine responses. Proc. Soc. exp. Biol. Med. (in press).

15 *Hermsmeyer, K.:* Contraction and membrane activation in several mammalian vascular muscles. Life Sci. *10:* 223–227 (1971).

16 *Holloway, E.T. and Bohr, D.F.:* Reactivity of vascular smooth muscle in hypertensive rats. Circulation Res. *33:* 678–685 (1973).
17 *Ishiyama, Y.; Yabu, H., and Miyazaki, E.:* Changes in contractility and calcium binding of guinea pig taenia coli by treatment with enzymes which hydrolyze sialic acid. Jap. J. Physiol. *25:* 719–732 (1975).
18 *Johansson, B.; Johsson, O.; Axelsson, J., et al.:* Electrical and mechanical characteristics of vascular smooth muscle response to norepinephrine and isoproterenol. Circulation Res. *21:* 619–633 (1967).
19 *Keene, J.J.; Seidel, C.L., and Bohr, D.F.:* Manganese on calcium flux and norepinephrine-induced tension in arterial smooth muscle. Proc. Soc. exp. Biol. Med. *139:* 1083–1085 (1972).
20 *Leonard, E.:* Alterations of contractile response of artery strip by a potassium-free solution, cardiac glycosides and changes in stimulation frequency. Am. J. Physiol. *189:* 185–190 (1957).
21 *Martonosi, A.N.:* The mechanism of calcium transport in sarcoplasmic reticulum; in Calcium transport in contraction and secretion, pp. 313–327 (North-Holland, Amsterdam 1975).
22 *Sigurdsson, S.B.; Uvelius, B., and Johansson, B.:* Relative contribution of superficially bound and extracellular calcium to activation of contraction in isolated rat portal vein. Acta physiol. scand. *95:* 263–269 (1975).
23 *Somlyo, A.P.; Somlyo, A.V.; Shuman, H., and Garfield, R.E.:* Calcium compartments in vascular smooth muscle: electron probe analysis; in Ionic actions on vascular smooth muscle with special regard to brain vessels, pp. 17–20 (Springer, Berlin 1976).
24 *Steinsland, O.S.; Furchgott, R.F., and Kirpekar, S.M.:* Biphasic vasoconstriction of the rabbit ear artery. Circulation Res. *32:* 49–58 (1973).
25 *Tada, M.; Kirchberger, M.A.; Li, H.-C., and Katz, A.M.:* Interrelationships between calcium and cyclic AMP in the mammalian heart; in Calcium transport in contraction and secretion, pp. 373–381 (North-Holland, Amsterdam 1975).
26 *Toda, N.:* Influence of cadmium ions on contractile response of isolated aortas to stimulatory agents. Am. J. Physiol. *225:* 350–355 (1973).
27 *Needleman, P. and Johnson, E.M., jr.:* Mechanism of tolerance development to organic nitrates. J. Pharmacol. exp. Ther. *184:* 709–715 (1973).
28 *Vallieres, J.; Scarpa, A., and Somlyo, A.P.:* Subcellular fractions of smooth muscle: isolation, substrate utilization and Ca^{++} transport by main pulmonary artery and mesenteric vein mitochondria. Archs Biochem. Biophys. *170:* 659–669 (1975).
29 *Webb, R.C. and Bohr, D.F.:* Potassium-induced relaxation as an indicator of Na^{+}–K^{+} ATPase activity in vascular smooth muscle. Blood Vessels (in press).
30 *Wei, J.-W.; Janis, R.A., and Daniel, E.E.:* Isolation and characterization of plasma membrane from rat mesenteric arteries. Blood Vessels *13:* 279–292 (1976).

D.F. Bohr, MD, Department of Physiology, University of Michigan Medical School, *Ann Arbor, MI 48109* (USA)

Mechanisms of Vasodilatation. Satellite Symp., 27th Int. Congr. Physiol. Sci., Wilrijk 1977, pp. 48–55 (Karger, Basel 1978)

Mechanics of Smooth Muscle Relaxation[1]

Börje Johansson

Department of Physiology and Biophysics, University of Lund, Lund

Vasoconstriction and vasodilatation can be due to changes in the chemo-mechanical energy transduction in the vascular smooth muscle or they can be passive results of altered transmural pressure. Active vasoconstriction is obviously caused by shortening of the contracting smooth muscle cells, but the response is seldom *isotonic* since the load on the muscle can change in the course of the event. For instance, if the shortening is associated with significant shift of force from parallel elastic structures to the contractile cells, the muscle response actually becomes *auxotonic*. Conversely, the decrease in vascular caliber may lead to gradual reduction in the load on the muscle due to the law of *Laplace*. A proper term for this kind of shortening would be *meiotonic* contraction. Passive narrowing of the vessels during a fall in blood pressure can of course involve an element of smooth muscle response since any active tone prevailing in the vascular wall will temporarily change its external manifestation from production of isometric force to some form of shortening. Lengthening of the smooth muscle during vasodilatation due to inhibition of contractile activity or increase in transmural pressure may be discussed in an analogous manner. Again, the load on the smooth muscle in the intact vessel can vary during the response.

1. Characteristics of Vascular Smooth Muscle Relaxation

In view of these complexities in intact cylindrical vessels, we shall at the present stage limit our discussion of the mechanics of relaxation to the simpler linear system of isolated vascular smooth muscle strips, mounted at low preload so that parallel elasticity can be neglected. Relaxation under these conditions appears as a decrease in isometric force or as isotonic lengthening. Its time

[1] Supported by grants from the Swedish Medical Research Council (14x-28), the Medical Faculty of the University of Lund, and AB Hässle, Göteborg.

course will be determined by a number of consecutive processes. Rate of removal of a stimulating agent or rate of appearance of an inhibitor at the receptor sites is the obvious primary determinant of the change in force or length. It is likely, however, that the course of relaxation may be influenced also by processes which couple the membrane receptors to the contractile system. The rate of relaxation is sometimes used in pharmacological studies as an indicator of the rate of agonist removal from the receptor sites. The rationale behind this procedure must be that the coupling processes are not engaged as rate limiting steps, a kind of assumption which may not always be justified.

2. Dissociation between Cell Membrane and Intracellular Mechanisms

Figure 1 illustrates an experiment in which a change in relaxation rate appeared to be due to an effect on intracellular mechanisms rather than on a change in membrane response. A portal vein preparation examined by the sucrose-gap method was first exposed to a Ca-free standard Krebs' solution

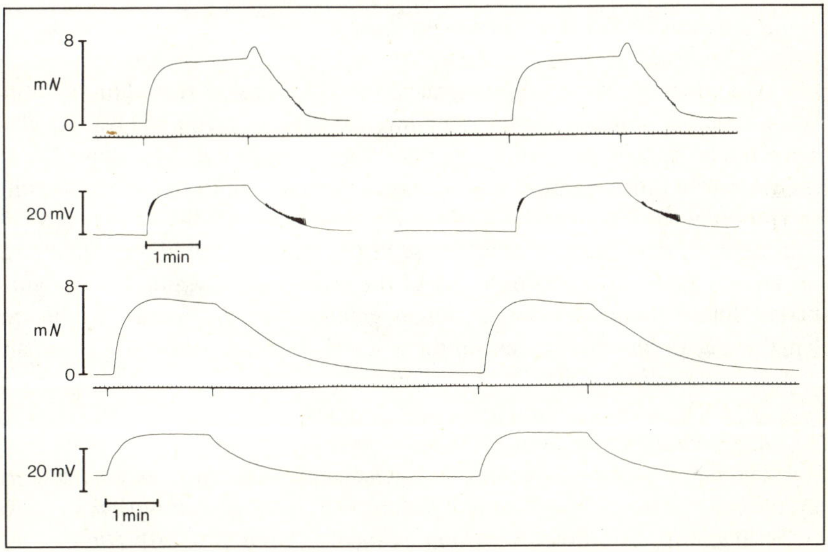

Fig. 1. Isometric active force (upper tracing) and electrical membrane response (lower tracing) of rat portal vein studied by the sucrose-gap method. Spontaneous activity abolished in Ca^{++}-free Krebs' solution. Upper part of the figure shows two responses at 37 °C to administration of high-K^{+} Krebs' with 2.5 m*M* Ca^{++} followed after 2 min by return to Ca^{++}-free standard Krebs'. Lower part shows responses to the same procedures at 13 °C. Note that rate of repolarization was not very different at the two temperatures while rate of relaxation was slowed down appreciably by the cooling.

which abolished the spontaneous electrical and mechanical activity. Switching to high K^+ solution with 2.5 m*M* Ca^{++} at 37 °C caused depolarization and development of a sustained contracture. After 2 min, the Ca-free standard medium was readministered so that the muscle repolarized and relaxed. These electrical and mechanical responses in the upper part of figure 1 seem to run in parallel, although the time courses are difficult to compare quantitatively due to the temporary occurrence of spikes and twitches. When the same stimulus and removal of stimulus were repeated at 13 °C (lower part of fig. 1) the electrical and mechanical responses were differently affected. The rate of repolarization as indicated by the electrical recording was not markedly changed by the decrease in temperature but relaxation was very much slower than at 37 °C. Figure 1 is not intended to give any precise information on the relation between rate of repolarization and rate of relaxation in the smooth muscle but these simple recordings may illustrate my point that changes in relaxation can occur as a result of predominant variations in intracellular processes. It is appropriate therefore to consider the mechanical principles by which intracellular events can be engaged as determinants of the rate of relaxation.

3. Mechanical Principles of Intracellular Events

According to *Hill*'s (3, 4) classical muscle model, a recording of isometric force does not indicate directly the level of activity in the contractile element since the latter will shorten or lengthen during rise or fall in tension due to the presence of a series-coupled passive elastic element. On the basis of this concept, relaxation would ultimately depend on the decay of the *active state* in the contractile system but the time course of the recorded fall in tension would also depend on the load-extension curve of the series elasticity and the configuration of the force-velocity relation at forces greater than P_o. Similarly, the rate of lengthening in an isotonic relaxation would be influenced by the external load and the *force-velocity* characteristics above P_o.

A. Decay of Active State

The decay of the active state during isometric relaxation may be determined by the controlled release method of *Ritchie* (12) or by measuring shortening velocity against a low afterload following isotonic releases (2, 5, 6). *Hill* (4) defined the intensity of the active state as the force that the contractile element bears when not changing its length. *Ritchie*'s method determines such force values directly while the isotonic release method gives a 'velocity-active state' which may or may not have the same time course as the 'force-active state' (cf. 8).

Figure 2 exemplifies the use of *Ritchie*'s method to study the decay of the active state in uniform phasic contractions of rat portal vein induced by ionic

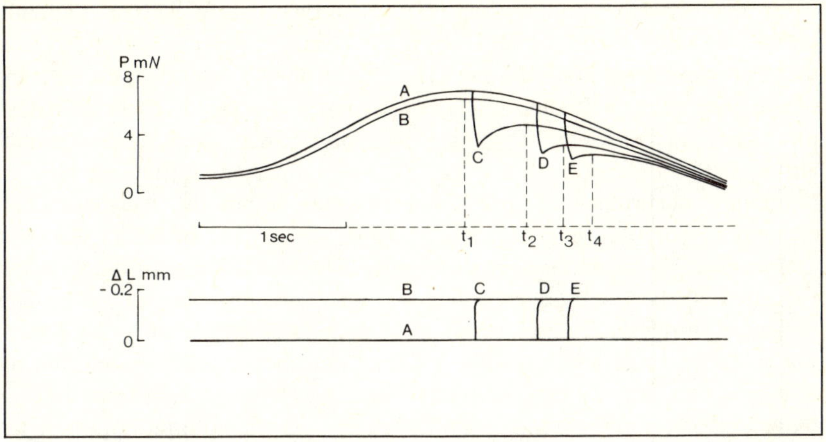

Fig. 2. Superimposed tracings of force and length during 5 contractions of isolated rat portal vein in a solution which made the spontaneous phasic responses regular and uniform (6). Curves A and B show full isometric responses obtained at about 5.0 and 4.8 mm muscle length, respectively. In C–E the muscle responded to a sudden release from the longer to the shorter length with an immediate drop in force followed by force recovery. The maximal force in curve B and in the recovery curves represent the active state intensity at times $t_1 - t_4$ for contraction at the shorter length.

changes in the extracellular fluid (6). The controlled release involves a step change in length which immediately reduces the series elastic force. If this drop is large enough, the contractile element, due to persisting chemomechanical transduction (or active state), will switch from lengthening to shortening so that redevelopment of force occurs. The peak value of the force recovery is a measure of the active state at the respective point in time as it would represent the P_0 value of the instantaneous force-velocity curve. The top values of force recovery after the releases (C, D, and E) in figure 2 thus give 3 points (at times t_2, t_3, and t_4) on the falling part of the active state curve. These points are all well below the corresponding force values recorded during the isometric relaxation (A) and this is true also if they are compared with the isometric response obtained from the muscle at the short length to which it was released (curve B in fig. 2).

The validity of *Ritchie*'s method for measuring active state has been questioned as the results have been shown to depend on the size of the length steps (1). Comparable objections may be raised against any of the methods designed to determine active state intensity. The mere concept of the active state has actually been under attack recently since work on skeletal muscle suggests that the time course of the isometric force does not so much reflect an influence of compliance in the series elastic element, but is rather a direct effect of the rate constants for cross-bridge formation and detachment (13). For vascular smooth

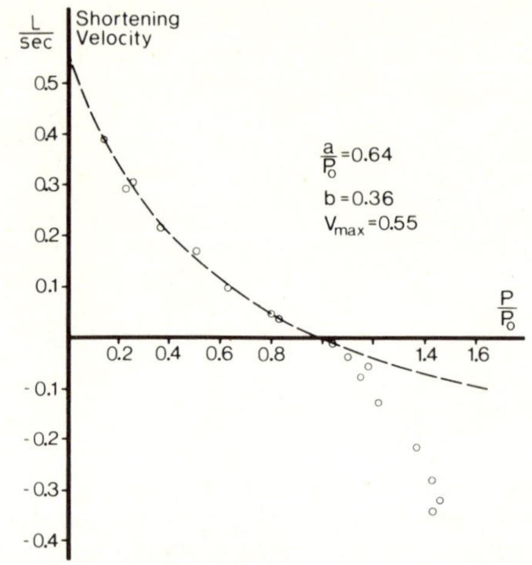

Fig. 3. Force-velocity data of portal vein obtained by isotonic quick release and quick stretch at the plateau of contractions evoked by AC stimulation. Shortening velocities against different loads $< P_o$ fitted to *Hill's* equation as indicated by broken line and values for a/P_o, b and V_{max}. From *Johansson et al.* (7).

muscles, which usually show much greater extension of series elasticity than striated muscle (10), it appears useful to retain the concept of active state even though its measurability in absolute terms may be questionable.

B. Force-Velocity Curve

The difference between the active state and the isometric force during relaxation in figure 2 indicates that the contractile element, when it lengthens, can carry a greater force than it would actually develop if it were itself under truly isometric conditions. Understanding of the mechanics of relaxation therefore requires that we consider the configuration of the force-velocity curve. At forces below P_o, the force-velocity relations of both striated and smooth muscles can be described by *Hill's* (3) hyperbolic equation. The rates at which a muscle or its contractile element *lengthens* under the influence of a load *greater* than the isometric force have been studied less extensively. In skeletal muscle, the force-velocity relation at loads $> P_o$ appears to deviate towards the force axis in comparison with *Hill's* hyperbola (9). This implies that the muscle lengthens very slowly and is therefore quite effective in sustaining the increased load.

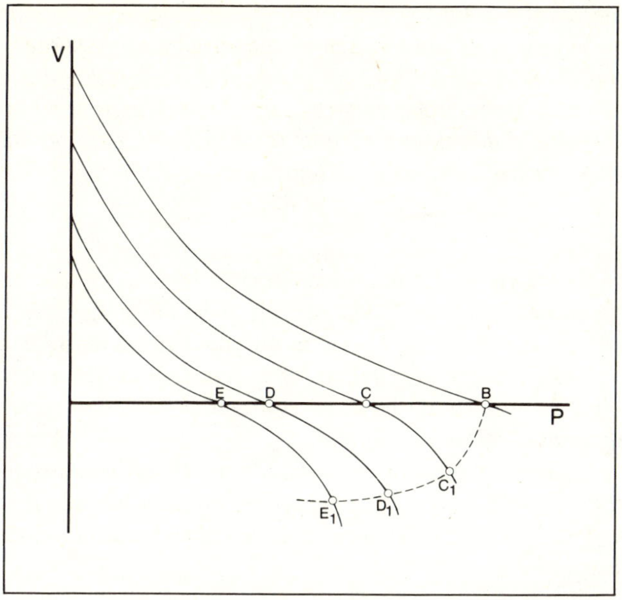

Fig. 4. Tentative diagram illustrating the mechanics of relaxation for the muscle in figure 2. Full lines indicate possible instantaneous force-velocity curves, corresponding to the active state levels obtained as maximum of isometric force or force recovery at times $t_1 - t_4$. These active state values (instantaneous P_0 values) are represented by points B, C, D, E on the force axis. Points B, C_1, D_1 and E_1 show, proportionately, the force recorded during isometric relaxation at the short length (curve B in fig. 2) and the possible relative rates of lengthening of the contractile element at $t_1 - t_4$. The broken line is the trajectory of this part of the isometric relaxation in the force-velocity dimensions.

However, should the load become very large, $>1.6 \times P_0$, there is quite an abrupt yield, or 'give', characterized by rapid lengthening of the skeletal muscle (9).

We have studied recently the force-velocity relation of smooth muscle from rat portal vein and rabbit urinary bladder including also measurements of the rates of lengthening at loads above isometric force (7). The data were obtained by isotonic quick-release and quick-stretch experiments. A detailed description and analysis of the recordings has been published elsewhere (7). Figure 3 shows force-velocity data obtained in a portal vein during the plateau of contractions evoked by AC stimulation (11). The values for contractile element shortening at different loads $< P_0$ could be fitted to *Hill*'s equation as indicated in figure 3. The length responses to loads greater than the isometric force showed a clear-cut series elastic component, but the subsequent isotonic lengthening was more difficult to characterize as its velocity decreased markedly with time, particularly at the highest loads. All the data in figure 3 including those above P_0 were taken at 100 msec after the change in load. Even if the absolute values of the rates of lengthening must be judged with caution due to the curvilinear responses, it is quite clear that the contractile element of the portal vein while it lengthens can carry a greater load than it develops isometrically. In comparison with the extrapolated hyperbola of *Hill*'s equation, the force-velocity relation of the smooth muscle above P_o deviates towards the velocity axis.

C. Isometric Relaxation

On the basis of this description of the force-velocity relation, we may return to the isometric relaxation in figure 2 and use the diagram in figure 4 for a discussion of its mechanics. Points B, C, D, E on the force axis in figure 4 represent the active state values (P_o values of instantaneous force-velocity curves) at times t_1–t_4. They illustrate how the contractile element is gradually inactivated over this period of time. Points B, C_1, D_1, E_1 in figure 4 give the isometric force at t_1–t_4 for the continuous isometric relaxation recorded at the shorter muscle length (curve B in fig. 2). This set of points also indicates the relative rates at which the contractile element may lengthen under the force of the recoiling series elastic element.

4. Conclusion

Four different relations are involved in the isometric relaxation: (1) isometric force as a function of time; (2) active state (or P_o) as a function of time; (3) force in the series elastic element as a function of length; (4) lengthening velocity of contractile element as a function of force above P_o. It should be evident from the foregoing that the mechanics of relaxation can be accounted for on the basis of the classical models only if experimental results described by these four relationships are all compatible with one another. We do not have sufficiently complete data on one and the same preparation to test whether such consistency prevails. It is a challenge for further work on smooth muscle to try to collect such complete information. Only then can we decide whether the classical muscle models can account for the mechanics of smooth muscle relaxation or whether other models must be introduced.

References

1 *Briden, K.L. and Alpert, N.R.:* Effect of shortening on the time-course of active state decay. J. gen. Physiol. *60:* 202–220 (1972).

2 *Edman, K.A.P. and Nilsson, E.:* Mechanical parameters of myocardial contraction studied at a constant length of the contractile element. Acta physiol. scand. *72:* 205–219 (1968).

3 *Hill, A.V.:* Heat of shortening and the dynamic constants of muscle. Proc. R. Soc. Lond. (Biol.) *126:* 136–195 (1938).

4 *Hill, A.V.:* Abrupt transition from rest to activity in muscle. Proc. R. Soc. Lond. (Biol.) *136:* 399–420 (1949).

5 *Jewell, B.R. and Wilkie, D.R.:* Analysis of the mechanical components in frog's striated muscle. J. Physiol., Lond. *143:* 515–540 (1958).

6 *Johansson, B.:* Active state in the smooth muscle of the rat portal vein in relation to electrical activity and isometric force. Circulation Res. *32:* 246–258 (1973).

7 *Johansson, B.; Hellstrand, P., and Uvelius, B.:* Responses of smooth muscle to quick load change studied at high time resolution. Blood Vessels (in press).

8 *Julian, F.J. and Moss, R.R.:* The concept of active state in striated muscle. Circulation Res. *38:* 53–59 (1976).

9 *Katz, B.:* The relation between force and speed in muscular contraction. J. Physiol., Lond. *96:* 45–64 (1939).

10 *Murphy, R.A.:* Contractile system function in mammalian smooth muscle. Blood Vessels *13:* 1–23 (1976).

11 *Peiper, U.; Laven, R., and Ehl, M.:* Force velocity relationships in vascular smooth muscle. Pflügers Arch. *356:* 33–45 (1975).

12 *Ritchie, J.M.:* Effect of nitrate on the active state of muscle. J. Physiol., Lond. *126:* 155–168 (1954).

13 *Simmons, R.M. and Jewell, B.R.:* Mechanics and models of muscular contraction; in *Linden* Recent advances in physiology, vol. 9, pp. 87–147 (Churchill/Livingstone, London 1974).

B. Johansson, MD, Department of Physiology and Biophysics, Sölvegatan 19,
S–223 62 Lund (Sweden)

Mechanisms of Vasodilatation. Satellite Symp., 27th Int. Congr. Physiol. Sci., Wilrijk 1977, pp. 56–72 (Karger, Basel 1978)

Membrane Potential and Relaxation in Vascular Smooth Muscle

G. Siegel, R. Ehehalt and H.P. Koepchen

Institute of Physiology, Biophysical Research Group, The Free University of Berlin, Berlin

Quantitative data on the relation between membrane potential and force development in smooth muscle, especially vascular smooth muscle, are scarce in the literature. One reason for this could be that even small potential changes in vascular muscle of the graded depolarization type could lead to considerable changes in mechanical tension (69, 71, 73), since at rest the cell membrane of this type of smooth muscle is relatively depolarized and approximates the threshold needed to induce mechanical activity (69). However, the validity of intracellular recordings of small potential changes in vascular smooth muscle cells is still doubtful due to the considerable technical difficulties involved in the measurement. Therefore, it could be easier to demonstrate a quantitative relation between membrane potential and mechanical force in smooth muscle cells in which contractions are triggered by spike potentials (52). The present discussion attempts, taking these problems into consideration, to explain some recent electrophysiological findings which link changes in membrane potential to the relaxation of vascular smooth muscle.

1. Electromechanical Coupling

If the concept of excitation-contraction and inhibition-relaxation coupling in vascular smooth muscle is valid, the obvious question then is in which conditions does the membrane of smooth muscle cells hyperpolarize and the cells relax as a consequence. It is established that autonomic neurotransmitters (5, 16, 17, 42, 74, 76, 79), ions (1, 7–9, 12, 13, 18, 19, 31, 49, 56, 64, 66, 67, 69, 71, 73, 74, 76, 81, 84, 85), hormones (3, 18, 31, 45, 47, 51, 56, 63, 64, 74, 76, 81), environmental (6, 26, 45, 54, 60, 74, 76) and cell metabolism (2, 11, 20, 24, 33, 66, 68, 69, 71, 73, 74, 76, 82) influence membrane potential and thus vascular tone. These factors can be referred to as effector influences on the instantaneous membrane potential (73). In the following discussion, we shall give an example of the influence of each of the effectors. We must mention here that the efficacy of these effectors to alter cellular activity in an oscillatory system such as the vascular smooth muscle is phase-dependent.

In addition to the classical concepts of *excitation-contraction* or *inhibition-relaxation coupling,* the work of *Somlyo* (74–76), *Casteels and Droogmans* (21) and others (5, 41, 47, 51, 80) has established that changes in mechanical activity of vascular smooth muscle can be induced without alteration in membrane potential through so-called *pharmacomechanical coupling* (74). Thus, *Bell* (5) has shown that polarized guinea-pig uterine arteries are relaxed by acetylcholine in the absence of detectable hyperpolarization, and *Casteels and Droogmans* (21) have given evidence that in the rabbit ear artery noradrenaline in a concentration range of 10^{-9} to 10^{-6} *M* causes a dose-dependent contraction without effect on the resting potential of the cells (cf. 47). Those are only two examples, and they are still controversial since other authors find that membrane depolarization always occurs in the presence of catecholamines (3, 28, 45, 63, 64, 74).

The picture is also complicated by the fact that at least two effector influences may interact. For example, it is known from the work of *Lorenz and Vanhoutte* (50) and *Paton* (55) that K ions influence the release of catecholamines from the sympathetic nerve terminals (cf. 31). A hyperpolarization of the cell membrane is possible with an increase in the K^+ permeability, decrease in the Na^+ permeability, an activation of an electrogenic Na^+ outward pump, an inhibition of an electrogenic K^+ inward pump or any combination of these actions; thus, cellular metabolism can influence the membrane potential directly. For more details, the reader is referred to the excellent review articles by *Somlyo and Somlyo* (74, 76) and *Speden* (80).

Figure 1 illustrates experiments supporting the theory of electromechanical coupling in vascular smooth muscle. It shows the already classical recordings by *Funaki* (28) (parts A and B) as well as by *Johansson* and his co-workers (3) (part C) in the portal vein. Parts A and C demonstrate beautifully that spontaneous spike discharges based on slow waves are accompanied by phasic contractions. Part B is interesting because here the slow wave activity without spikes was recorded at a higher temperature (39 °C). These graded depolarizations were also followed by a contraction but with a reduced amplitude, whereby the tension curve takes the form of a sine wave. In this connection, it should be mentioned that *Mekata and Niu*'s (51) experiments on the carotid artery showed that changes in the vascular tone are synchronous with the recorded small-amplitude oscillations of the membrane potential, even during the exposure to adrenaline (cf. 51). These authors write that with adrenaline (10^{-8} g/ml) depolarizations up to 3 mV occur, their average is only 0.3 mV. The force developed was approximately 0.1 p. Figure 2 shows the mechanical force developed in a dog carotid artery as a function of the membrane potential (71). The

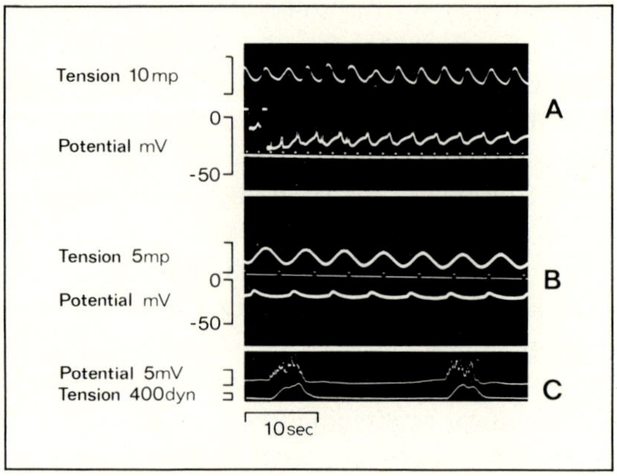

Fig. 1. A Spontaneous spike discharges based on slow waves in the portal vein of the rat at temperature of 36 °C, followed by phasic contractions (time marks = 0.5 sec). *B* Repeated slow potentials without spikes at higher temperature (39 °C), also accompanied by contractions but of reduced amplitude. Time marks = 1 sec. *A* and *B* from *Funaki* (28). *C* Spontaneous pattern of the activity in the portal vein of the rat consisting of bursts of spikes at intervals, closely correlated to development of tension. From *Axelsson et al.* (3).

variable here was the extracellular K^+ concentration. The tension was only slightly reduced at high external K^+ concentrations after α-receptor blockade (cf. 50, 55). In the linear part of the curve, a potential change of only 4.4 mV causes a change of 1 p in the force developed (71, 73). Thus, for a force development of 0.1 p, a depolarization of only 0.4 mV is required, which agrees well with *Mekata and Niu*'s (51) findings for the addition of adrenaline in the same potential range and in the same vessel. A similar relation between contraction and membrane potential was found by *Mironneau* in voltage-clamped uterine (52) and vascular smooth muscle (53).

2. *Electrical and Mechanical Responses to Ions*

Under which conditions does the membrane of vascular smooth muscle cells hyperpolarize, causing the cells to relax (33)? Here we would like to mention only the effects of the ions H^+, K^+, Mg^{++} and Ca^{++}. If one varies $[K^+]_o$ in the range 0–100 m*M* in the carotid artery (66, 67, 69, 71, 73), then the membrane potential changes relatively little (cf. 12, 13, 33, 74). The maximal hyperpolarization is obtained around 2.5 m*M* $[K^+]_o$. Other external K^+ concentrations lead to a depolarization. However, we have observed a hyperpolarization of about 2.6 ± 0.7 mV in 45% of the transitions from 4.7 to 11.7 m*M* $[K^+]_o$, and in a few cases a hyperpolarization of about 2.3 ± 0.8 mV during the transition to

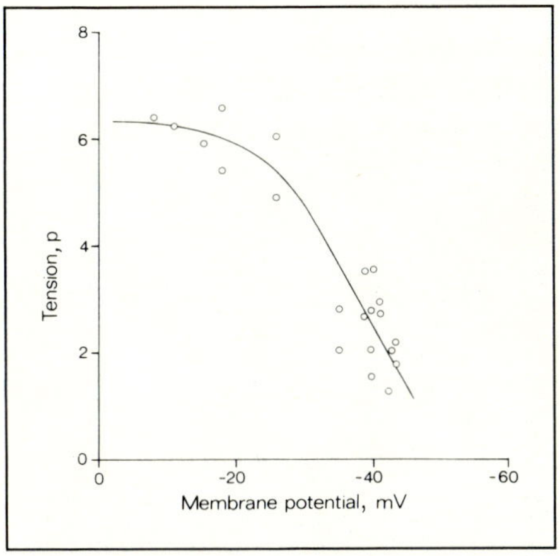

Fig. 2. The dependency of mechanical tension on the membrane potential in isolated carotid segments. The membrane potential and mechanical force were altered by variations of the extracellular K^+ concentration. The steepness of the curve can be taken as a measure of the degree of electromechanical coupling.

23.5 m*M* $[K^+]_o$ (67, 71). Besides a possible stimulation of an electrogenic Na^+ pump (12, 13), our results indicate that these opposing effects depend on the instantaneous value of the membrane potential when the hyperkalemic solution is introduced during a slow wave (71). This underscores the phase dependency of the effector action in this self-oscillating system. We will see later from figure 7 that the reaction of a vessel, whether it be dilatation or constriction, depends on the time of the intervention in the phase of its physiological membrane oscillation rhythm.

In contrast to the membrane potential, the tension changes by a factor of 1:5 during variations of $[K^+]_o$ in the same range 0–100 m*M* (71, 73). The mechanical force increases both above and below 6 m*M* $[K^+]_o$, which means that the same tension can appear at different external K^+ concentrations (cf. 13, 19, 31, 33, 49, 71, 73, 74). This is true for the stationary tension course. Maximum hyperpolarization and relaxation were registered at different external K^+ concentrations because the curve for mechanical force development was displaced to ever higher, nonstationary $[K^+]_o$ values as the equilibration time of specimens in the various Krebs' solutions continuously increased. The tension reaches equilibrium after 1.5–2 h in the various stages of external K^+ concentration. Then the maximum relaxation was found at 6 m*M* $[K^+]_o$ (71, 73). In contrast, the membrane potentials were measured during stable intracellular penetrations about 10–20 min after the application of each of the test solutions with differing external K^+ concentrations.

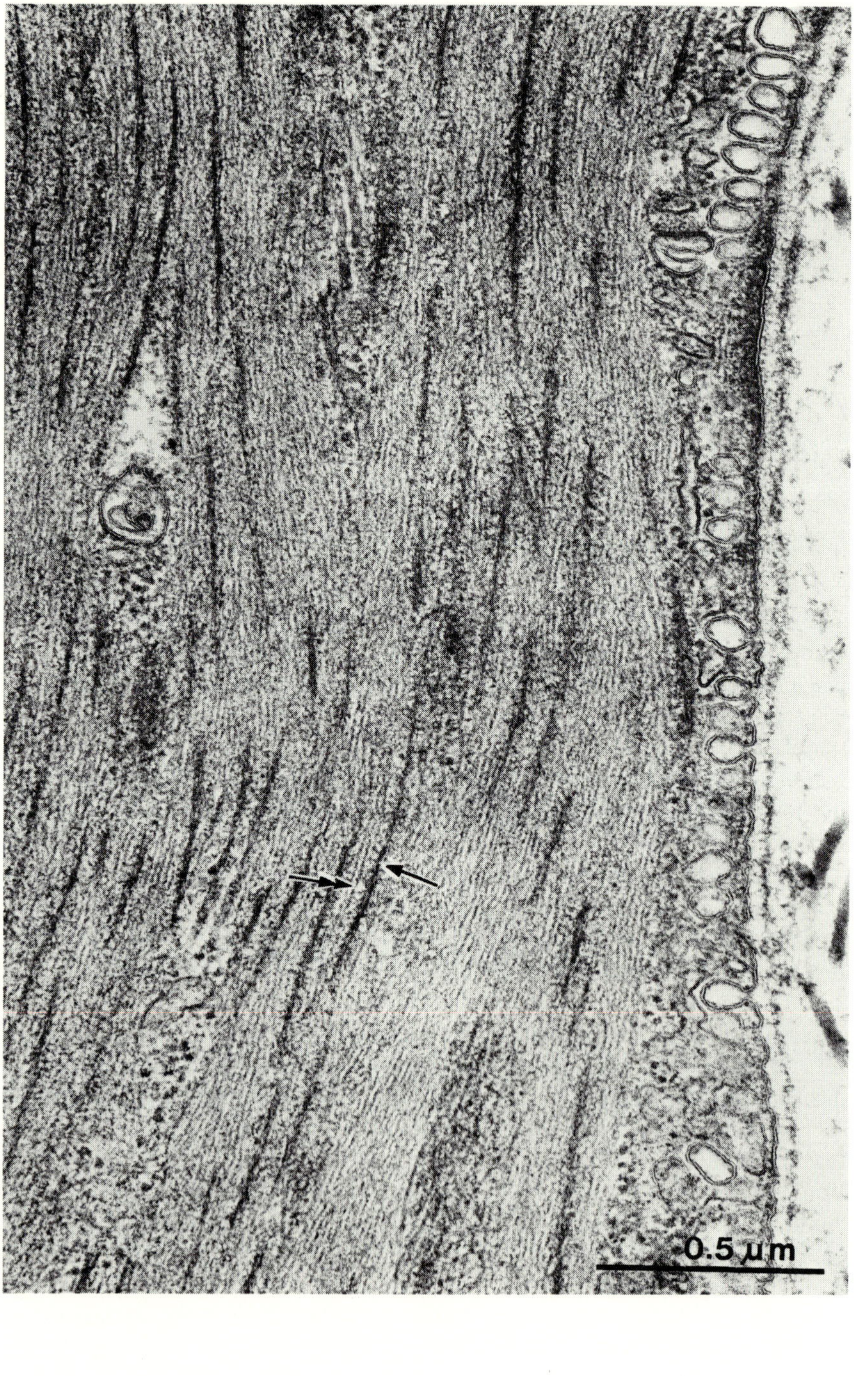
0.5 µm

To test the coupling of membrane potential, which is the integrated result of the various effector influences, to the mechanical force development, we changed the pH value of the Krebs' solution. A displacement of the pH in the range 6.8–7.8 causes a linear change in the membrane potential (66, 69, 71, 73). Alkalosis causes a depolarization and contraction, and acidosis results in hyperpolarization and dilatation (cf. 1, 7, 18, 56, 84). Equal hyperpolarizations produced by changes in external H^+ *or* K^+ concentrations cause approximately equal relaxations. If we compare the tension change as a function of the alteration of membrane potential during pH variation to the curve for electromechanical coupling found by $[K^+]_o$ changes, this curve is confirmed.

The change in the membrane potential with pH variations could be due to a conformational change in the membrane phase or a change in ion concentrations in the immediate neighborhood of the membrane. *Somlyo et al.* (74, 77) have pointed out that increased or decreased cation concentrations in the narrow tissue clefts between the basement membrane and the cell membrane could be important. Figure 3 shows a longitudinal section of a smooth muscle cell from the rabbit portal-anterior mesenteric vein (78). The cell membrane together with the basement membrane, and its filamentous projections, surround small pockets or narrow tissue clefts of extracellular space. In electron micrographs of intestinal (29, 30) and vascular smooth muscles (78), the outer aspect of the cell membrane and the inner aspect of the basal lamina are separated by an electron-lucent gap of about 8–15 nm. We studied the binding characteristics of mono- and divalent cations in isolated adventitial connective tissue as a function of the pH value. Figure 4 shows the K binding. Starting at pH = 4, the K binding increases with increasing pH. It attains a maximum of 25 mmole/kg dry weight at pH = 6.8 and decreases with basic pH. This decrease of the K binding with increasing pH is an extremely unusual result for a pH-dependent cation-binding curve, which in principle is similar to a pH titration curve. An explanation for this behavior is possible when one considers the strong increase in Ca binding in the basic pH range (cf. fig. 5). Clearly, the K and Ca ions compete for the same binding sites on connective tissue structures (cf. 59). Because of the higher affinity of Ca ions, the K ions are expelled from the binding sites. The actual K-binding curve shows that with a displacement of the pH from 7.4 towards acidity, the K ions are bound to connective tissue; with a displacement towards basicity they are released. The physiological meaning of this mechanism could be

Fig. 3. Longitudinal section of a smooth muscle cell from rabbit portal-anterior mesenteric vein showing thick (arrow) and thin (double arrow) filaments. Cross-bridges can be seen on the thick filaments. Note the prominent basement membrane with filamentous projections. Specimen fixed in glutaraldehyde, post-fixed in osmium followed by tannic acid and subsequently block-stained with uranyl acetate. From *Somlyo* (78) – unpublished micrograph.

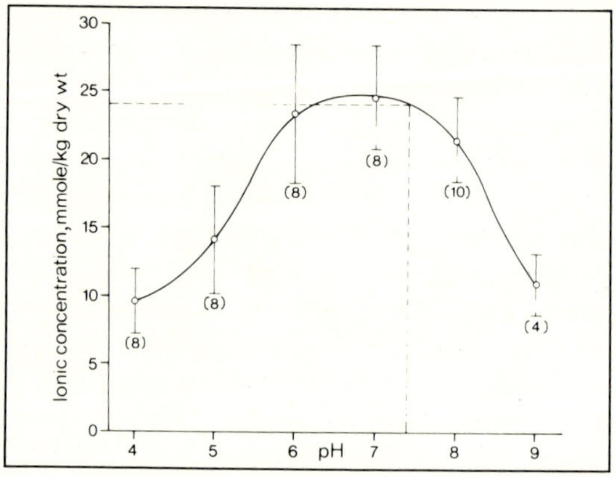

Fig. 4. K binding of isolated adventitial connective tissue from the dog carotid artery as a function of the pH of the Krebs solution. Dashed lines indicate characteristic physiological pH and concentration values.

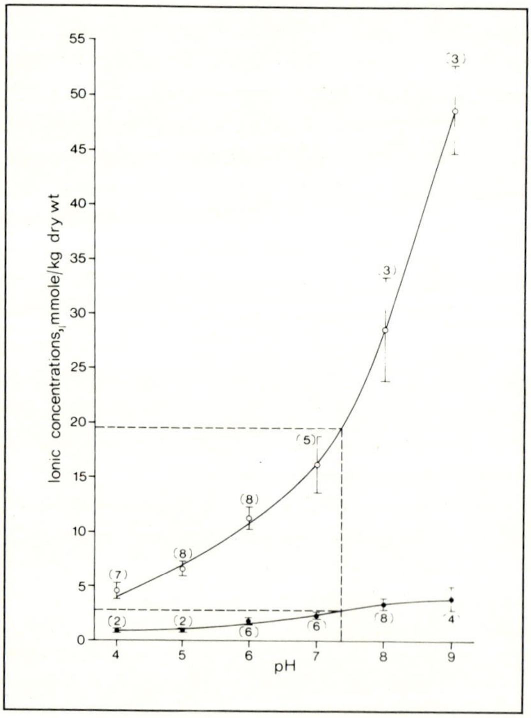

Fig. 5. Ca (○) and Mg binding (●) of isolated adventitial connective tissue from the dog carotid artery as a function of the pH of the Krebs solution.

that alkalosis increases the extracellular K^+ concentration in the neighborhood of the cell membrane, while acidosis has the opposite effect. The first would lead to membrane depolarization with contraction, the second to hyperpolarization with relaxation (66, 69, 73).

In an earlier paper, it was speculated that changing the pH to tissue acidosis leads to a release of Ca ions from connective tissue proteoglycans (73), especially from the basal lamina of the vascular smooth muscle cells (22, 74, 77). The released Ca^{++} increases the Ca^{++} concentration in the narrow clefts between the cell membrane and the basal lamina, which could possibly cause membrane stabilization with hyperpolarization and/or a shift of the threshold for initiation of contraction to a greater depolarization (9, 19). The dual effect of Ca^{++} in vascular smooth muscle is well established by the work of many laboratories (e.g., 9, 74, 77). *Bohr* (10) observed that an increase in the extracellular Ca^{++} concentration leads to hyperpolarization and relaxation in the rat tail artery, and that this effect does not occur in the presence of ouabain. In these experiments, hyperpolarization was probably achieved by activation of the Na^+-K^+ pump (2, 58). Figure 5 shows the Ca and Mg binding in isolated adventitial connective tissue. The very steep Ca-binding curve confirms that even a very small pH change in the physiological range can lead to a considerable release or binding of Ca ions. In principle, the same argument holds for Mg ions, but their binding curve is flatter. However, it must be noted that the Mg^{++} concentration in our modified Krebs' solution was only 1/10 that of normal blood plasma (19). Despite their antagonistic competition for binding sites in vascular connective tissue, K and Ca ions can work synergistically in their physiological effects: tissue acidosis could lead to a membrane hyperpolarization and relaxation in vascular smooth muscle cells, tissue alkalosis to depolarization and contraction.

The processes of a contraction-relaxation cycle in phasic, spike-producing smooth muscle can be easily explained, but those in tonic, graded depolarization-type smooth muscle are still hypothetical (74). *Holman* (41) has summarized the different mechanisms which may cause an increase of Ca ion concentration leading to contraction of smooth muscle (15, 43, 59, 62, 74–77, 83). A repolarization or hyperpolarization in the graded depolarization type of smooth muscle could lead to a decrease of intracellular Ca ion concentration and thus to relaxation via the following mechanisms: (1) At rest under equilibrium conditions, the Ca^{++} influx and efflux are identical. Hyperpolarization would cause a decrease in P_{Ca}, thus a predominance of the Ca^{++} efflux results (58). (2) During hyperpolarization, the Na electrochemical gradient increases. An increase in the Na^+ influx would lead to increased Ca^{++} extrusion via the Na-Ca exchange mechanism (86). If $[Na^+]_o$, $[Na^+]_i$ and $[Ca^{++}]_o$ are held constant, then hyperpolarization leads directly to a decrease in $[Ca^{++}]_i$ (8). *Blaustein* (8) demonstrated further that vascular smooth muscle fibers are especially sensitive to changes in the Na electrochemical gradient; slight changes in $[Na^+]_i$

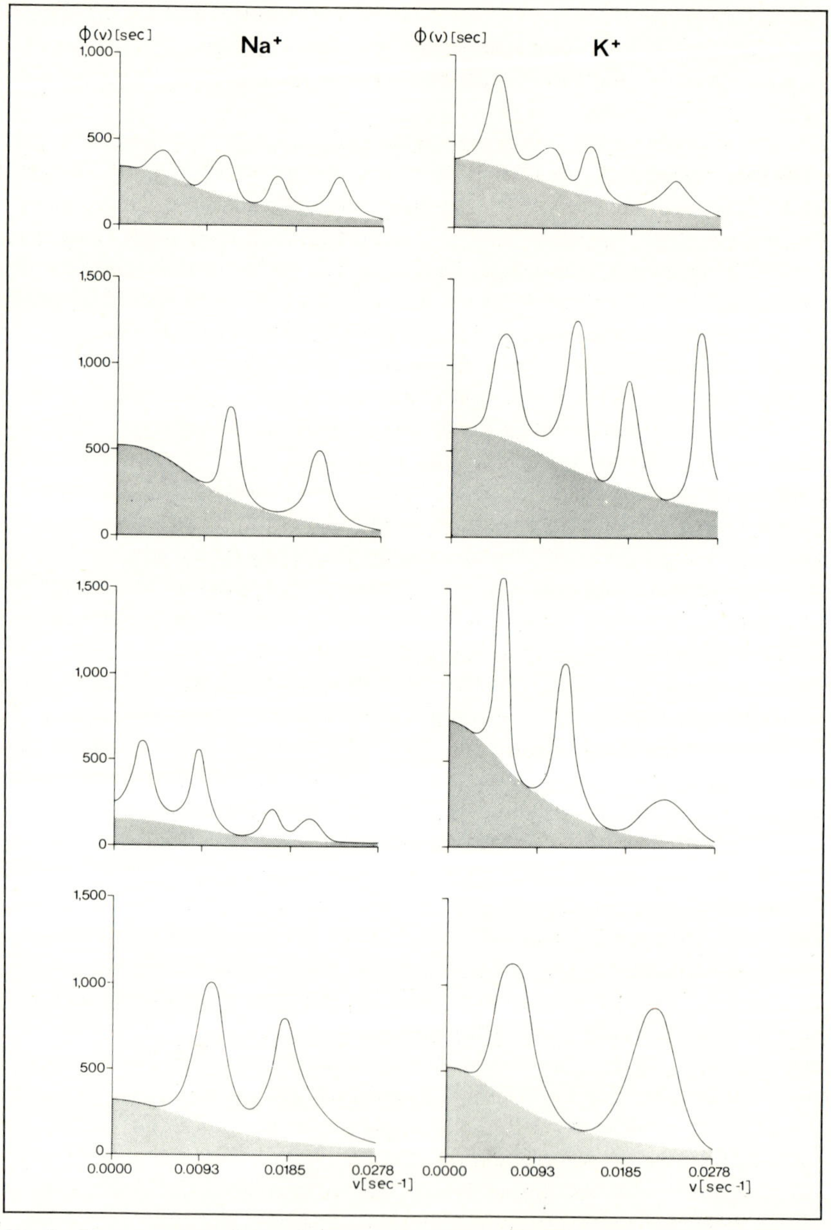

Fig. 6. Power spectra from four separate efflux experiments with simultaneous measurement of $^{24}Na^+$ and $^{42}K^+$ in the dog carotid artery by the indirect method. Distinct periodicities of the Na^+ and K^+ efflux are recognizable. Some frequency ranges were preferred by the individual preparations.

have drastic effects on $[Ca^{++}]_i$ and mechanical tension. This depends partly on the fact that $[Ca^{++}]_i$ is normally held at a suprathreshold level (cf. 69). This means that the contractile tissue is balanced on the rising phase of the $[Ca^{++}]_i$-tension curve and even a small change in $[Ca^{++}]_i$ will manifest itself as a change in tension. (3) A hyperpolarization could cause Ca ions to bind to specific and/or unspecific intracellular and/or membrane sites (14, 41, 43, 59, 74, 77, 83, 85). Whether one or more of these mechanisms play a role has to be determined.

3. Influence of Cell Metabolism and Nerves on Membrane Potential and Vascular Tone

There is another possible regulatory mechanism which could lead to vasodilatation or vasoconstriction. It has long been known that the peripheral, functioning vascular system shows electrical and mechanical oscillations (3, 23–25, 27, 28, 32, 40, 42, 44, 48, 61). We could demonstrate that the slow wave activity of vascular smooth muscle is apparently due to a rhythmical transmembrane ion exchange (65, 66, 68–72). In figure 6, the power spectra from four indirect Na^+–K^+ efflux experiments are given one below the other as a function of frequency. The left half of the figure is derived from the Na^+ efflux, the right half from the K^+ efflux. It is obvious that a periodicity exists for both fluxes. The flux oscillations occur often at integral multiples of one frequency. An analysis of the amplitudes of the flux oscillations and of the phase shifts between the Na^+ and K^+ flux oscillations strongly suggested that the activity of the electrogenic Na^+ pump determines the membrane potential oscillations in smooth muscle cells (68, 70). Through this mechanism, the energy metabolism directly generates the membrane oscillations (cf. 24, 72).

In figure 7, fluctuations in resistance are illustrated in an isolated, perfused skin area of a dog (57, 61). Figure 7A shows spontaneous oscillations. In figure 7B, various phases of increasing resistance were modified by a *single* stimulation of the vasoconstrictor nerve to the area. The excitability, as measured by the constricting effect of the same nerve stimulation, depends on the phase of the spontaneous constriction-dilatation cycle. Thus, for example, a stimulus applied at the peak of the constriction phase has no effect. By contrast, if the stimulus is applied later, in the dilatory phase, the vascular resistance suddenly (within approximately 5 sec) changes in opposite direction and a vasoconstriction follows. In figure 7C, a short train of impulses was applied to the vasoconstrictor nerve, during the spontaneous, rhythmic constrictions. After an initial increase in resistance, vasodilatation occurred regularly. This relaxation could have two explanations: (1) The stimulation of the skin nerve causes release of kinins from the sweat glands (cf. 45). This mechanism would

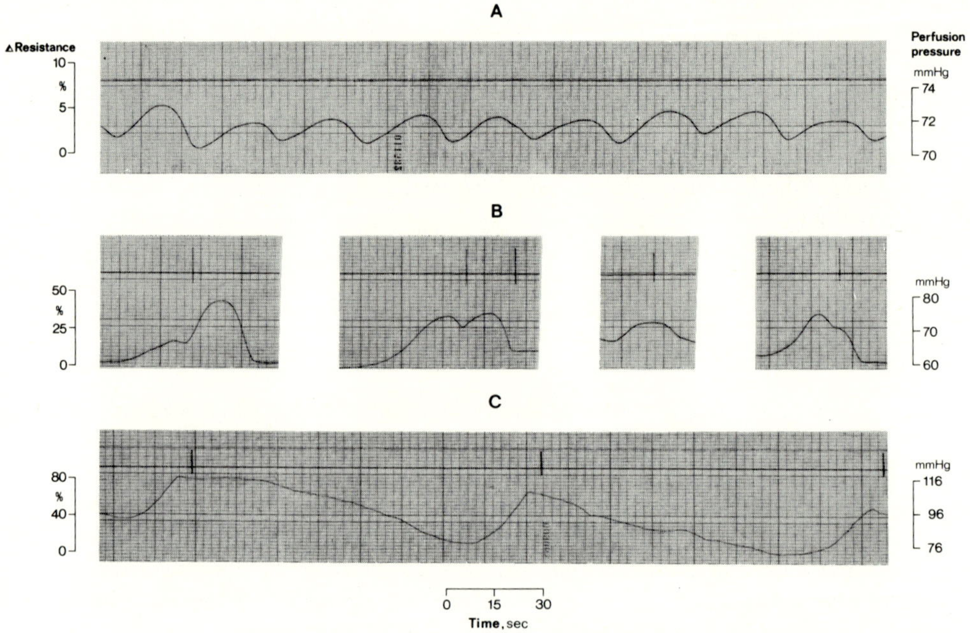

Fig. 7. Oscillations of resistance in an isolated perfused skin area of a dog. *A* Spontaneous fluctuations. *B* During single stimulation of the vasoconstrictor nerve at various phases of resistance increase. *C* During a short impulse series to the vasoconstrictor nerve of a spontaneously rhythmic vessel.

explain the time lag between nerve stimulation and the start of vasodilatation, since the kinins would have to diffuse in the vicinity of the vascular smooth muscle cells. (2) Stimulation of dilatory purinergic nerve fibers running in the mixed efferent skin nerve (17).

It has long been known that the vascular smooth muscle possesses a high glycolytic capacity (4, 46, 66, 70, 72, 74). Auto-oscillations in the glycolytic metabolism were described by *Hess et al.* (34–36). The enzyme phosphofructokinase (PFK), which can be influenced by multiple allosteric effectors, was identified as the primary oscillator (37–39, 70, 72). Figure 8 shows one of *Hess*'s (34–36) original recordings demonstrating the control of glycolytic oscillations by titration with ADP. ADP is a dynamic control variable and exerts a positive feedback on the PFK activity. The concentration of NADH as a function of time was registered. From this picture, one sees that titration with ADP can reverse the phase of the oscillation without essentially changing the

Mechanisms of Vasodilatation (Karger, Basel 1978)

Erratum

Dear Reader,

Because of a mistake in the final production stage, parts of figure 7 of the article by *G. Siegel, R. Ehehalt and H.P. Koepchen* on page 66 of this volume have been wrongly reproduced. The correct position of the figure is shown below.

Thanking you for your understanding,
S. Karger AG, Basel

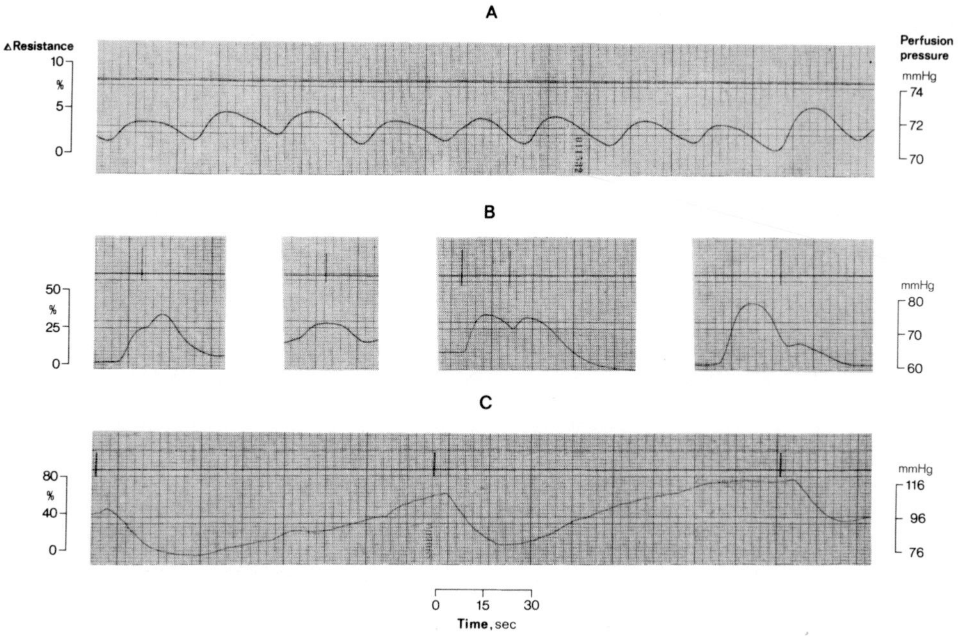

Fig. 7. Oscillations of resistance in an isolated perfused skin area of a dog. *A* Spontaneous fluctuations. *B* During single stimulation of the vasoconstrictor nerve at various phases of resistance increase. *C* During a short impulse series to the vasoconstrictor nerve of a spontaneously rhythmic vessel.

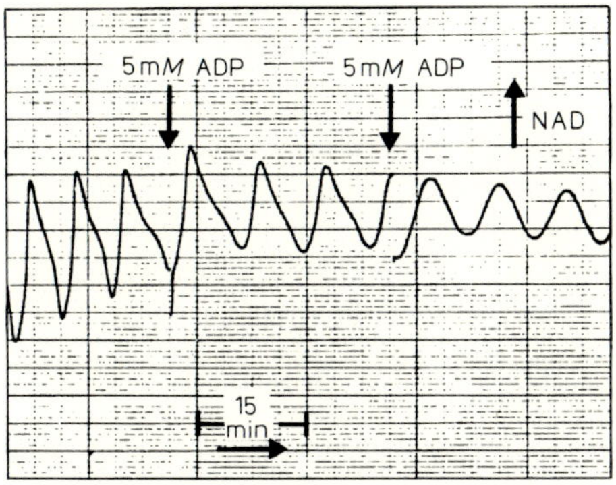

Fig. 8. Titration of glycolytic oscillations with ADP. Recording of NADH fluorescence in an extract of *S. carlsbergensis.* From *Hess* (34).

frequency and amplitude. However, this only occurs in the quadrant of the oscillation phase in which the metabolite ADP controls the state of glycolysis via an interaction with a primary element of the oscillating system (34). The similarity in phase dependency of the actions of the three different effectors (introduction of hyperkalemic solution during a slow wave of the membrane potential, sympathetic nerve stimulation on the spontaneous rhythm of vascular resistance, and ADP on the oscillation of NADH concentration) is striking. Even the time durations of the phase jumps are similar (cf. fig. 7B, 8). A coupling of these three systems is of course speculative, but a sudden change in the energy metabolism, as indicated by a phase jump, could alter the activity of electrogenic ion pumps quickly and drastically. The effect of electrogenic ion pumps on membrane potential of vascular smooth muscle cells is well established (2, 12, 13, 33, 66, 68–71, 73, 74, 82).

4. Conclusion

For the discussion of the validity of the concept of electromechanical coupling for vascular smooth muscle, the influence of the effector nerves, ions, hormones, and metabolism on the cellular membrane potential and vascular tone was studied. Hyperpolarization and relaxation of the vascular smooth muscle cells can be caused by a modest increase or decrease of the external K^+ concentration or an increase of the extracellular H^+ or Ca^{++} concentrations. The effect seems to depend on the instantaneous membrane potential during a slow wave. The result of a vasoconstrictor nerve stimulation is similarly dependent on the phase of the physiological resistance oscillations. It is possible that with alterations in the local metabolic conditions, the intermediary metabolism can vary the vascular lumen rapidly by making phase jumps in oscillating enzyme systems and thus changing the energy supply of the electrogenic ion pumps. Such a local change in the physical state functions of the tissue can lead to a specific change in the bound ion fraction of vascular connective tissue substances. An alteration of the ionic milieu in the immediate neighborhood of the cell membrane would influence the membrane potential of vascular smooth muscle cells.

Acknowledgements

The authors thank Mrs. *C. Fuhrmann* for her excellent technical assistance and Dr. *L. Seldon* for his aid and advice at all stages in the preparation of this manuscript. We are grateful to Miss *M. Schindler* for her outstanding work in preparing the illustrations and to Mr. *P. Holzner* for the excellent photographic work. We would also like to thank the staff of the *Grossrechenzentrum für die Wissenschaft in Berlin* for their cooperation and for generously making computer time available.

This work was supported by the Deutsche Forschungsgemeinschaft (Si 182/1–4) within its overall program *Structure and Function of Biological Membranes*, the Stiftung Volkswagenwerk (Az. 111 327) and the Erwin Riesch-Stiftung (ERS 7/8).

References

1 *Aaberg, G.; Mohme-Lundholm, E., and Vamos, N.:* The effect of H^+ and lactate ions on the electrical activity and content of high energy phosphate compounds of taenia coli from the guinea pig. Acta physiol. scand. *69:* 129–133 (1967).

2 *Anderson, D.K.:* Cell potential and the sodium-potassium pump in vascular smooth muscle. Fed. Proc. Fed. Am. Socs exp. Biol. *35:* 1294–1297 (1976).

3 *Axelsson, J.; Johansson, B.; Jonsson, O., and Wahlström, B.:* The effects of adrenergic drugs on electrical and mechanical activity of the portal vein; in *Keatinge* Symp. Electrical Activity and Innervation of Blood Vessels. Biblthca anat., No. 8, pp. 16–20 (Karger, Basel 1967).

4 *Bass, A.; Brdiczka, D.; Eyer, P.; Hofer, S., and Pette, D.:* Metabolic differentiation of distinct muscle types at the level of enzymatic organization. Eur. J. Biochem. *10:* 198–206 (1969).

5 *Bell, C.:* Transmission from vasoconstrictor and vasodilator nerves to single smooth muscle cells of the guinea-pig uterine artery. J. Physiol., Lond. *205:* 695–708 (1969).

6 *Berne, R.M.:* Cardiac nucleotides in hypoxia: possible role in regulation of coronary blood flow. Am. J. Physiol. *204:* 317–322 (1963).

7 *Betz, E.; Enzenross, H.G., and Vlahov, V.:* Interaction of H^+ and Ca^{++} in the regulation of local pial vascular resistance. Pflügers Arch. *343:* 79–88 (1973).

8 *Blaustein, M-P.:* Sodium ions, calcium ions, blood pressure regulation, and hypertension: a reassessment and a hypothesis. Am. J. Physiol. *232:* C165–C173 (1977).

9 *Bohr, D.F.:* Vascular smooth muscle: dual effect of calcium. Science, N.Y. *139:* 597–599 (1963).

10 *Bohr, D.F.:* Personal commun. (1977).

11 *Bolton, T.B.:* Effects of electrogenic sodium pumping on the membrane potential of longitudinal smooth muscle from terminal ileum of guinea-pig. J. Physiol., Lond. *228:* 693–712 (1973).

12 *Brace, R.A. and Anderson, D.K.:* Predicting transient and steady-state changes in resting membrane potential. J. appl. Physiol. *35:* 90–94 (1973).

13 *Brace, R.A.; Anderson, D.K.; Chen, W.-T.; Scott, J.B., and Haddy, F.J.:* Local effects of hypokalemia on coronary resistance and myocardial contractile force. Am. J. Physiol. *227:* 590–597 (1974).

14 *Brauer, E.; Kupka, K.D., and Rudloff, V.:* Conformational changes of spectrin as a result of calcium binding. Bioelectrochem. Bioenerget. *3:* 509–518 (1976).

15 *Bülbring, E. and Kuriyama, H.:* The action of catecholamines on guinea-pig taenia coli. Phil. Trans. R. Soc. Lond. B *265:* 115–121 (1973).

16 *Burnstock, G. and Holman, M.E.:* The transmission of excitation from autonomic nerve to smooth muscle. J. Physiol., Lond. *155:* 115–133 (1961).

17 *Burnstock, G.:* Purinergic transmission; in *Iversen, Iversen, and Snyder* Handbook of psychopharmacology, vol. 5, pp. 131–194 (Plenum Publishing, New York 1975).

18 *Bygdeman, S. and Euler, U.S.v.:* The effect of respiratory acidosis upon the peripheral vascular reactivity to noradrenaline in the cat. Acta physiol. scand. *54:* 138–146 (1962).

19 *Carrier, O., jr.; Hester, R.K.; Jurevics, H.A., and Tenner, T.E., jr.:* Influence of magnesium on calcium- and potassium-related responses in vascular smooth muscle. Blood Vessels *13:* 321–337 (1976).

20 *Casteels, R.; Droogmans, G., and Hendrickx, H.:* Electrogenic sodium pump in smooth muscle cells of the guinea-pig's taenia coli. J. Physiol., Lond. *217:* 297–313 (1971).

21 *Casteels, R. and Droogmans, G.:* Membrane potential and excitation-contraction coupling in the smooth muscle cells of the rabbit ear artery. J. Physiol., Lond. *263:* 163P–164P (1976).

22 *Catchpole, H.R.; Joseph, N.R., and Engel, M.B.:* Homeostasis of connective tissues. Archs Path. *61:* 503–511 (1956).

23 *Connor, J.A.; Prosser, C.L., and Weems, W.A.:* A study of pacemaker activity in intestinal smooth muscle. J. Physiol., Lond. *240:* 671–701 (1974).

24 *Connor, J.A.; Kreulen, D.L., and Prosser, C.L.:* Relation between oxidative metabolism and slow rhythmic potentials in mammalian intestinal muscle. Proc. natn. Acad. Sci. USA *73:* 4239–4243 (1976).

25 *Cow, D.:* Some reactions of surviving arteries. J. Physiol., Lond. *42:* 125–143 (1911).

26 *Detar, R. and Bohr, D.F.:* Oxygen and vascular smooth muscle contraction. Am. J. Physiol. *214:* 241–244 (1968).

27 *Friedmann, H.:* Über Spontankontraktionen überlebender Arterien. I. Mitteilung. Pflügers Arch. ges. Physiol. *181:* 206–212 (1920).

28 *Funaki, S.:* Electrical and mechanical activity of isolated smooth muscle from the portal vein of the rat; in *Keatinge* Symp. Electrical Activity and Innervation of Blood Vessels. Biblthca anat., No. 8, pp. 5–10 (Karger, Basel 1967).

29 *Gabella, G.:* Quantitative morphological study of smooth muscle cells of the guinea-pig taenia coli. Cell Tissue Res. *170:* 161–186 (1976).

30 *Gabella, G.:* Personal commun. (1977).

31 *Gebert, G. and Piechowiak, H.:* Effect of potassium and norepinephrine on the tone of the isolated artery: changes by ouabain pretreatment. Experientia *30:* 46–47 (1974).

32 *Golenhofen, K.:* Slow rhythms in smooth muscle (minute-rhythm); in *Bülbring, Brading, Jones and Tomita* Smooth muscle, pp. 316–342 (Arnold, London 1970).

33 *Hendrickx, H. and Casteels, R.:* Electrogenic sodium pump in arterial smooth muscle cells. Pflügers Arch. *346:* 299–306 (1974).

34 *Hess, B.:* Modelle enzymatischer Prozesse; in *Scharf und Bruns* Biologische Modelle, vol. 33, pp. 195–230 (Barth, Leipzig 1968).

35 *Hess, B. and Boiteux, A.:* Control of glycolysis; in *Järnefelt* Regulatory functions of biological membranes, vol. 11, pp. 148–162 (Elsevier, Amsterdam 1968).

36 *Hess, B.; Boiteux, A., and Krüger, J.:* Cooperation of glycolytic enzymes; in *Weber* Advances in enzyme regulation, vol. 7, pp. 149–167 (Pergamon Press, Oxford 1969).

37 *Hofer, H.W. und Pette, D.:* Wirkungen und Wechselwirkungen von Substraten und Effektoren an der Phosphofructokinase des Kaninchen-Skeletmuskels. Hoppe-Seyler's Z. physiol. Chem. *349:* 1378–1392 (1968).

38 *Hofer, H.W.:* Influence of enzyme concentration on the kinetic behaviour of rabbit muscle phosphofructokinase. Hoppe-Seyler's Z. physiol. Chem. *352:* 997–1004 (1971).

39 *Hofmann, E.:* The significance of phosphofructokinase to the regulation of carbohydrate metabolism. Rev. Physiol. Biochem. Pharmacol. *75:* 1–68 (1976).

40 *Holman, M.E.; Kasby, C.B.; Suthers, M.B., and Wilson, J.A.F.:* Some properties of the smooth muscle of rabbit portal vein. J. Physiol., Lond. *196:* 111–132 (1968).

41 *Holman, M.E.:* A survey of the new findings presented and the discussions arising during sessions I, II and III. Phil. Trans. R. Soc. Lond. B *265:* 157–165 (1973).

42 *Huizinga, D.:* Untersuchungen über die Innervation der Gefässe in der Schwimmhaut des Frosches. Pflügers Arch. ges. Physiol. *11:* 207–221 (1875).

43 *Janis, R.A. and Daniel, E.E.:* Ca^{2+} transport by subcellular fractions from smooth muscle; in *Stephens* The biochemistry of smooth muscle, pp. 653–671 (University Park Press, Baltimore 1977).

44 *Johansson, B. and Bohr, D.F.:* Rhythmic activity in smooth muscle from small subcutaneous arteries. Am. J. Physiol. *210:* 801–806 (1966).

45 *Keatinge, W.R.:* Electrical and mechanical responses of vascular smooth muscle to vasodilator agents and vasoactive polypeptides. Circulation Res. *18:* 641–649 (1966).

46 *Kirk, J.E.:* Enzyme activities and coenzyme concentrations of normal human vascular tissue; in *Comèl and Laszt* Biochemie der Gefässwand, part II, pp. 145–161 (Karger, Basel 1969).

47 *Kitamura, K.; Kuriyama, H., and Suzuki, H.:* Effect of noradrenaline on the membrane properties and on the contraction of the rabbit main pulmonary artery. J. Physiol., Lond. *263:* 164P–165P (1976).

48 *Koepchen, H.P.:* Die Blutdruckrhythmik (Steinkopff, Darmstadt 1962).

49 *Konold, P.; Gebert, G., and Brecht, K.:* The effect of potassium on the tone of isolated arteries. Pflügers Arch. ges. Physiol. *301:* 285–291 (1968).

50 *Lorenz, R.R. and Vanhoutte, P.M.:* Inhibition of adrenergic neurotransmission in isolated veins of the dog by potassium ions. J. Physiol., Lond. *246:* 479–500 (1975).

51 *Mekata, F. and Niu, H.:* Biophysical effects of adrenaline on the smooth muscle of the rabbit common carotid artery. J. gen. Physiol. *59:* 92–102 (1972).

52 *Mironneau, J.:* Excitation-contraction coupling in voltage clamped uterine smooth muscle. J. Physiol., Lond. *233:* 127–141 (1973).

53 *Mironneau, J. and Savineau, J.-P.:* Ionic currents and contraction in vascular smooth muscle. Proc. Int. Union Physiol. Sci., vol. XIII, Abstr. Volunteer Papers, p. 515. Secr. 27e Congr. Int. Sci. Physiol., Paris 1977.

54 *Olsson, R.A.; Davis, C.J.; Khouri, E.M., and Patterson, R.E.:* Evidence for an adenosine receptor on the surface of dog coronary myocytes. Circulation Res. *39:* 93–98 (1976).
55 *Paton, D.M.:* Mechanism of K^+-induced release of cytoplasmic noradrenaline from adrenergic nerves. Br. J. Pharmacol. *55:* 247P (1975).
56 *Peiper, U. und Wende, W.:* Wirkung der extracellulären Wasserstoffionen-Konzentration auf die adrenerge Aktivierung des Gefässmuskels. Pflügers Arch. ges. Physiol. *314:* 14–26 (1970).
57 *Polster, J.; Seller, H.; Langhorst, P. und Koepchen, H.P.:* Zeitliche Eigenschaften der Vasomotorik. I. Über den Verlauf von Widerstandsänderungen an Hautgefässen bei indirekter Reizung. Pflügers Arch. ges. Physiol. *296:* 95–109 (1967).
58 *Racker, E.:* Structure and function of ATP-driven ion pumps. Trends Biochem. Sci. *1:* 244–247 (1976).
59 *Romero, P.J.:* Role of membrane-bound Ca in ghost permeability to Na and K. J. Membr. Biol. *29:* 329–343 (1976).
60 *Schrader, J.; Haddy, F.J., and Gerlach, E.:* Release of adenosine, inosine and hypoxanthine from the isolated guinea pig heart during hypoxia, flow-autoregulation and reactive hyperemia. Pflügers Arch. *369:* 1–6 (1977).
61 *Seller, H.; Langhorst, P.; Polster, J. und Koepchen, H.P.:* Zeitliche Eigenschaften der Vasomotorik. II. Erscheinungsformen und Entstehung spontaner und nervös induzierter Gefässrhythmen. Pflügers Arch. ges. Physiol. *296:* 110–132 (1967).
62 *Shoenberg, C.F. and Needham, D.M.:* A study of the mechanism of contraction in vertebrate smooth muscle. Biol. Rev. *51:* 53–104 (1976).
63 *Shuba, M.F.; Gurkovskaya, A.V.; Klevetz, M.J.; Kochemasova, N.G., and Taranenko, V.M.:* Mechanism of the excitatory and inhibitory actions of catecholamines on the membrane of smooth muscle cells; in *Bülbring and Shuba* Physiology of smooth muscle, pp. 347–355 (Raven Press, New York 1976).
64 *Shuba, M.F.:* Electrophysiological properties of vascular smooth muscle cells. Proc. Int. Union Physiol. Sci., vol. XII, Abstr. Round Tables and Lectures, p. 337. Secr. 27e Congr. Int. Sci. Physiol., Paris 1977.
65 *Siegel, G.; Koepchen, H.P., and Roedel, H.:* Slow oscillations of transmembrane Na and K fluxes in vascular smooth muscle; in *Betz* Vascular smooth muscle, pp. 3–6 (Springer, Berlin 1972).
66 *Siegel, G.; Jäger, R.; Nolte, J.; Bertsche, O.; Roedel, H., and Schröter, R.:* Ionic concentrations and membrane potential in cerebral and extracerebral arteries; in *Cervós-Navarro* Pathology of cerebral microcirculation, pp. 96–120 (de Gruyter, Berlin 1974).
67 *Siegel, G.; Roedel, H.; Jäger, R., and Bertsche, O.:* Relationship between membrane potential of vascular smooth muscle and external K^+ concentration. Pflügers Arch. *347:* R14 (1974).
68 *Siegel, G. and Roedel, H.:* Active ion transport in vascular smooth muscle; in *Leniger-Follert and Lübbers* Regulation of microcirculation. Drug Res. *25:* 1667 (1975).
69 *Siegel, G.; Roedel, H.; Nolte, J.; Hofer, H.W., and Bertsche, O.:* Ionic composition and ion exchange in vascular smooth muscle; in *Bülbring and Shuba* Physiology of smooth muscle, pp. 19–39 (Raven Press, New York 1976).
70 *Siegel, G.; Roedel, H., and Hofer, H.W.:* Basic rhythms in vascular smooth muscle; in *Worcel and Vassort* Smooth muscle pharmacology and physiology. INSERM Colloque, vol. 50, pp. 215–232 (INSERM, Paris 1976).
71 *Siegel, G.; Niesert, G.; Ehehalt, R., and Bertsche, O.:* Membrane basis of vascular regulation; in *Betz* Ionic actions on vascular smooth muscle, pp. 48–55 (Springer, Berlin 1976).

72 *Siegel, G.; Hofer, H.W.; Ehehalt, R., and Nolte, J.:* Oscillatory phenomena in vascular smooth muscle. Proc. Int. Union Physiol. Sci., vol. XII, Abstr. Round Tables and Lectures, p. 336. Secr. 27e Congr. Int. Sci. Physiol., Paris 1977.

73 *Siegel, G.; Gustavsson, H.; Ehehalt, R., and Lindman, B.:* The role of membrane potential in the regulation of vascular tone; in *Lewis* Recent advances in basic microcirculatory research. Biblthca anat., No. 15, pp. 126–135 (Karger, Basel 1977).

74 *Somlyo, A.P. and Somlyo, A.V.:* Vascular smooth muscle. I. Normal structure, pathology, biochemistry, and biophysics. Pharmacol. Rev. *20:* 197–272 (1968).

75 *Somlyo, A.V. and Somlyo, A.P.:* Electromechanical and pharmacomechanical coupling in vascular smooth muscle. J. Pharmac. exp. Ther. *159:* 129–145 (1968).

76 *Somlyo, A.P. and Somlyo, A.V.:* Vascular smooth muscle. II. Pharmacology of normal and hypertensive vessels. Pharmacol. Rev. *22:* 249–353 (1970).

77 *Somlyo, A.P.; Somlyo, A.V.; Shuman, H., and Garfield, R.E.:* Calcium compartments in vascular smooth muscle: electron probe analysis; in *Betz* Ionic actions on vascular smooth muscle, pp. 17–20 (Springer, Berlin 1976).

78 *Somlyo, A.V.:* Personal commun. (1977).

79 *Speden, R.N.:* Electrical activity of single smooth muscle cells of the mesenteric artery produced by splanchnic nerve stimulation in the guinea pig. Nature, Lond. *202:* 193–194 (1964).

80 *Speden, R.N.:* Excitation of vascular smooth muscle; in *Bülbring, Brading, Jones and Tomita* Smooth muscle, pp. 558–588 (Arnold, London 1970).

81 *Sutter, M.C.:* Quantitative effects of external calcium concentration on contraction of rat portal vein compared to thoracic aorta. Acta physiol. scand. *98:* 266–268 (1976).

82 *Thomas, R.C.:* Electrogenic sodium pump in nerve and muscle cells. Physiol. Rev. *52:* 563–594 (1972).

83 *Tomita, T. and Watanabe, H.:* Factors controlling myogenic activity in smooth muscle. Phil. Trans. R. Soc. Lond. B *265:* 73–85 (1973).

84 *Vanhoutte, P.:* Influence de variations du pH sur la réactivité de fragments des veines isolées du chien. C. r. Séanc. Soc. Biol. *161:* 2318–2320 (1967).

85 *Wei, J.-W.; Janis, R.A., and Daniel, E.E.:* Alterations in calcium transport and binding by the plasma membrane of mesenteric arteries from spontaneously hypertensive rats. Blood Vessels *14:* 55–64 (1977).

86 *Wilbrandt, W. und Koller, H.:* Die Calciumwirkung am Froschherzen als Funktion des Ionengleichgewichts zwischen Zellmembran und Umgebung. Helv. physiol. pharmac. Acta *6:* 208–221 (1948).

Dr. *G. Siegel,* Institute of Physiology, Biophysical Research Group, The Free University of Berlin, Arnimallee 22, *D–1000 Berlin 33* (FRG)

Modulation by Local and Humoral Events

Mechanisms of Vasodilatation. Satellite Symp., 27th Int. Congr. Physiol. Sci., Wilrijk 1977, pp. 73–78 (Karger, Basel 1978)

Myogenic Tone in Resistance Vessels[1]

Paul C. Johnson

Department of Physiology, University of Arizona College of Medicine, Tucson, Ariz.

The mechanisms which can act upon vascular smooth muscle to alter blood vessel caliber are numerous. The vasodilator influences on the blood vessel are many, as discussed throughout this volume. There are substantially fewer vasoconstrictor mechanisms, even on a hypothetical basis. Moreover, the existence of some of the proposed vasoconstrictor mechanisms is doubtful.

In reality, all peripheral vascular beds have some degree of vascular tone under resting conditions. This tone is due in part to the extrinsic nerves. However, the importance of this influence is variable and in virtually all beds most of the tone persists after denervation. Several mechanisms have been proposed for this non-neural tone.

1. Metabolic Regulation

Circulating vasoconstrictor substances have been considered since the time of *August Krogh,* but the evidence is not convincing. If such substances are present in plasma, they are quite long-lasting since basal tone is found in vascular beds perfused for a period of hours with separated oxygenated plasma. Overall, the evidence for a circulating humoral substance is not strong. A second suggestion is that oxygen in the arterial blood or surrounding tissues acts directly on vascular smooth muscle, stimulating it to contract. Recent evidence from direct measurement of oxygen tension in the vicinity of the arteriole does not support such a mechanism – see *Johnson* (5) for references.

[1] This investigation was supported by NIH grants HL 15390, HL 17421, and RRO 5675.

2. *Myogenic Regulation*

By exclusion of external influences, such as those described above, attention has been directed to the vascular smooth muscle cell itself, and intrinsic mechanisms. *Bohr* and co-workers, cited by *Johnson* (5), found that isolated segments of small arteries developed basal tone during perfusion with a balanced salt solution over a period of some hours. They proposed that the mechanism might be an increased permeability of the vascular smooth muscle to calcium.

The possibility has also been considered that intravascular pressure constitutes a stimulus to the smooth muscle cells to contract. This hypothesis was first proposed by *Bayliss* (3) in 1902 on the basis of very indirect evidence. Over the years, several lines of evidence (principally indirect) have been developed which suggest that intravascular pressure is a stimulus. First, it has been noted in reactive hyperemia studies that very brief periods of arterial occlusion lead to substantial vasodilatation. It seems unlikely that there is sufficient reduction in nutrient levels or in oxygen during this period of time to cause substantial metabolic changes in the tissue. This observation implies that pressure reduction probably causes the fall in vascular tone.

The second line of evidence in favor of the myogenic hypothesis is the effect of venous pressure elevation on vascular resistance and capillary filtration coefficient. It has been observed that elevated venous pressure causes a sustained increase in precapillary resistance and capillary filtration coefficient in a variety of tissues and organs including intestine, liver, and certain skeletal muscle beds although not all skeletal muscles show this response. It can be shown that this vasoconstriction is not due to reflex nerve activity or tissue pressure changes (4). In addition, venous pressure elevation leads to a decrease in blood flow which should cause vasodilatation through metabolite accumulation. A related type of experiment involves simultaneous elevation of arterial and venous pressure by the same increment. In such studies, flow decreases and calculated vascular resistance rises, in accordance with the myogenic hypothesis.

A third type of study is reduction of ambient pressure around an organ. This leads to elevated vascular transmural pressure and sustained vasoconstriction. By elimination of other possibilities, the myogenic mechanism appears to be the most probable explanation.

At the *microcirculatory level* a number of observations support the myogenic hypothesis. Elevation of intravascular pressure increases the frequency of vasomotion in venules and in lymphatic channels. This response is seen whether flow is increased or decreased at the same time as pressure is raised. However, the increased intravascular pressure is accompanied by an increase in mean diameter of the venule or lymphatic vessel. Thus, the stimulus in this instance for vasomotion might be elongation of the smooth muscle cell. A mechanism by which stretch of the smooth muscle cell leads to depolarization of the cell membrane and excitation of the contractile machinery is not difficult to envision. However, the data on the resistance vessels implies a sustained constriction of the blood vessel to less than its control diameter. There are microcirculatory data which show that such responses do in fact occur. Such data are obtained from three types of experiments as described below.

First, reduction of arterial pressure leads to dilatation of the arterioles in mesentery and in sartorius muscle. Such dilatation is consistent with the myogenic hypothesis, but could also be explained by accumulation of vasoactive metabolites as blood flow is reduced. More recently, we have also measured red cell velocity in these arterioles and determined that volume flow does not fall and in many instances actually increases as arterial pressure is reduced (6). Thus, it does not seem probable that there is a decrease in supply of oxygen and other nutrients to the tissue when arterial pressure (within limits) is reduced.

A second line of microcirculatory evidence has to do with the effect of venous pressure elevation. Studies of rat mesoappendix have shown that elevation of venous pressure causes constriction of the smallest arteriolar branches (metarterioles and precapillary sphincters) while small arterioles constrict only moderately and large arterioles do not constrict at all (2). Simultaneous flow measurements were not made in this study but it can be safely assumed that flow decreased. This point has been well documented in a variety of gross flow studies mentioned earlier. The graded effect on the vessels is to be expected from the myogenic theory. The smallest vessels, being nearest the venous side, would experience the greatest rise in internal pressure when venous pressure is elevated. The larger vessels would also experience a pressure increase, but this would not be as great. Moreover, the concomitant decrease in flow would be expected to have some counteracting influence. Indeed, in similar studies on cat sartorius muscle, we have observed a dilatation in the largest arterioles even while the smallest arterioles constrict with venous pressure elevation. In these studies on sartorius muscle, flow was measured in individual arterioles and decreased uniformly at all levels.

Measurement of capillary flow (red cell velocity) in mesentery has also provided data of interest relating to the myogenic mechanism in the microcirculation. Periodic flow is prominent in mesenteric capillaries in certain preparations. This periodicity can be traced to vasomotion of the terminal vessels of the arteriolar network and often to precapillary sphincters. When arterial pressure is reduced, vasomotion ceases and capillary red cell velocity increases on the average (7). Vasomotion can be restored at this time either by elevating arterial pressure or by elevating venous pressure. Elevation of venous pressure reduces flow by half but restores vasomotion. Thus, vasomotion appears to depend upon intravascular pressure rather than upon flow.

Arteriolar diameter measurements under no-flow conditions also provide evidence that the arterioles are sensitive to intravascular pressure. In studies of mesenteric arterioles, we occluded the vessel downstream from the site of diameter measurement so as to stop flow in the arteriole completely. We then reduced arterial pressure to the isolated, perfused mesenteric preparation. Under these conditions the arteriole dilated with pressure reduction in a majority of instances (6). This finding is certainly strong evidence in favor of a myogenic mechanism. The only other explanation which might be advanced is that flow in nearby arterioles might have decreased as arterial pressure was reduced, causing a change in the metabolite level in the surrounding tissue. This is unlikely considering the substantial distance (100–200 μm) between arterioles and the rapid change in diameter which occurs upon pressure reduction. A second set of experiments provides apparently incontrovertible evidence that the arterioles are pressure sensitive. In these studies, arterial and venous circuits were clamped and as a result flow in the preparation stopped. The two circuits were then connected to a standing reservoir so that the pressure in the two circuits was fixed at the pressure in the reservoir. Reservoir pressure was set at 10–15 mm Hg and arteriolar diameter allowed to stabilize. Reservoir pressure was then increased to 40 mm Hg for 2–3 min. In most instances, the arteriole constricted with elevation of intravascular pressure (6). When pressure was reduced, the arteriole dilated again. Since there was no flow in the preparation during the experiment, flow effects can be eliminated as a possible cause. Similar observations were reported earlier by *Baez* (1) on rat mesoappendix so this behavior is not limited to cat mesentery.

3. Myogenic versus Metabolic Regulation

Given that there is a pressure-sensitive mechanism in the arterioles, a key question arises: To what extent does this mechanism contribute to local regulatory phenomena such as reactive hyperemia, the venous-arteriolar response, and autoregulation? It seems evident, as noted earlier, that the myogenic response could be responsible in part for reactive hyperemia, at least for dilatation which follows the initial pressure reduction. With longer periods of occlusion, the hyperemic phase is prolonged and this probably represents an accumulation of vasodilator metabolites in the tissues. To date, the key experiments have not yet been done which would determine the relative importance of metabolites and the myogenic mechanism in the initial dilatation. The venous-arteriolar response seems entirely explainable on the basis of the myogenic mechanism. This behavior is not explainable as a flow-dependent phenomenon and it is not abolished by nerve-blocking agents (4). Moreover, the response is greater in the smaller vessels, as the myogenic theory predicts. Autoregulation of blood flow can be explained on the basis of myogenic or metabolic theory but each mechanism would appear to have some limitations. On the one hand, the myogenic mechanism is essentially a pressure-sensitive rather than a flow-sensitive system. Thus, it is not clear how flow could be so perfectly regulated as it appears to be in certain vascular beds. The perfect flow regulation suggests that flow is the controlled variable and thus a metabolic explanation might be the one of choice. However, this mechanism requires a certain error signal. In the autoregulating range, vascular resistance may change by a factor of 2 or more, without a detectable change in flow. This implies exquisite sensitivity of the resistance vessels to metabolite levels. Recent evidence from our laboratory suggests that the myogenic mechanism may be responsible in fact for autoregulation of blood flow. When arterial pressure to the isolated cat sartorius muscle is reduced from 120 to 100 or 80 mm Hg, dilatation is seen primarily in the larger arterioles. No dilatation is seen in the most peripheral arteriolar branches. The dilatation is sufficient to keep flow constant and in fact flow increases on the average as pressure is reduced. Under these conditions, it seems likely that the dilatation of the larger arterioles fully offsets the effect of arterial pressure reduction so that the distending pressure in the small arterioles is not decreased. Thus, these vessels do not dilate. When arterial pressure is decreased further, no further dilatation is seen in the large arterioles but now the smaller arterioles dilate. This response is sufficient to maintain constant flow to a pressure of 60 mm Hg, but when pressure is reduced to 40 mm Hg, flow decreases. We believe that these findings can be explained in terms of a sequential myogenic dilatation which involves progressively smaller, more peripheral vessels as arterial pressure is reduced. The mechanism is tied to intravascular pressure. Upstream vessels dilate as arterial pressure falls and this dilatation tends to restore pressure in the smaller vessels downstream. The level to which this dilatation proceeds in the vasculature depends upon the magnitude of the pressure drop. If the fall is modest, pressure in the downstream vessels can be returned to normal by dilatation of the first one or two series of arteriolar branches. If the fall is greater, intermediate-sized vessels will also dilate. If the fall is large the pressure in the smaller vessels decreases also and these vessels will dilate as well. In effect the arteriolar network is a series-coupled arrangement of myogenic sensor-effector units. Each reacts in such a way as to keep intravascular pressure constant. The effect is to keep blood flow constant also.

References

1 *Baez, S.:* Bayliss response in the microcirculation. Fed. Proc. Fed. Am. Socs exp. Biol. *27:* 1410–1415 (1968).

2 *Baez, S.; Laidlaw, Z., and Orkin, L.R.:* Localization and measurement of microcirculatory responses to venous pressure elevation in the rat. Blood Vessels *11:* 260–276 (1974).

3 *Bayliss, W.M.:* On the local reactions of the arterial wall to changes in internal pressure. J. Physiol., Lond. *28:* 220–231 (1902).

4 *Johnson, P.C.:* Myogenic nature of increase in intestinal vascular resistance with venous pressure elevation. Circulation Res. *6:* 992–999 (1958).

5 *Johnson, P.C.:* The microcirculation and local and humoral control of the circulation; in *Guyton and Jones* Cardiovascular physiology. MTP Review of Physiology (Series One), vol. 1, pp. 163–195 (Butterworth, London 1974).

6 *Johnson, P.C. and Intaglietta, M.:* Contributions of pressure and flow sensitivity to autoregulation in mesenteric arterioles. Am. J. Physiol. *231:* 1686–1698 (1976).

7 *Johnson, P.C. and Wayland, H.:* Regulation of blood flow in single capillaries. Am. J. Physiol. *212:* 1405–1415 (1967).

P.C. Johnson, PhD, Department of Physiology, University of Arizona, College of Medicine, *Tucson, AZ 85724* (USA)

Mechanisms of Vasodilatation. Satellite Symp., 27th Int. Congr. Physiol. Sci., Wilrijk 1977, pp. 79–88 (Karger, Basel 1978)

Relation of Stiffness, Energy Metabolism, and Isometric Tension in a Vascular Smooth Muscle

John W. Peterson

II. Physiologisches Institut, Universität Heidelberg, Heidelberg

Measurements of muscle stiffness have long been parameters of interest to muscle physiologists. In skeletal muscle studies, its original designation as 'series elastic component' (SEC) arose largely from studies of contractile velocity, culminating in the two-(or more)-component and sliding filament model of contraction, in which an active element shortens against an internal elastic element, producing isometric force and a rapid elastic recoil upon sudden release to tensions less than isometric. In the 1950s (13), the SEC was described as large and non-linear. Over the next 15 years, with various technical improvements (primarily the speed with which force transients could be resolved), the SEC became progressively smaller and the force-extension relation more linear. In gross over-simplification, these improvements coupled with a better picture of the structural basis for contractility, led to the 1971 Huxley-Simmons formulation, and the SEC became the 'T1 curve'. This historical development is a topic of frequent review, a recent and thorough example being that of *White and Thorson* (32).

In smooth muscle, a somewhat similar progression is ongoing. Earlier studies of smooth muscle stiffness focused primarily on the concept of 'viscoelasticity' and were often confused by spontaneous activity or basal tone, and a basic lack of information regarding the fundamental origin of force in smooth muscle. It was generally found, however, that smooth muscle became stiffer and more viscous with increases in tonus (4, 27), but it remained unclear whether these observations reflected primarily on contractile properties or on cellular control of passive resistive properties.

The essence of the Huxley-Simmons 1971 formulation (11) for our purposes is that 'instantaneous stiffness' (that is, force changes occurring in concert with very rapid length changes) reflects primarily an elastic component that resides within the myosin cross-bridges themselves. Structural studies (30, 31) have revealed that the actin-myosin cross-bridge is also the fundamental contractile unit in vascular smooth muscle (VSM). The studies to be described demonstrate that measurements of 'instantaneous stiffness' in vascular muscle appear also to reflect the elastic properties of cross-bridges themselves, and therefore quantitatively measure the number of actin-myosin cross-bridges in linkage at a given moment in time.

Studies of the tension-dependent ATPase in VSM, on the other hand, measure the total rate at which cross-bridges hydrolyze ATP, thus the net rate at which cross-bridges are going through the attachment/force generation/detachment cycle. While stiffness measures the number of active cross-bridges, actomyosin ATPase measures the time-average product of number of active bridges and the rate at which they cycle. By measuring both stiffness and tension-dependent ATPase, it is possible to resolve the relative number of active bridges and the relative speeds of operation under a variety of mechanical conditions. For the present, we focus on the isometric state, and especially the role of calcium in the regulation of these parameters.

1. Methods

Tension-dependent (or actomyosin) ATPase, which we equate in the steady state to tension-dependent metabolic energy production, is measured by the steady-state rates of O_2 consumption and lactate production in intact preparations of VSM under well-defined mechanical conditions. This methodology has recently been reviewed (22) and its general validity substantiated (19, 20). Our own work has shown clearly that, when appropriate pains are taken to subtract or eliminate the influences of pharmacology and activation on vascular energy metabolism *not directly* involved with the contractile process, one is left with a method of measuring parameters that, at a minimum, reflect directly on the total actomyosin ATPase under rather physiological conditions (21, 23, 24).

The techniques of the quick stretch studies and more detailed results are presented elsewhere (25). Except when noted, all studies are on a preparation of swine carotid artery, dissected in the method described by *Herlihy and Murphy* (8), at 37 °C, pH 7.4, in a bicarbonate-buffered Krebs-Henseleit physiological saline solution (PSS) at 40% O_2 tension and 5% CO_2 with 10 m*M* glucose added. For the quick stretch studies, we found that a 50% K^+-substituted Na^+-PSS plus histamine (70 μM; K^+-Hist solution) produced a maximal stable contraction, and has been used throughout as the standard activating solution. Length steps (rise time about 1.5 msec) are given as a percentage of resting length, taken as that length where passive tension is zero in order to minimize the necessary passive stiffness corrections.

2. Stiffness Measurements

Figure 1 illustrates the time course of a normal isometric contraction in K^+-Hist activating solution and a series of length steps as described in the legend.

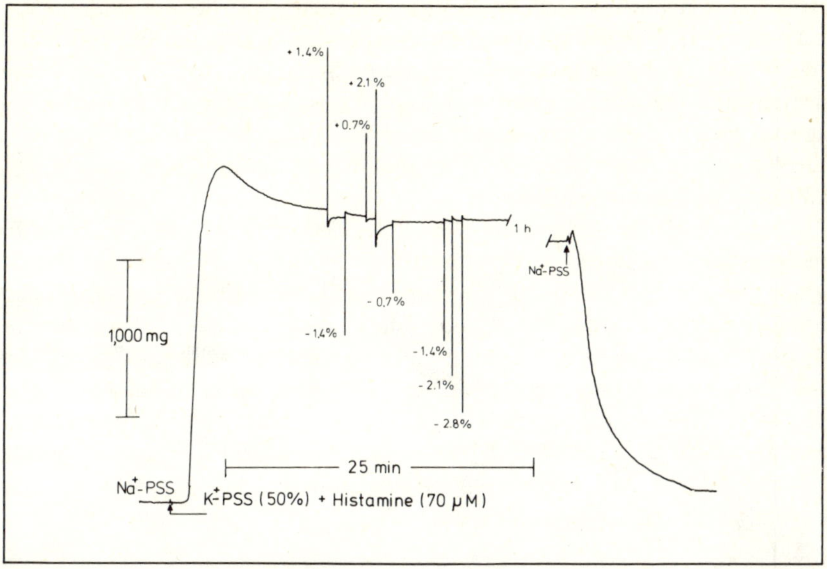

Fig. 1. Time-course of normal isometric contraction with length steps of varying sizes during stable tension, and relaxation upon return to Na^+-PSS. At this length, force/unit area was typically 0.7 kg/cm^2. The duration of the length steps was 200 msec, thus appearing simply as vertical lines on the slow recording shown. Force responses were measured on a recording oscilloscope.

A plot of such stiffness measurements is shown in figure 2, in which the peak force changes are added to the active isometric tension at the instant of the length change, and then normalized to active isometric tension to allow for a time-dependent fade of contractile strength in subsequent contractions and for comparison to other muscles. The plot represents the force-extension curve of the active muscle, while the slope ($\Delta P/\Delta L$) measures the instantaneous stiffness.

There are three principal features to note in figure 2, and an important but implicit fourth:

(1) In the region of stretch, the stiffness curve is quite linear. Non-linearities less than 10% are sometimes seen at the largest stretch, but have been both above or below the linear value, and are probably within experimental error or perhaps reflect a yield phenomenon as noted by *Murphy and Seidel* (18).

(2) A non-linear tailing-off is seen in release. Although increasingly rapid machines have minimized this effect in skeletal muscle, a non-linearity is still seen (15) and is perhaps irreducible.

(3) The linear region extrapolates to a total discharge value of about 1% (1.5% average for 9 arteries).

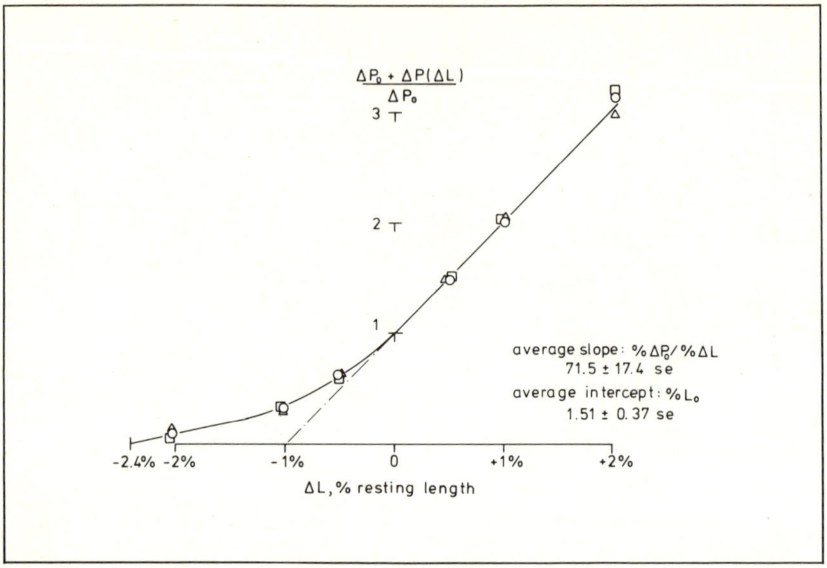

Fig. 2. Instantaneous force changes (corrected by previous measurements of passive stiffness) are added or subtracted from active isometric tension at instant of length step. Notation used throughout figures: ΔP_O = active isometric force; $\Delta P(\Delta L)$ = instantaneous force change; ΔL = % resting length. Data for 3 series of measurements during the 5-hour experiment in one arterial segment. Average values given are for similar experiments with 9 different arteries.

The implicit principal feature is that the relation is quantitatively nearly identical to that for skeletal muscle (12), which forces us, in the absence of new insights, to interpret it in a like manner. The linear slope, which is representative of cross-bridge elasticity, averaged a 70% increase in force for a 1% stretch. In amphibian skeletal muscle, this number is nearer 100%. The average linear extrapolated intercept is nearly double the value for frog skeletal muscle. This observation may fit in interestingly with the fact that quasi-structural units, more or less analogous to sarcomeres but of greater lengths, can be identified in electron micrographs (3). The much broader force-length relation in VSM is also indicative of longer functional sarcomere lengths, as are plausible explanations of reduced contractile velocities and greater holding economies (28).

We can readily hypothesize several explanations for this non-linear tail in VSM: (1) substantial tension recovery actually occurs during the length step itself (1.5 msec); (2) a functionally and structurally significant non-bridge, non-linear series elastic component is present; or (3) additional, but artifactual, series elastic elements are present (for example, the presence of dead or non-responsive

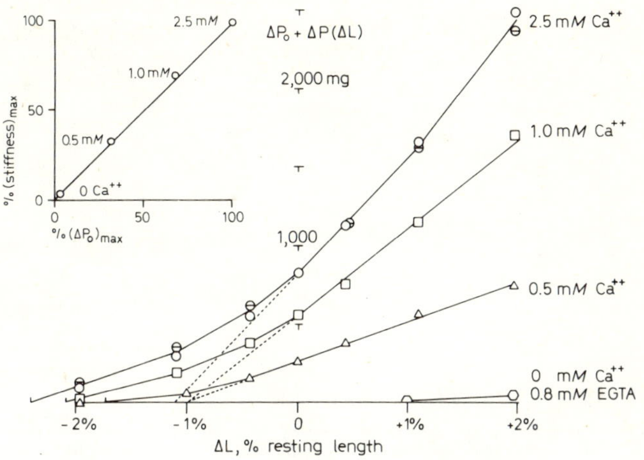

Fig. 3. Un-normalized stiffness at various Ca^{++} levels in K^{+}-Hist activating solution. Sequence was 2.5, 0+EGTA, 0.5, 1.0 and 2.5 m*M* Ca^{++}. Isometric tension values at $\Delta L = 0$.

cells in the tissue). While the ATP-dependent cycling of cross-bridges in VSM is surely orders of magnitude slower than in skeletal muscle (28), we cannot rule out very rapid re-equilibration processes occurring prior to bridge detachment. While non-artifactual *intra*cellular series elasticities are very likely present (e.g. dense body to actin linkages), *Mulvany and Halpern* (17) have presented evidence indicating that *inter*cellular compliances can be quite small.

In this preparation, as is well-known for many vascular muscles, it is possible to produce stable sub-maximal isometric contractions during K^{+} activation simply by varying the external calcium (9, 10). We have therefore measured stiffness curves at various graded-Ca^{++} tension levels, as illustrated in figure 3. The principal features described earlier are evident in all three curves, which extrapolate to very nearly the same place, and are almost superimposable when normalized to active isometric tension. This argues strongly that the linear regime is primarily representative of more or less quantitized units of cross-bridge elasticity. That the non-linear tails go to progressively smaller values with reduced isometric tensions argues weakly in favor of an additional series elastic component. In the inset, the relative stiffness is plotted against the relative active tension at the 4 calcium concentrations. The values lie very nearly on the unity response line. That stiffness increases with Ca^{++}-dependent tension (but observed

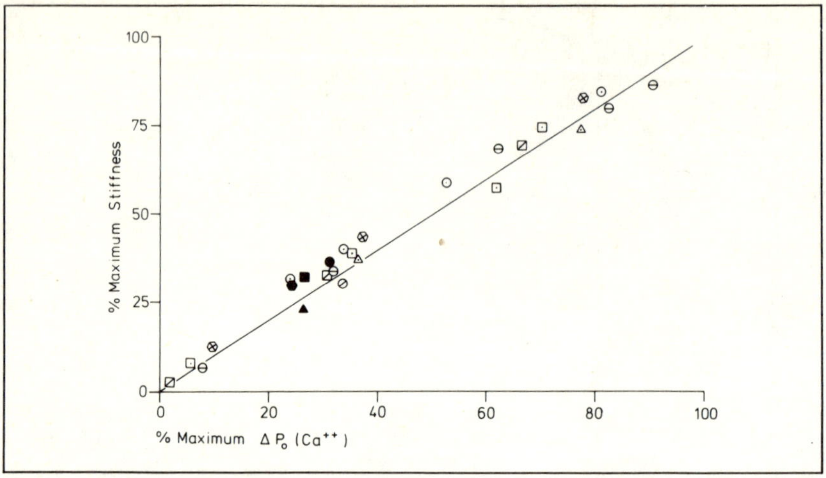

Fig. 4. Plot of relative stiffness versus relative tension for graded-Ca^{++} contractures in K^+-Hist (seven arteries). Solid symbols show stiffness measured for K^+ contractions. The line shown is the identity line (0,0–100,100).

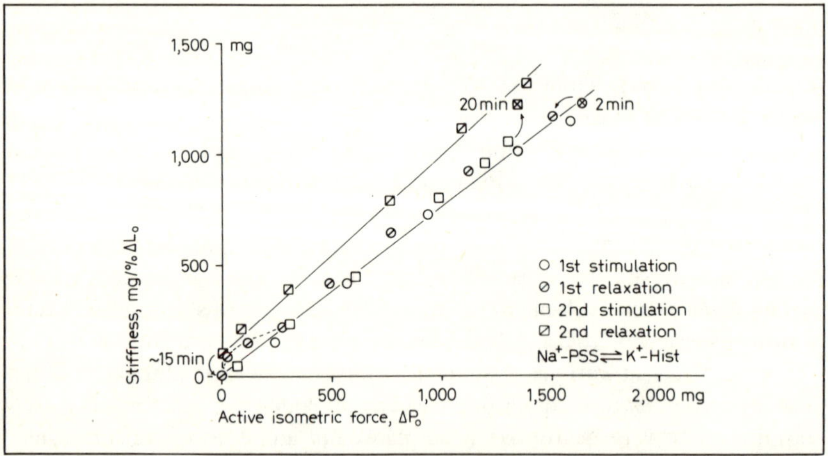

Fig. 5. A plot of absolute stiffness and tension during two series of stimulation/relaxation cycles are shown. In the first series (circles), tension and stiffness rose in unison ($T_{1/2}$ approximately 40 sec). 2 min after attaining peak stable tension, the tissue was transferred to Na^+-PSS, and stiffness falls with tension during relaxation along the same line. For the second series (squares), stiffness and tension during activation again follow the initial line. Allowing now 20 min at peak tension before relaxation, stiffness is increased at all tensions during relaxation. In both cases, there is an indication that at very low tensions in relaxation, there remains a remnant of stiffness without consequent tension (cf. 29).

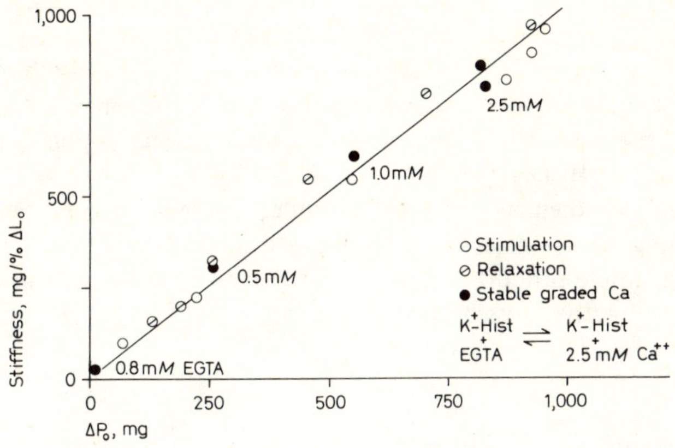

Fig. 6. Plot of absolute stiffness and tension during activation by addition and removal of Ca^{++}. Solid circles measured for stable graded-Ca^{++} contractions as in figure 3.

in a different context) is evident in some earlier work (1). Data from such measurements in seven arteries is shown in figure 4. Clearly, when varying Ca^{++} produces varying isometric tension, stiffness is directly related. This argues that intracellular Ca^{++} levels determine the number of cross-bridges in operation. The figure also suggests that K^+-activation (which gives only 20–30% of maximum tension) simply results in fewer active cross-bridges. Basic correlations between stiffness and tension have been observed in a variety of preparations (2, 5, 14, 18).

In addition, we have measured stiffness during the rise and fall of isometric tension due to K^+-Hist activation and return to Na^+-PSS. A plot of stiffness against isometric tension during the time-dependent stimulation/relaxation cycle is shown in figure 5, with one especially notable feature. We find that during relaxation, the artery is slightly stiffer. In repeated experiments, this increase varied between 5 and 20%, and is clearly dependent upon allowing sufficient time for some process to occur. We can offer tentative explanations at a very fundamental level, relating to bridges rotating into non-force generating configurations without detachment during relaxation (16, 29); or at a more artifactual level, relating to a difference in the uniformity of the spreads of activation and relaxation among the population of cells in the tissue.

We have also looked at the stimulation/relaxation cycle in a way that depends directly upon Ca^{++}, simply by adding Ca^{++} to a Ca^{++}-depleted tissue maintained in activating solution, and subsequently removing it with added

EGTA. Except for a 3-fold slowing down of the process, the results are identical. The data in figure 6 show only a slight increase in stiffness during relaxation, but the principal point is that the relation between tension and stiffness during the influx/outflux of Ca^{++} is exactly the same as that for stable graded-Ca^{++} contractures, both of which are exactly the same as the tension-stiffness relation during the rise and fall of tension either by electrical depolarization (K^+) or pharmacological (Hist) activation.

In summary then, we have found under a variety of conditions, especially those relating to the role of Ca^{++}, that stiffness and tension are inextricably related in a simple proportional way, and that the primary role of Ca^{++} is in the recruitment of active cross-bridges.

3. Metabolic Measurements

At the moment, the metabolic arguments reinforcing this conclusion are equally strong. In the bovine mesenteric vein (22, 23), we have found a remarkably linear relation between tension-dependent ATPase (as measured by O_2 consumption and lactate production) and isometric tension under a variety of conditions of graded pharmacological stimulation and due to varying muscle length at maximal activation (a comparison which, in addition, allows an estimate of the energy requirements of maintaining activation). When activation energy is accounted for, a simple linear proportionality between energy utilization and isometric tension was found over the full range of contractile lengths (21).

In the hog carotid artery, based on similar measurements, again a rather linear proportionality is found for graded K^+ or histamine contractions (6). This study is rendered more interesting metabolically by the observation of inversely compensatory roles of aerobic and anaerobic metabolism, where the net computed ATP production of both is considerably more linear and less variable than either alone. A similar relation was observed in bovine mesenteric vein (24). More recently, we have measured O_2 consumption rates for series of smoothly graded Ca^{++}-dependent contractions, and for graded contractions where the K^+/Na^+ ratio in activating solution is varied. In both cases, a simple linear relation is indicated.

Hellstrand (7), following nearly identical protocols, has recently demonstrated a similar linear response of metabolism to isometric tension maintenance with varying calcium and varying muscle length in the rat portal vein. It is interesting to note that in this spontaneously active and rather rapid preparation, the cost of tension maintenance and the relative cost of activation are both much higher than in the more tonic preparations (28).

4. Conclusion

These studies all lead to the conclusion that variations in isometric tension, whether due to graded degrees of activation or alterations in muscle length, are accompanied by simple proportional changes in the rates of tension-dependent actomyosin ATPase. Our more recent studies of stiffness in the hog carotid artery lead to the conclusion that increased tension is the result of a simple linear increase in the number of active actomyosin cross-bridges. A similar 3-fold linear relationship between tension, ATPase, and stiffness has also recently been demonstrated in cardiac muscle (26).

Acknowledgements

The author thanks *J.C. Rüegg, R.J. Paul, E. Glück* and *G. Pfitzer* for their assistance and discussions in this work. Supported by the Deutsche Forschungsgemeinschaft.

References

1 *Alexander, R.S.:* Role of calcium in the plasticity of venous smooth muscle. Am. J. Physiol. *213:* 287–294 (1967).

2 *Alexander, R.S.:* Series elasticity of urinary bladder smooth muscle. Am. J. Physiol. *231:* 1337–1342 (1976).

3 *Ashton, F.T.; Somlyo, A.V., and Somlyo, A.P.:* The contractile apparatus of vascular smooth muscle: intermediate high voltage stereo electron microscopy. J. molec. Biol. *98:* 17–29 (1975).

4 *Burton, A.C.:* Relation of structure to function of the tissues of the wall of blood vessels. Physiol. Rev. *34:* 613–642 (1954).

5 *Dobrin, P.B. and Canfield, T.R.:* Series elastic and contractile elements in vascular smooth muscle. Circulation Res. *33:* 454–464 (1973).

6 *Glück, E. and Paul, R.J.:* The aerobic metabolism of porcine carotid artery and its relationship to isometric force. Pflügers Arch. *370:* 9–18 (1977).

7 *Hellstrand, P.:* Oxygen consumption and lactate production of the rat portal vein in relation to its contractile activity. Acta physiol. scand. *100:* 91–106 (1977).

8 *Herlihy, J.T. and Murphy, R.A.:* Length-tension relationship of smooth muscle of the hog carotid artery. Circulation Res. *33:* 275–283 (1973).

9 *Hinke, J.A.M.:* Calcium requirements for noradrenaline and high potassium ion contraction in arterial smooth muscle; in *Paul, Daniel, Kay and Monckton* Muscle, pp. 269–285 (Pergamon, Oxford 1965).

10 *Hudgins, P.M. and Weiss, G.B.:* Differential effects of calcium removal upon vascular smooth muscle contraction induced by norepinephrine, histamine and potassium. J. Pharmac. exp. Ther. *159:* 91–97 (1968).

11 *Huxley, A.F. and Simmons, R.M.:* Proposed mechanism of force generation in striated muscle. Nature, Lond. *233:* 533–538 (1971).

12 *Huxley, A.F. and Simmons, R.M.:* Mechanical transients and the origin of muscular force. Cold Spring Harb. Symp. quant. Biol. *37:* 669–683 (1972).

13 *Jewell, B.R. and Wilkie, D.R.:* An analysis of the mechanical components in frog's striated muscle. J. Physiol., Lond. *143:* 515–540 (1958).

14 *Johansson, B.; Hellstrand, P., and Uvelius, B.:* Responses of smooth muscle to quick load change at high time resolution. Hawaii Conf. on Hypertension (in press, 1977).

15 *Julian, F.J. and Sollins, M.R.:* Variation of muscle stiffness with force at increasing speeds of shortening. J. gen. Physiol. *66:* 287–302 (1975).

16 *Julian, F.J.; Sollins, K.R., and Sollins, M.R.:* A model for muscle contraction in which cross-bridge attachment and force generation are distinct. Cold Spring Harb. Symp. quant. Biol. *37:* 685–688 (1972).

17 *Mulvany, M.J. and Halpern, W.:* Circulation Res. (in press, 1977).

18 *Murphy, R.A. and Seidel, C.L.:* Chemomechanical transduction in vascular smooth muscle; in *Bevan, Burnstock, Johansson, Maxwell and Nedergaard* Vascular neuroeffector mechanisms, pp. 47–57 (Karger, Basel 1976).

19 *Namm, D.H.:* Control of energy metabolism in vascular smooth muscle; in *Bevan, Burnstock, Johansson, Maxwell and Nedergaard* Vascular neuroeffector mechanisms, pp. 86–97 (Karger, Basel 1976).

20 *Paul, R.J.:* The chemical energetics of vascular smooth muscle: intermediary metabolism and its relation to contractility. Handbook of circulation (in press, 1977).

21 *Paul, R.J. and Peterson, J.W.:* Relation between length, isometric force, and O_2 consumption in vascular smooth muscle. Am. J. Physiol. *228:* 915–922 (1975).

22 *Paul, R.J. and Peterson, J.W.:* The mechanochemistry of smooth muscle; in *Stephens* The biochemistry of smooth muscle, pp. 15–39 (University Park Press, Baltimore 1977).

23 *Paul, R.J.; Peterson, J.W., and Caplan, S.W.:* A nonequilibrium thermodynamic description of vascular smooth muscle mechanochemistry. J. Mechanochem. Cell Motility *3:* 19–32 (1974).

24 *Peterson, J.W. and Paul, R.J.:* Aerobic glycolysis in vascular smooth muscle: relation to isometric tension. Biochim. biophys. Acta *357:* 167–176 (1974).

25 *Peterson, J.W.; Pfitzer, G., and Rüegg, J.C.:* Dependence of instantaneous stiffness and energy metabolism on calcium-dependent K^+ contractures in vascular smooth muscle. Pflügers Arch. (in press, 1977).

26 *Reiermann, H.J.; Herzig, J.W., and Rüegg, J.C.:* Ca^{++} activation of ATPase activity, ATP-P_i exchange and tension in briefly glycerinated heart muscle. Basic Res. Cardiol. *72:* 133–139 (1977).

27 *Remington, J.W. and Alexander, R.S.:* Relation of tissue extensibility to smooth muscle tone. Am. J. Physiol. *135:* 302–308 (1956).

28 *Rüegg, J.C.:* Smooth muscle tone. Physiol. Rev. *51:* 201–248 (1971).

29 *Siegman, M.J.; Butler, T.M.; Mooers, S.U., and Davies, R.E.:* Calcium-dependent resistance to stretch and stress-relaxation in resting smooth muscles. Am. J. Physiol. *231:* 1501–1503 (1976).

30 *Somlyo, A.P.; Devine, C.E.; Somlyo, A.V., and Rice, R.V.:* Filament organization in vertebrate smooth muscle. Phil. Trans. R. Soc. Lond. B *265:* 223–229 (1973).

31 *Somlyo, A.P. and Somlyo, A.V.:* Ultrastructure of smooth muscle; in *Daniel and Paton* Methods of pharmacology, vol. 3, pp. 3–45 (Plenum Publishing, New York 1975).

32 *White, D.C.S. and Thorson, J.:* The kinetics of muscle contraction. Prog. Biophys. molec. Biol. *27:* 175–237 (1973).

Dr. *J.W. Peterson*, II. Physiologisches Institut, Universität Heidelberg, *D–6900 Heidelberg 1* (FRG)

Mechanisms of Vasodilatation. Satellite Symp., 27th Int. Congr. Physiol. Sci., Wilrijk 1977, pp. 89–97 (Karger, Basel 1978)

Disposition of Vasoconstrictor Agonists

W. Osswald

Laboratório de Farmacologia, Faculdade de Medicina, Porto

As stated elsewhere in this book, vasodilatation may be due to an altered disposition of vasoconstrictor agonists, resulting in a decrease of tone. It is therefore justified, in the scope of a symposium dedicated to mechanisms involved in vasodilatation, to briefly review the processes by which blood vessels may dispose of vasoconstrictor agonists, especially of those of proven or putative physiological significance. As a group, these substances have rather evanescent effects, suggesting that they are rapidly removed from the biophase and/or subject to metabolism.

1. Catecholamines

Since existing information refers predominantly to catecholamines and since noradrenaline is the main transmitter in mammalian blood vessels, most of this review will be dedicated to catecholamines (fig. 1).

1.1. Noradrenaline Released from Sympathetic Nerve Terminals

Disposition of the transmitter released by electric stimulation of blood vessels has been described in more detail in an earlier review (32). Although methodological difficulties still hamper with this kind of investigation, it is clear that the fate of the transmitter depends mainly on the structural characteristics of the vessel under study. Thus, in the saphenous vein of the dog, a vessel with dense adrenergic innervation distributed throughout the media and characterized by relatively narrow synaptic clefts (10), neuronal reuptake is an important mechanism of disposition (6–8); by contrast in elastic arteries with predominantly large neuromuscular gaps and adventitio-medial innervation (e.g. the rabbit aorta and pulmonary artery), diffusion through the adventitia and *O*-methylation in extraneuronally located sites contribute in a significant way to

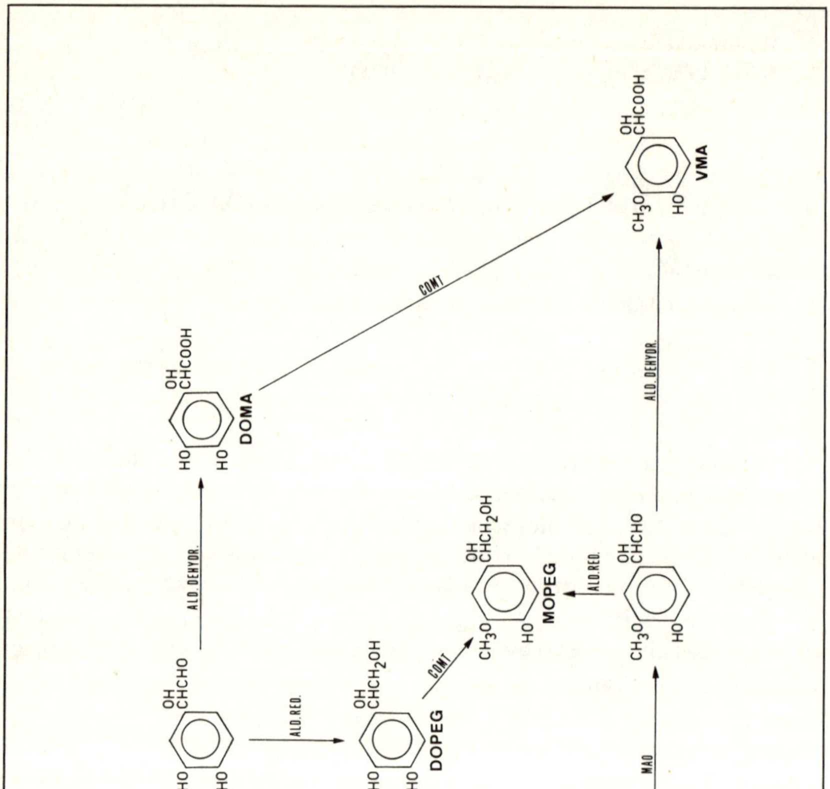
NA
MAO
COMT
NM
ALD.DEHYDR.
ALD.RED.
DOMA
DOPEG
COMT
MOPEG
ALD.RED.
MAO
ALD.DEHYDR.
COMT
VMA

the disposition of the adrenergic transmitter (3, 6, 18, 42). By contrast, oxidative deamination does not appear to play an important role in the termination of the action of released noradrenaline (6, 18, 47, 48), although monoamine oxidase (MAO) is present and functional in vascular tissue (12, 29, 30, 34, 39, 40, 45, 46). The reason for this apparent discrepancy may be that MAO has a predominantly intraneuronal location, and thus acts on the transmitter when it has already been removed from the biophase by neuronal reuptake, and therefore is physiologically 'inactive'. A recent report (7) lends support to this hypothesis: in the isolated saphenous vein of the dog, electric stimulation causes a marked increase in the outflow of noradrenaline and dihydroxyphenylglycol (DOPEG); blockade of neuronal uptake by cocaine increases the noradrenaline fraction and abolishes the formation of DOPEG, whereas methylated (normetanephrine, NMN) and deaminated-and-methylated metabolites (OMDA) increase significantly. The use of inhibitors of extraneuronal uptake (corticosteroids) or of *O*-methylation (U-0521) allows to attribute an exclusively extraneuronal origin to methylated and methylated-and-deaminated metabolites. Thus, at least in the saphenous vein, most of the transmitter is subject to neuronal reuptake and subsequent deamination; however, a fraction escapes metabolism and appears in the outflow or is taken up by extraneuronal sites and subsequently *O*-methylated or methylated and deaminated. Although extraneuronal disposition plays a role even in the presence of an intact function of the adrenergic nerves it may function as an alternative pathway when neuronal uptake is blocked (7, 8, 25, 26, 35). Using the rabbit pulmonary artery preparation, *Endo et al.* (14) reached similar conclusions; the differences in the pattern of metabolization they observed may be attributed to the structural characteristics of the pulmonary artery, which favour diffusion and extraneuronal disposition of released noradrenaline.

The meaning of these results, obtained after incubation with ^{3}H-noradrenaline depends of course on the source of the overflow of tritiated material evoked by electric stimulation; although *Schrold and Nedergaard* (37) have found that in ring preparations of rabbit pulmonary artery preloaded with tritiated noradrenaline an important fraction of the tritium overflow stems from extraneuronal structures, there is ample evidence that, under usual experimental conditions, the stimulation evoked overflow of tritium is exclusively neural in origin (14, 41, 42).

Fig. 1. Schematic representation of the metabolism of noradrenaline. NA = Noradrenaline; NM = normetanephrine; DOMA = dihydroxymandelic acid; MOPEG = methoxyhydroxyphenylglycol; DOPEG = dihydroxyphenylglycol; VMA = vanylmandelic acid; COMT = catechol-*O*-methyltransferase; MAO = monoamineoxidase; Ald. dehydr. = aldehydedehydrogenase; Ald. red. = aldehyde reductase.

1.2. Exogenous Catecholamines

The disposition of exogenous catecholamines by blood vessels has been the object of a large number of investigations, since *Tatum* (43) described the capacity of the wall of large blood vessels to destroy added adrenaline. The mechanisms of inactivation present in blood vessels appear not to differ from those found in other tissues: neuronal and extraneuronal uptake, enzymatic degradation by catechol-*O*-methyl transferase (COMT) and MAO, diffusion and binding to tissue constituents (collagen, elastin, endothelium, fibroblasts). These mechanisms have been repeatedly reviewed (4, 27, 35, 38, 40).

Perfusion experiments conducted with *noradrenaline* (9, 19, 33) have shown that the amine is rapidly and efficiently removed by peripheral vascular beds. The removal capacity is even higher for adrenaline: the perfused hind-limb of the dog removes about 8 times more adrenaline than noradrenaline (44). The effect of drugs, and of denervation, on the removal and the accumulation of these amines allows one to conclude that both neuronal and extraneuronal mechanisms are involved; increasing the concentration of the catecholamine results in saturation of accumulation, but not of removal. There are some important differences between noradrenaline and *adrenaline,* the latter being more extensively metabolized by extraneuronal mechanisms (especially *O*-methylation) and less prone to be accumulated in free form. For noradrenaline, extraneuronal mechanisms play an important role only when neuronal uptake and deamination are blocked or when higher concentrations are used, whereas for adrenaline, both neuronal and extraneuronal mechanisms appear to be operating at all concentrations studied (33, 44).

Using the oil immersion technique, *Kalsner and Nickerson* (29) found that in the rabbit aorta *O*-methylation was the major mechanisms of inactivation for added noradrenaline, followed by oxidative deamination and neuronal uptake. This was confirmed by *Wyse* (47) who, however, found that cocaine had a greater effect than that reported in the original work. In the dog saphenous vein (34) and the rat tail artery (48), inhibition of neuronal uptake and of oxidative deamination were found to play a more important role than *O*-methylation. In the case of adrenaline, *O*-methylation is relatively more important (29, 34).

Incubation of vascular tissues with ^{3}H-noradrenaline has contributed to increase our knowledge on the disposition of the amine. In the rabbit aorta, *O*-methylation is predominant, but topographic differences are evident, since DOPEG is formed primarily in the innervated adventitia and NMN in the media (30); blockade of deamination increases the formation of methylated metabolites and vice-versa (31). In the rabbit ear artery, neuronal uptake is very pronounced, and NMN is the major metabolite; cocaine and denervation greatly reduce the formation of deaminated metabolites (22). In the mesenteric artery of the dog, a vessel with a high metabolizing capacity, neuronal uptake and deamination represent the main inactivation mechanisms. DOPEG is the major metabolite, followed by the OMDA fraction, DOMA and NMN. Here again, blockade of an inactivation mechanism may favour an alternative one (15). Experiments conducted on the saphenous vein of the dog (21) led to similar results, neuronal uptake being the single most important factor of inactivation; however, methylated and deaminated metabolites were the major metabolites, followed by DOPEG, DOMA and NMN. In these experiments, adrenaline was also used; accumulation was less marked and the OMDA fraction and MN accounted for more than 90% of the metabolites. This confirms the notion that adrenaline has a greater affinity for extraneuronal uptake than noradrenaline, when used in equiactive concentrations, and that increasing the concentration favours extraneuronal mechanisms (20, 33, 44). On the other hand, the influence of drugs known to inhibit transport systems allows to conclude that DOPEG is the major metabolite of catecholamines taken up into nerve terminals, whereas NMN and MN are predominantly formed in extraneuronal sites; the OMDA fraction appears to be also formed outside the nerve terminals, but not necessarily in the same compartment as NMN.

For the rabbit aorta, elaborate compartmental analysis of efflux curves (after exposure to ^{3}H-noradrenaline of strips where MAO and COMT were inhibited) has shown that there are 3 important distribution spaces: extracellular (I and II), extraneuronal (III and IV) and neuronal (IV and V) compartments (23). The use of cocaine confirmed the neuronal nature of the compartments identified by the preceding analysis (24). However, it is not yet possible to ascribe a morphologic equivalent to the compartments referred to above, as shown by the fact that compartment IV has the characteristics of both neuronal and extraneuronal sites. This work has also demonstrated that redistribution of noradrenaline does occur during relaxation (due to wash out), as well as that COMT is the major noradrenaline-metabolizing enzyme of rabbit aorta (13). This last finding is in good agreement with the earlier work referred to above.

Binding of catecholamines to 'silent' receptor sites has been repeatedly described (2, 16, 17), and it has been suggested that such binding plays a role in termination of action of the transmitter and may modify the response of the effector cells to catecholamines (36).

Although it is difficult to compare results obtained with different isolated blood vessels, due to the variety of experimental approaches used, it is clear that blood vessels are able to remove and inactivate considerable amounts of noradrenaline and adrenaline. Therefore, the hypothesis has been advanced that blood vessels may be the most important site of disposition both for circulating catecholamines and those added to perfusing fluids (5, 11, 33, 35).

The third physiological catecholamine, *dopamine,* appears to be mainly metabolized (in the saphenous vein) by MAO (unpublished results).

2. *Non-Adrenergic Agonists*

Histamine, which has both vasodilator and vasoconstrictor properties, is degraded in the rabbit aorta by both oxidative deamination (through diamine oxidase, DAO) and methylation (catalyzed by imidazole *N*-methyltransferase). According to *Kalsner* (28), these catabolic routes are approximately of the same importance, whereas *Adams and Hudgins* (1) found oxidative deamination to represent the major pathway.

In the saphenous vein, acetylcholine, as could be expected, is mainly hydrolyzed, as shown by experiments with physostigmine. In the same preparation, oxidative deamination represents the major inactivation mechanism for 5-hydroxytryptamine (unpublished results).

Angiotensin and *prostaglandins* are rapidly inactivated in a single passage through perfused organs, but there is little information available regarding the role played by the vessels in this inactivation process; preliminary results (unpublished) appear to indicate that this role is not important.

3. *Conclusion*

Blood vessels are equipped with enzymatic and non-enzymatic mechanisms which dispose in an efficient way of most of the vasoconstrictor agonists of physiological significance. There is ample evidence, at least for the catecholamines, that blood vessels represent the most important site of inactivation of these agonists.

References

1 *Adams, M.D. and Hudgins, P.M.:* Characteristics of histamine disposition in vascular smooth muscle. Pharmacology *14:* 330–338 (1976).

2 *Avakian, O.V. and Gillespie, J.S.:* Uptake of noradrenaline by adrenergic nerves, smooth muscle and connective tissue in isolated perfused arteries and its correlation with the vasoconstrictor response. Br. J. Pharmacol. *32:* 168–184 (1968).

3 *Bevan, J.A.; Bevan, R.D.; Purdy, R.E.; Robinson, C.P.; Su, C., and Waterson, J.G.:* Comparison of adrenergic mechanisms in an elastic and a muscular artery of the rabbit. Circulation Res. *30:* 541–548 (1972).

4 *Bevan, J.A. and Su, C.:* Sympathetic mechanisms in blood vessels: nerve and muscle relationships. A. Rev. Pharmac. *13:* 269–285 (1973).

5 *Bonisch, H.; Uhlig, W., and Trendelenburg, U.:* Analysis of the compartments involved in the extraneuronal storage and metabolism of isoprenaline in the perfused heart. Arch. Pharmakol. *283:* 223–244 (1974).

6 *Brandão, F.:* A comparative study of the role played by some inactivation pathways in the disposition of the transmitter in the rabbit aorta and the saphenous vein of the dog. Blood Vessels *13:* 309–318 (1976).

7 *Brandão, F.:* Inactivation of norepinephrine in an isolated vein. J. Pharmac. exp. Ther. *203:* 23–29 (1977).
8 *Brandão, F. and Guimarães, S.:* Inactivation of endogenous noradrenaline released by electrical stimulation *in vitro* of dog saphenous vein. Blood Vessels *11:* 45–54 (1974).
9 *Cellander, O. and Mellander, S.:* Elimination of adrenaline and noradrenaline from the circulating blood. Nature, Lond. *176:* 973–974 (1955).
10 *Coimbra, A.; Ribeiro-Silva, A., and Osswald, W.:* Fine structural and autoradiographic study of the adrenergic innervation of the dog lateral saphenous vein. Blood Vessels *11:* 128–144 (1974).
11 *Lande, I.S. de la; Harvey, J.A., and Holt, S.:* Response of the rabbit coronary arteries to autonomic agents. Blood Vessels *11:* 319–337 (1974).
12 *Lande, I.S. de la and Jellett, L.B.:* Relationship between the roles of monoamine oxidase and sympathetic nerves in the vasoconstrictor response of the rabbit ear artery to norepinephrine. J. Pharmac. exp. Ther. *180:* 47–55 (1972).
13 *Eckert, E.; Henseling, M., and Trendelenburg, U.:* The effect of inhibitors of extraneuronal uptake on the distribution of $^3H(\pm)$ noradrenaline in nerve-free rabbit aortic strips. Arch. Pharmakol. *293:* 115–127 (1976).
14 *Endo, T.; Starke, K.; Bangerter, A., and Taube, H.D.:* Presynaptic receptor systems on the noradrenergic neurons of the rabbit pulmonary artery. Arch. Pharmakol. *296:* 229–247 (1977).
15 *Garrett, J. and Branco, D.:* Uptake and metabolism of noradrenaline by the mesenteric arteries of the dog. Blood Vessels *14:* 43–54 (1977).
16 *Gillespie, J.S.:* Tissue binding of noradrenaline. Proc. R. Soc. B *166:* 1–10 (1966).
17 *Gillespie, J.S.; Hamilton, D.N.H., and Hosie, R.J.A.:* The relationship between development of tissue fluorescence and the retention of nfused noradrenaline. J. Physiol., Lond. *190:* 38P–39P (1967).
18 *Gillis, C.N.:* Inactivation of norepinephrine released by electrical stimulation of rabbit aorta in a gaseous medium; in *Bevan, Furchgott, Maxwell and Somlyo* Physiology and pharmacology of vascular neuroeffector systems, pp. 47–52 (Karger, Basel 1971).
19 *Gryglewski, R. and Vane, J.R.:* The inactivation of noradrenaline and isoprenaline in dogs. Br. J. Pharmacol. *39:* 573–584 (1970).
20 *Guimarães, S. and Paiva, M.Q.:* The role played by the extraneuronal system in the disposition of noradrenaline and adrenaline in vessels. Arch. Pharmakol. *296:* 279–287 (1977).
21 *Guimarães, S.; Paiva, M.Q., and Osswald, W.:* A comparative study of the disposition of adrenaline and noradrenaline in an isolated venous structure. Abstr. Joint Meeting Germ. Ital. Pharmacol., Venezia 1977, p. 370.
22 *Head, R.J.; Johnson, S.M.; Berry, D., and Lande, I.S. de la:* Denervation and *O*-methylation of noradrenaline in the rabbit ear artery. Clin. exp. Pharmacol. Physiol. *2:* suppl., pp. 39–42 (1975).
23 *Henseling, M.; Eckert, E., and Trendelenburg, U.:* The distribution of $^3H \pm$ noradrenaline in rabbit aortic strips after inhibition of the noradrenaline-metabolizing enzymes. Arch. Pharmakol. *292:* 205–217 (1976).
24 *Henseling, M.; Eckert, E., and Trendelenburg, U.:* The effect of cocaine on the distribution of labelled noradrenaline in rabbit aortic strips and on efflux of radioactivity from the strips. Arch. Pharmakol. *292:* 231–241 (1976).
25 *Hughes, J.:* Evaluation of mechanisms controlling the release and inactivation of the adrenergic transmitter in the rabbit portal vein vas deferens. Br. J. Pharmacol. *44:* 472–491 (1972).
26 *Hughes, J. and Roth, R.H.:* Variation in noradrenaline output with changes in stimulus

frequency and train length: role of different noradrenaline pools. Br. J. Pharmacol. *51:* 373–381 (1974).

27 *Iversen, L.L.:* Role of transmitter uptake mechanisms in synaptic neurotransmission. Br. J. Pharmacol. *41:* 571–591 (1971).

28 *Kalsner, S.:* Potentiation of responses to histamine in aortic strips by steroids. Can. J. Physiol. Pharmacol. *48:* 443–449 (1970).

29 *Kalsner, S. and Nickerson, M.:* Disposition of norepinephrine and epinephrine in vascular tissue, determined by the technique of oil immersion. J. Pharmac. exp. Ther. *165:* 152–165 (1969).

30 *Levin, J.A.:* The uptake and metabolism of ^{3}H-*l*- and ^{3}H-*dl*-norepinephrine by intact rabbit aorta and by isolated adventitia and media. J. Pharmac. exp. Ther. *190:* 210–226 (1974).

31 *Levin, J.A.:* Extraneuronal uptake and metabolism of ^{3}H-*l*-noradrenaline in rabbit aorta. Proc. 6th Int. Congr. Pharmacol., Helsinki 1975, vol. 2, pp. 139–148.

32 *Osswald, W.:* Transmitter disposition mechanisms. Vascular Neuroeffector Mechanisms 2nd Int. Symp., Odense 1975, pp. 123–130 (Karger, Basel 1976).

33 *Osswald, W. and Branco, D.:* The effects of drugs and denervation on removal and accumulation of noradrenaline in the perfused hind-limb of the dog. Arch. Pharmakol. *277:* 175–190 (1973).

34 *Osswald, W.; Guimarães, S., and Coimbra, A.:* The termination of action of catecholamines in the isolated venous tissue of the dog. Arch. Pharmakol. *269:* 15–31 (1971).

35 *Osswald, W.; Garrett, J., and Guimarães, S.:* Extraneuronal uptake and metabolism in dog vascular structures. Proc. 6th Int. Congr. Pharmacol., Helsinki 1975, vol. 2, pp. 149–158.

36 *Powis, G.:* The binding of noradrenaline to connective and elastic tissue. Proc. 6th Int. Congr. Pharmacol., Helsinki 1975, vol. 2, pp. 131–137.

37 *Schrold, J. and Nedergaard, O.A.:* ^{3}H-noradrenaline outflow induced from isolated adventitia and intima-media of rabbit aorta by electrical field stimulation. Eur. J. Pharmacol. *39:* 423–427 (1976).

38 *Somlyo, A.P. and Somlyo, A.V.:* Vascular smooth muscle. II. Pharmacology of normal and hypertensive vessels. Pharmacol. Rev. *22:* 249–353 (1970).

39 *Spector, S.; Tarver, J., and Berkowitz, B.:* Catecholamine biosynthesis and metabolism in the vasculature of normotensive and hypertensive rats; in *Okamoto* Spontaneous hypertension. Its pathogenesis and complications, pp. 41–45 (Ikagu Shoin, Tokyo 1972).

40 *Spector, S.; Tarver, J., and Berkowitz, B.:* Effects of drugs and physiological factors in the disposition of catecholamines in blood vessels. Pharmacol. Rev. *24:* 191–202 (1972).

41 *Starke, K.; Montel, H.; Gayk, W., and Merker, R.:* Comparison of the effects of clonidine on pre- and postsynaptic adrenoceptors in the rabbit pulmonary artery. Arch. Pharmakol. *285:* 133–150 (1974).

42 *Su, C. and Bevan, J.A.:* The release of H^{3}-norepinephrine in arterial strips studied by the technique of superfusion and transmural stimulation. J. Pharmac. exp. Ther. *172:* 62–68 (1970).

43 *Tatum, A.L.:* On the destruction of epinephrine and constrictor substances of serum by oxygenation in the presence of blood vessel walls. J. Pharmac. exp. Ther. *4:* 151–156 (1912).

44 *Teixeira, F.:* The effects of drugs and denervation on removal and accumulation of adrenaline in the perfused hind-limb of the dog. Archs int. Pharmacodyn. Thér. *225:* 221–231 (1977).

45 *Vanhoutte, P.M. and Shepherd, J.T.:* Activity and thermosensitivity of canine cutaneous veins after inhibition of monoamine oxidase and catechol-*O*-methyl transferase. Circulation Res. *25:* 607–616 (1969).

46 *Verity, M.A.; Su, C., and Bevan, J.A.:* Transmural and subcellular localization of monoamine oxydase and catechol-*O*-methyl transferase in rabbit aorta. Biochem. Pharmacol. *21:* 193–201 (1972).

47 *Wyse, D.G.:* On the role of neuronal uptake (uptake$_1$) in the inactivation of noradrenaline by aortic strips. Can. J. Physiol. Pharmacol. *52:* 1102–1109 (1974).

48 *Wyse, D.G.:* Inactivation of neural and exogenous norepinephrine in rat tail artery studied by the oil immersion technique. J. Pharmac. exp. Ther. *198:* 102–111 (1976).

Prof. Dr. *W. Osswald,* Laboratório de Farmacologia, Faculdade de Medicina, *Porto* (Portugal)

Mechanisms of Vasodilatation. Satellite Symp., 27th Int. Congr. Physiol. Sci., Wilrijk 1977, pp. 98–106 (Karger, Basel 1978)

Interactions of Locally Produced Humoral Substances in Regulation of the Microcirculation

Burton M. Altura and Bella T. Altura

Department of Physiology, SUNY Downstate Medical Center, Brooklyn, N.Y.

Local regulation of microcirculatory blood flow is thought to be largely dependent upon the presence or liberation of chemical and humoral mediators in the immediate vicinity of the small microscopic blood vessels; these mediators are either blood-borne, produced in the parenchymal tissues, or possibly formed in the blood vessel wall itself (1, 13, 16, 33). Thus, the local regulation of blood flow is dependent upon the balance or the interaction between vasodilator and vasoconstrictor factors which modulate the myogenic activity of the vascular smooth muscle cells (1–2, 3, 13, 33).

1. Natural Physiologic Antagonism between Circulating Constrictors and Dilators

Naturally occurring humoral or locally produced substances which can induce vasodilatation include prostaglandins (PGs) (8, 14, 17–19, 22, 29, 31), histamine (2, 5, 11, 20, 25, 26), kinins (1–3, 5, 14, 20, 25, 29, 32), serotonin (20, 27), and possibly acetylcholine (2, 3, 20, 26, 29). Most of these substances initiate vasodilatation by depressing the myogenic activity of the microvascular smooth muscle cells (1–3, 5, 31). In addition, some of these substances may induce vasodilatation indirectly by inhibiting the release of, or by antagonizing the constrictor action of neuro-humoral agents such as catecholamines, angiotensin, neurohypophyseal peptides, etc. (1–3, 5, 8, 9, 12, 14, 19, 21–23, 29). Of particular interest are the findings that such antagonism can be demonstrated for histamine, acetylcholine, and bradykinin with direct, quantitative *in vivo* microscopy on intact muscular microvessels, including the smallest components of the microcirculation, the precapillary sphincters (fig. 1–3). These dilators induce a parallel, rightward shift of the dose-response curves for a variety of constrictor agonists. Although it is known that some of these dilators (e.g. acetylcholine, serotonin, histamine) can inhibit adrenergic neurotransmission in certain blood vessels (23, 26, 29), it is unlikely that this could account for the marked attenuation of constrictions induced by vasopressin, angiotensin, and barium ions. It is more tempting to speculate that these vasodilators induce inhibition by interfering with the release or the storage of activator calcium ions which are necessary for the constrictor responses.

2. Vasodilator Hormone Interactions and Modification by Steroids

Certain tissue hormones (e.g. kinins, angiotensins) might owe part of their vasoactivity to the synthesis and the release of other agents (e.g. PGs) (14, 15, 21, 28). A good example of this at the microcirculatory level is illustrated by the data shown in figure 3. Intra-arterial infusion of the PG-synthetase inhibitor,

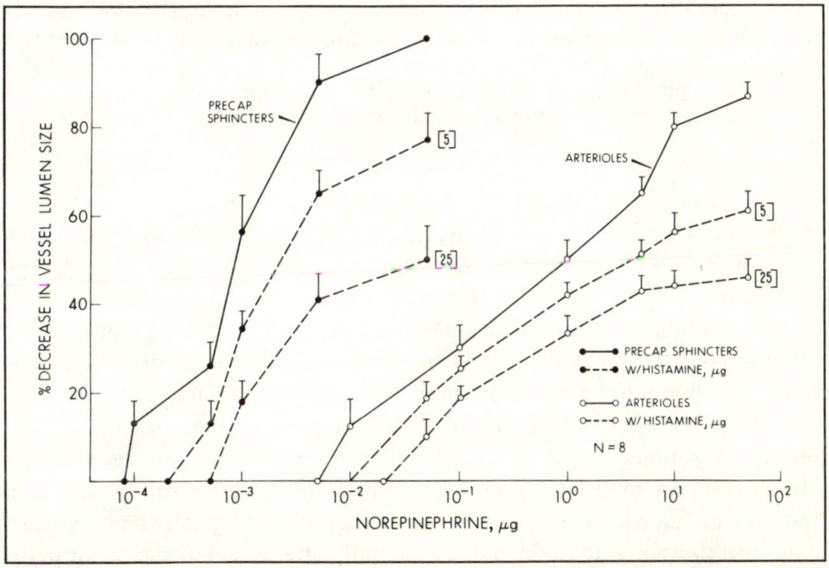

Fig. 1. Non-relaxant doses of histamine antagonize microvascular constrictions induced by norepinephrine. An image-splitting TV microscope recording system – see work by *Altura* (4, 7) for details – was employed for these *in vivo* measurements made on rat mesenteric precapillary sphincters (•) (5.6 ± SE 0.4 μm) and terminal arterioles (○) (20.4 ± SE 1.1 μm) here as well as those in figures 2–6. Histamine (broken lines) was applied locally for 3–5 min prior to application of norepinephrine (significantly different from controls, $p < 0.05$). Bars here and in other figures are SEMs.

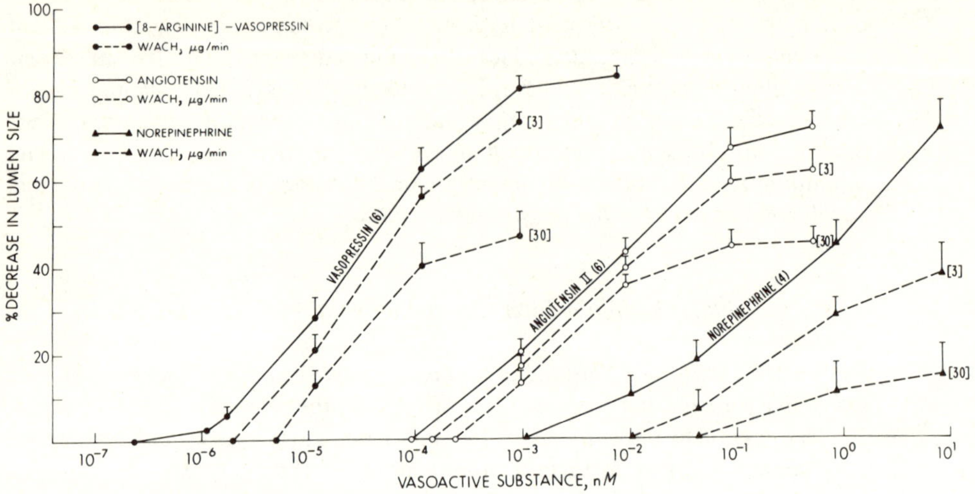

Fig. 2. Influence of intra-arterial administration (via branch of ileo-colic artery) of acetylcholine (broken lines) on terminal arteriolar constrictions induced by topical [8-arginine]-vasopressin, angiotensin II, and norepinephrine. Acetylcholine chloride infusion was started 10 min prior to reapplication of test doses of the constrictor agonists and continued throughout administration of the latter substances (significantly different from controls, $p < 0.05$). Figures in parentheses represent numbers of different animals that were utilized.

indomethacin, into a branch of the ileo-colic artery, counteracts the bradykinin-induced non-specific attenuation of the constrictor response to norepinephrine, angiotensin and, to some extent, to vasopressin. Such findings suggest that some of the vasoinhibitory properties of substances such as bradykinin are probably due to PG-like compounds. Of particular interest is that all of these actions can be produced by physiological amounts of these humoral substances.

Many reports in the literature conclude or infer that glucocorticoids have important actions on peripheral blood flow, endothelial cells, and the formed elements of the blood. Despite this common impression, it should be stated that there is no convincing evidence that either physiological or pharmacological concentrations of glucocorticoids can directly alter vessel diameters or peripheral blood flow – see references cited by *Altura and Altura* (6) and *Shepherd and Vanhoutte* (26). A more likely interpretation is that although physiological concentrations of the glucocorticoid hormones by themselves have little effect on the intrinsic activity of vascular smooth muscle, they alter blood-vessel caliber and flow in the microcirculation by affecting the response to circulating or locally released catecholamines, histamine, and kinins. Although glucocor-

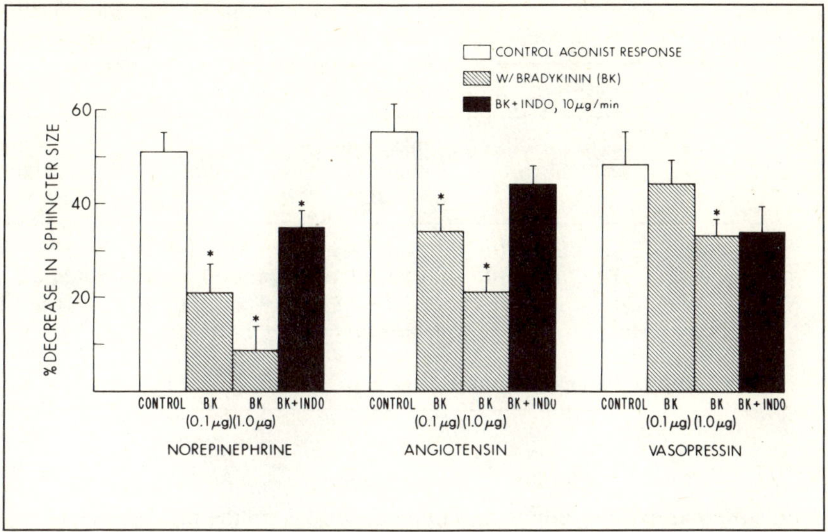

Fig. 3. Bradykinin-induced inhibition of precapillary sphincter constrictor responses to norepinephrine, angiotensin, and vasopressin: reversal by intra-arterial infusion of indomethacin (INDO, 10 μg/min). Bradykinin (BK) was applied locally 2–5 min prior to topical application of the constrictor agonists. INDO infusion was started 20 min prior to reapplication of BK + agonists and continued during evaluation of the latters' responses. Values with an asterisk are significantly different from controls ($p < 0.05$).

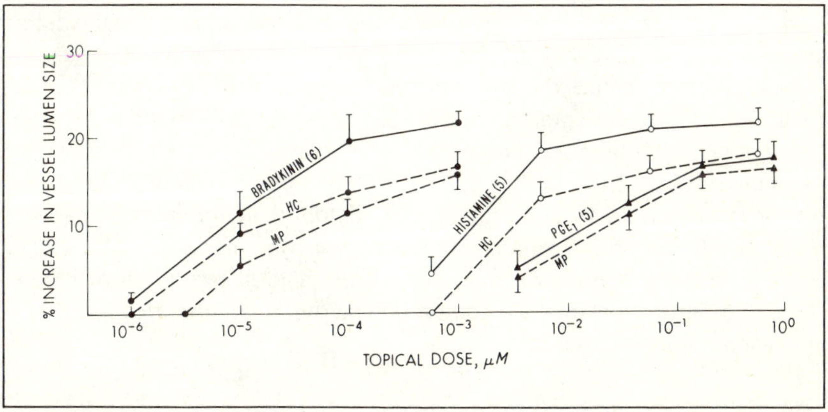

Fig. 4. Inhibition of bradykinin and histamine vasodilatation of precapillary sphincters by intra-arterial administration of hydrocortisone sodium succinate (HC; 20 μg/min) and methylprednisolone sodium succinate (MP; 20 μg/min). Histamine and bradykinin dose-response curves in presence of HC and MP are significantly different from controls ($p < 0.05$).

Table I. Influence of sex on reactivity of mesenteric metarterioles to bradykinin, histamine and PGE_1 in young, adult male and female rats

Dilator	Male		Female	
	ED_{50} $\mu M \pm SE$	maximum dilatation % ± SE	ED_{50} $\mu M \pm SE$	maximum dilatation % ± SE
Bradykinin	$1.2 \pm 0.3 \times 10^{-5}$	24.6 ± 3.4	$2.5 \pm 0.3 \times 10^{-5}$*	15.8 ± 2.7*
Histamine	$1.1 \pm 0.2 \times 10^{-3}$	30.4 ± 2.2	$1.9 \pm 0.2 \times 10^{-3}$*	24.2 ± 2.4*
PGE_1	$1.3 \pm 0.3 \times 10^{-2}$	16.2 ± 1.8	$1.5 \pm 0.3 \times 10^{-2}$	10.2 ± 1.2*

N = 6–9 each.
* Significantly different from values in male ($p < 0.03$).

ticoids can exert other important, indirect actions on the microcirculation – see *Altura and Altura* (6) for review – the latter effects on histamine and kinins are particularly germane to the present discussion. Earlier semi-quantitative *in vivo* microcirculatory studies reveal that these steroids can attenuate the microvascular dilator effects induced by histamine and kinins (1, 24). More recently, using closed circuit high-resolution television microscopy, we have confirmed that low doses of steroids can, indeed, diminish the microcirculatory vasodilator actions of both histamine and bradykinin but not those induced by PGs (fig. 4). The shift of the dilator agonist dose-response curves might indicate that these steroids interfere with relaxation by acting at a common cellular site; recent evidence suggests that they affect calcium translocation (5, 7).

Sex hormones, like the glucocorticoids, apparently do not affect myogenic tone in microvascular smooth muscle (10); however, they modify the actions of endogenous neurohumoral constrictor and dilator agents (3–5, 10). It is of particular interest to note that microvessels of female mammals may ordinarily be less responsive to local hormonal vasodilators than are those of male animals (table I). Overall, these findings reinforce the hypothesis that glucocorticoids and sex steroidal hormones may play a role in local regulation of blood flow through modification of the response of microvascular smooth muscle to other endogenous humoral agents.

3. Synergism Among Certain Local Hormonal Vasodilators

Zweifach (32) suggested that certain local tissue hormones interact to produce unusual reactions at the microcirculatory level. In particular, he

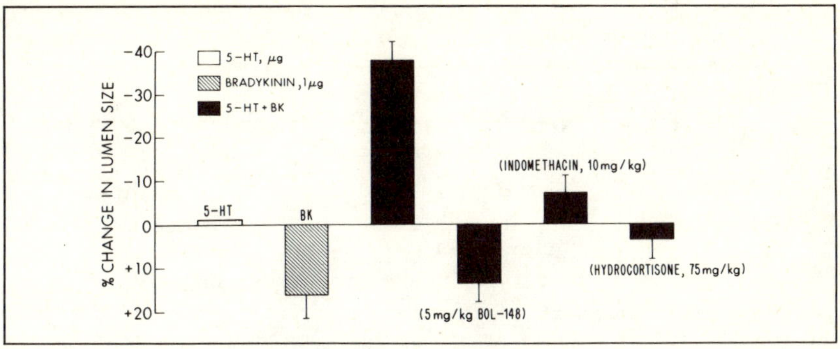

Fig. 5. Synergistic interaction between topically applied 5-HT and bradykinin in muscular venules. BOL-148, indomethacin, and hydrocortisone were administered i.v. to different animals 10–20 min prior to challenge of vessels with 5-HT and BK.

reported that bradykinin and serotonin (5-HT) have a synergistic action in the rat mesenteric microcirculation. This is characterized by intense venular spasm followed by capillary stasis and extravasation of red blood cells through defects in the walls of the postcapillary venules. These observations were confirmed by two other laboratories (24, 30). *Weiner and Altura* (30) suggested that rhexis of the postcapillary collecting venules is brought about by a local increase of intravascular pressure as a consequence of severe venular constriction. More recent findings support this hypothesis. Thus, the experiments shown in figure 5 illustrate that low concentrations of 5-HT, when combined with vasodilator doses of bradykinin, produce intense constriction of muscular venules, which can be completely blocked by a serotonin antagonist, BOL-148, or hydrocortisone. Indomethacin can markedly attenuate this synergistic reaction, suggesting that the association of bradykinin and 5-HT promotes the synthesis and the release of PG-like material. These findings could help explain not only why in man bradykinin, which normally produces itching in skin, potentiates the action of 5-HT on pain receptors (27), but may also aid in defining the painful phenomena associated with a variety of vascular disorders such as migraine, angina pectoris, and myocardial infarction.

4. Role of Prostaglandins and Histamine in Reactive Hyperemia

While PG-like compounds and histamine have been proposed as important mediators of reactive or post-occlusion hyperemia (17, 18, 25), few studies have examined this possibility *in vivo* at the microcirculatory level (11, 22). One such experiment, illustrated in figure 6, provides good evidence, at least for the rat

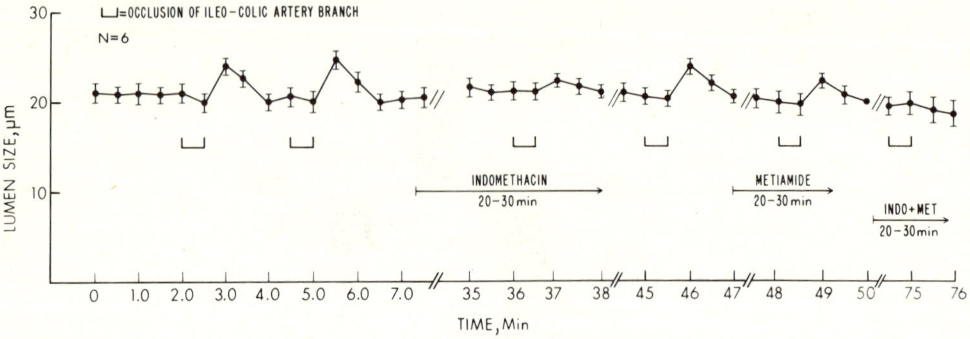

Fig. 6. Inhibition of mesenteric arteriolar post-occlusion vasodilatation by superfusion with indomethacin (20 μg/min) and metiamide (10–30 μg/min). A branch of ileo-colic artery was temporarily occluded for 30-sec periods. Values are means ± SDs.

mesenteric arterioles, that both PG-like compounds and histamine may indeed be mediators of reactive hyperemia. Metiamide, a specific H_2-histamine receptor antagonist, markedly attenuates the post-occlusion arteriolar hyperemia, which suggests that the participation of histamine may be mediated by H_2-receptors.

5. Conclusion

Many endogenous and locally synthesized dilator substances have been suggested to participate in the moment-to-moment regulation of microvascular tone. The impression derived from the literature, and from our own experiments, suggests that no single humoral substance is responsible for the control of blood flow through the microcirculation, but that combinations of humoral and chemical substances, which probably differ in each vascular bed, may interact with one another to ultimately control nutritive blood flow in the microcirculation. Evidence can be provided that the vasodilator effect of humoral substances, such as histamine, kinins, PGs, acetylcholine, and serotonin, can be counterbalanced or modified by combinations of other hormones such as catecholamines, angiotensin, vasopressin, glucocorticoids and sex steroids.

References

1 *Altura, B.M.:* Role of glucocorticoids in local regulation of blood flow. Am. J. Physiol. *211:* 1393–1397 (1966).

2 *Altura, B.M.:* Evaluation of neurohumoral substances in local regulation of blood flow. Am. J. Physiol. *212:* 1447–1454 (1967).

3 *Altura, B.M.:* Chemical and humoral regulation of blood flow through the precapillary sphincter. Microvasc. Res. *3:* 361–384 (1971).

4 *Altura, B.M.:* Sex and estrogens and responsiveness of terminal arterioles to neurohypophyseal hormones and catecholamines. J. Pharmac. exp. Ther. *193:* 403–412 (1975).

5 *Altura, B.M.:* Pharmacology of venular smooth muscle: new insights. Microvasc. Res. (in press).

6 *Altura, B.M. and Altura, B.T.:* Effects of local anesthetics, antihistamines, and glucocorticoids on peripheral blood flow and vascular smooth muscle. Anesthesiology *41:* 197–214 (1974).

7 *Altura, B.M. and Altura, B.T.:* Peripheral vascular actions of glucocorticoids and their relationship to protection in circulatory shock. J. Pharmac. exp. Ther. *190:* 300–315 (1974).

8 *Altura, B.M. and Altura, B.T.:* Vascular smooth muscle and prostaglandins. Fed. Proc. Fed. Am. Socs exp. Biol. *35:* 2360–2366 (1976).

9 *Altura, B.M. and Altura, B.T.:* Vascular smooth muscle and neurohypophyseal hormones. Fed. Proc. Fed. Am. Socs exp. Biol. *36:* 1853–1861 (1977).

10 *Altura, B.M. and Altura, B.T.:* Influence of sex hormones, oral contraceptives and pregnancy on vascular muscle and its reactivity; in *Carrier and Shibata* Factors influencing reactivity in vascular smooth muscle (Igaku-Shoin, Tokyo, in press).

11 *Altura, B.M. and Halevy, S.:* Cardiovascular actions of histamine; in *Rocha e Silva* Handbook of experimental pharmacology. Histamine and anti-histaminics, vol. XVIII, part 2, pp. 1–39 (Springer, Berlin, in press).

12 *Baez, S.; Orkin, L.R., and Lorenzo-Lagisquet, J.A.:* Antagonism of some vascular smooth muscle agonists by histamine and betahistine. Microvasc. Res. *3:* 170–182 (1971).

13 *Berne, R.M.:* Regulation of coronary flow. Physiol. Rev. *44:* 1–29 (1964).

14 *Blumberg, A.L.; Denny, S.E.; Marshall, G., and Needleman, P.:* Blood-vessel hormone interactions: angiotensin, bradykinin and prostaglandins. Am. J. Physiol. *232:* H305–H310 (1977).

15 *Gimbrone, M.A., jr. and Alexander, R.W.:* Angiotensin II stimulation of prostaglandin production in cultured human vascular endothelium. Science *189:* 219–220 (1975).

16 *Haddy, F.J. and Scott, J.B.:* Metabolic factors in peripheral circulatory regulation. Fed. Proc. Fed. Am. Socs exp. Biol. *34:* 2006–2011 (1975).

17 *Herbaczynska-Cedro, K.; Staszewska-Barczak, J., and Janczewska, H.:* The release of prostaglandin-like substances during reactive and functional hyperemia in the hindleg of the dog. Pol. J. Pharmacol. Pharm. *26:* 167–170 (1974).

18 *Herbaczynska-Cedro, K. and Vane, J.R.:* Prostaglandins as mediators of reactive hyperemia in kidney. Nature, Lond. *247:* 492 (1974).

19 *Kadowitz, P.J.; Joiner, P.D., and Hyman, A.L.:* Physiological and pharmacological roles of prostaglandins. Annu. Rev. Pharmacol. *15:* 285–306 (1975).

20 *Lanciault, G. and Jacobson, E.D.:* The gastrointestinal circulation. Gastroenterology *71:* 851–873 (1976).

21 *Messina, E.J.; Weiner, R., and Kaley, G.:* Inhibition of bradykinin and potentiation of norepinephrine and angiotensin vasoconstriction by inhibitors of prostaglandin synthesis in skeletal muscle of the rat. Circulation Res. *37:* 430–437 (1975).

22 *Messina, E.J.; Weiner, R., and Kaley, G.:* Prostaglandins and local circulatory control. Fed. Proc. Fed. Am. Socs exp. Biol. *35:* 2367–2375 (1976).

23 *McGrath, M.A. and Shepherd, J.T.:* Inhibition of adrenergic neurotransmission in canine vascular smooth muscle by histamine. Circulation Res. *39:* 566–573 (1976).

24 *Northover, A.M. and Northover, B.J.:* The effects of histamine, 5-hydroxytryptamine and bradykinin on rat mesenteric blood vessels. J. Path. *98:* 265–276 (1969).

25 *Powell, J.R. and Brody, M.J.:* Identification and blockade of vascular H_2-receptors. Fed. Proc. Fed. Am. Socs exp. Biol. *35:* 1935–1941 (1976).

26 *Shepherd, J.T. and Vanhoutte, P.M.:* Veins and their control (Saunders, Philadelphia 1975).

27 *Sicuteri, F.; Fanciullacci, M.; Franchi, G.; Bianco, P.L. del:* Serotonin-bradykinin potentiation on the pain receptors in man. Life Sci. *4:* 309–316 (1965).

28 *Vane, J.R. and McGiff, J.C.:* Possible contributions of endogenous prostaglandins to the control of blood pressure. Circulation Res. *36/37:* suppl. I, pp. I-68–1-75 (1975).

29 *Vanhoutte, P.M.; Lorenz, R.R., and Tyce, G.E.:* Inhibition of norepinephrine-^{3}H release from sympathetic nerve endings in veins by acetylcholine. J. Pharmac. exp. Ther. *185:* 386–394 (1973).

30 *Weiner, R. and Altura, B.M.:* Serotonin-bradykinin synergism in the mammalian capillary bed. Proc. Soc. exp. Biol. Med. *124:* 494–497 (1967).

31 *Weiner, R. and Kaley, G.:* Influence of prostaglandin E_1 on the terminal vascular bed. Am. J. Physiol. *217:* 563–566 (1969).

32 *Zweifach, B.W.:* The behavior of the vascular barrier during tissue injury. Biblthca anat., No. 4, pp. 21–31 (Karger, Basel 1964).

33 *Zweifach, B.W.:* Perspectives in microcirculation; in *Kaley and Altura* Microcirculation, vol. 1, pp. 1–19 (University Park Press, Baltimore 1977).

Prof. *B.M. Altura*, Department of Physiology, Box 31, SUNY Downstate Medical Center, 450 Clarkson Avenue, *Brooklyn NY 11203* (USA)

Mechanisms of Vasodilatation. Satellite Symp., 27th Int. Congr. Physiol. Sci., Wilrijk 1977, pp. 107–121 (Karger, Basel 1978)

Prostacyclin (PGI_2), the Vascular Wall and Vasodilatation

S. Moncada and J.R. Vane

Wellcome Research Laboratories, Langley Court, Beckenham, Kent

1. Background: Vascular Effects of Prostaglandins

Prostaglandins (PGs) are potent vasoactive agents with a wide variety of actions depending on the species tested, the vascular bed and the PG used (6, 9, 52). PGs of the F series constrict mainly capacitance vessels whereas PGs of the E series dilate arterioles, metarterioles, precapillaries and venules (24, 41).

PGE_2 constricts renal blood vessels of the rat and potentiates vascular responses to constrictor agents in different vascular beds of the rat (61, 62, 79). In most other species, including man, PGE_2 relaxes vascular smooth muscle, counteracts vasoconstriction induced by catecholamines, angiotensin II and vasopressin (80) and modulates sympathetic neurotransmission (47).

The effects of intravenous and intra-arterial injections of PGs are usually brief, passing off within a few minutes (81). One of the reasons for such a fleeting effect is pulmonary metabolism. Since E and F type PGs are actively removed by the lungs (34), it is unlikely that endogenous circulating PGs will affect vascular tone. The regulatory role of PGs will depend on their action at the site of synthesis, i.e. in the vascular wall.

PGs have potent actions on the micro-circulation where, in contrast to their effect when given intravenously, they have a long-lasting effect (80) and then antagonize the responses to pressor agents for periods which outlast their own vasodilator effect or their presence in the external bathing fluid (106).

Ferreira (30) showed that erythema induced by intradermal PGE_1 in man lasts for up to 10 h. The role of PGs in the mediation or modulation of reactive and functional hyperemia, in regulation of blood flow and in the development of the inflammatory response has recently been the subject of many reviews (63, 64, 97) and a recent symposium (57).

Arachidonic acid, the precursor of PGE_2 and $PGF_{2\alpha}$, produces vasodilatation when administered intravenously to cats (55), rats (17), rabbits (53, 59) and dogs (91), or locally into the coronary circulation of the dog's heart (50) or to isolated preparations *in vitro,* such as the Langendorff perfused rabbit heart (82). Since vasodilatation induced by arachidonic acid can be inhibited by aspirin-like drugs, which inhibit PG biosynthesis (105), it has been assumed that this effect is due to intramural conversion of arachidonic acid into PGE_2, its main recognized vasodilator metabolite.

Until recently, there seemed substantial evidence to suggest that vascular tissue mainly synthesized PGs of the E_2 type. Homogenates of rabbit aorta converted only 1% of its radioactive precursor into PGE_1, which represented a low conversion when compared with 41% converted by the rabbit renal medulla (15). Low rates of conversion in aortas were reported by *Hollander et al.* (51). Moreover, incubations of human umbilical artery produce an immunoreactive $PGF_{2\alpha}$ – like material at a rate of 15 ng/g tissue wet weight in 2 h (104) and cultured human endothelial cells harvested from umbilical veins produce PGE as detected by radio-immunoassay (38). *Terragno et al.* (100) reported that slices of bovine mesenteric arteries and veins released into the incubating medium a material which was biologically and chromatographically characterized as PGE_2 and $PGF_{2\alpha}$. They found a high rate of conversion of arachidonic acid (20%) in bovine vessels.

Various types of excitation initiate tissue PG synthesis, including changes in the chemical and physical environment. Some tissues, like the cat and dog spleen *in vitro* or *in vivo* (20, 31–33, 37), dog kidney (48), the rabbit heart *in vitro* and the dog heart *in vivo* (5, 12) and gravid uterus (103) continuously release a PG into the venous effluent. This release can be enhanced by a variety of stimuli including trauma, vasoconstriction, vasodilatation and anoxia. Although it is not clear which cell types generate PGs, it is very likely that some come from the vasculature.

Indirect evidence for the participation of PGs in vascular reactivity has been obtained by the use of PG synthetase inhibitors (31, 96, 105). In the first experiments using indomethacin, it was found that the vasoconstrictor effects of adrenaline were enhanced after inhibition of PG synthesis in the dog isolated spleen (31). The responses to noradrenaline, angiotensin II and nerve stimulation were also enhanced after PG synthesis inhibition in the cat isolated spleen (33).

The release of PGs into renal venous blood has been well documented (19, 25, 65, 66), and indomethacin potentiates the vasoconstrictor responses to angiotensin and noradrenaline and nerve stimulation in autoperfused dog kidney (4) and pump-perfused rat kidney (84).

Similar results have been obtained using the blood-perfused dog's paw and the perfused ear of the rabbit (1, 54). Indomethacin or eicosatetraynoic acid (EYA), another inhibitor of PG biosynthesis (1, 23), increase the perfusion pressure and enhance responses to nerve stimulation and noradrenaline (107). Administration of PG synthetase inhibitors to vascular areas where no basal release of PGs was observed under resting conditions (4, 32, 60) produced no increase in resting perfusion pressure (4, 66) or potentiation of vasoconstrictor responses to angiotensin or noradrenaline. However, in the cremaster muscle of the rat, indomethacin and EYA potentiated the vasoconstrictor effects of noradrenaline and angiotensin (68, 69).

Aiken (2) showed that relaxation induced by bradykinin in strips of coeliac and mesenteric arteries was abolished by indomethacin. More recently, *McGiff* (67, 101, 102) and colleagues suggested that vascular PGs may modulate not only vasoconstrictor but also vasodilator responses. They demonstrated that in the canine kidney bradykinin elicits vasodilatation and release of a PGE-like substance (characterized by bioassay and thin-layer chromatography – t.l.c.) and that indomethacin reduces bradykinin vasodilatation. Similar results were obtained by *Messina et al.* (69) and by *Needleman* (82) using the rabbit-isolated heart. *Gryglewski and Korbut* (44) and *Grodzinska et al.* (42) found that noradrenaline infusions released PGE-like activity into the perfusate from the rabbit mesenteric vascular beds *in vitro.* This release was inhibited by treatment of the preparation with indomethacin or fenamate. They observed that 'tolerance' to noradrenaline was related to PG release and that this was abolished by inhibition of PG biosynthesis. *Aiken* (3) demonstrated that indomethacin or meclofenamate reversed angiotensin tachyphylaxis in isolated strips of rabbit coeliac and mesenteric arteries. He attributed this effect to the release of PGE_2.

PG synthetase inhibitors potentiate responses to low frequency stimulation of strips of rabbit portal vein (40) and perfused rabbit arteries (45) suggesting the existence of a PG-mediated negative feedback mechanism. There were several discrepancies between arachidonic acid effects and PGE_2 effects. On rabbit isolated heart where arachidonic acid dilated the coronary vasculature, PGE_2 had little or no effect (11, 82). In dogs, doses of arachidonic acid which produced equivalent hypotension to a concentration of PGE_2 did not have the same positive inotropic effects (91). Finally, the bovine coronary artery, which relaxed to arachidonic acid, contracted to PGE_2 (58).

With the isolation and availability of PG endoperoxides, PGG_2 and PGH_2 (46, 85) and the demonstration of their powerful vasoactive properties both *in vivo* and *in vitro* (72), the proposition was made that at least in some preparations, such as the isolated human or bovine coronary artery, the PG endoperoxides were the main biologically active products, since these substances differed from PGE_2 in relaxing this tissue (83).

The same suggestion was made by *Hintze and Kaley* (50) who observed coronary dilatation to arachidonic acid in the canine heart *in vivo,* and by *Fitzpatrick et al.* (35) who showed arachidonic acid induced vasoconstriction in the perfused hind limb of the dog.

2. The Discovery of Prostacyclin

Working with a microsomal enzyme extracted from blood vessels, we recently found that PG endoperoxides were transformed into an unstable substance which is a potent vasodilator and an inhibitor of platelet aggregation (75). This compound, which we originally called PGX (43, 74, 75), was later chemically identified as an unstable intermediate in the formation of 6-oxo-$PGF_{1\alpha}$, a compound described earlier (21, 86). PGX was then renamed prostacyclin or PGI_2 (56).

PGI_2 is formed by other vascular tissues from several species including rabbit, ox and human (13, 26, 77), and has been established as the main metabolic product of arachidonic acid in vascular tissue (56, 74, 75). In view of this, all the evidence suggesting the participation of PGE_2 or $PGF_{2\alpha}$ or their unstable intermediates in the regulation of vascular tone has to be reinterpreted.

PGI_2 is a vasodilator and relaxes strips of coeliac and mesenteric arteries of the rabbit (13) and bovine coronary arteries (26). *Dusting et al.* (26) have shown that PGI_2 is the metabolite of arachidonic acid responsible for the paradoxical coronary artery relaxation described by *Kulkarni et al.* (58). The bovine coronary artery strip is contracted by PGE_2, PGD_2 and $PGF_{2\alpha}$ but relaxed by arachidonate, PGH_2 (26, 58), and PGI_2 (26). Unlike the bovine coronary artery, the porcine coronary artery contracts to PGH_2 (28, 98), and we have shown that PGI_2 is a weak constrictor of this tissue (27). Moreover, PGI_2 is a potent coronary vasodilator in the Langendorff perfused guinea-pig and rabbit heart, where it represents the major metabolic product of arachidonic acid (94, 95) and is released by vasodilators such as bradykinin (93). In the rabbit heart, others have reported the release of 6-oxo-$PGF_{1\alpha}$, the metabolite of PGI_2, and PGI_2 itself (22, 90). In the canine coronary circulation, PGI_2 is a vasodilator in doses ranging from 50–1,000 ng, which makes it more potent than PGE_2 (7). The effects of PGI_2 are enhanced after indomethacin treatment. In rats and rabbits, PGI_2 given intravenously is 4–8 times more potent than PGE_2 in producing hypotension and more than 100 times more active than its stable end-product 6-oxo-$PGF_{1\alpha}$ (fig. 1) (8).

Interestingly, in the rat and rabbit, PGI_2 is equipotent as a vasodilator when given intra-arterially or intravenously (8). In this respect, PGI_2 is different from PGE_2 which is much less active intravenously due to pulmonary inactivation (34); PGI_2 is not metabolized by passage through the lungs (unpublished observations).

Further evidence for the formation of PGI_2 in blood vessel walls has been obtained by the use of 15-hydroperoxy arachidonic acid (15-HPAA), a potent and selective inhibitor of its formation in microsomal fractions and fresh rings of vascular tissue (43, 74). In coeliac and mesenteric artery strips of the rabbit, the PG endoperoxide PGH_2 produced a biphasic response, consisting of a sharp

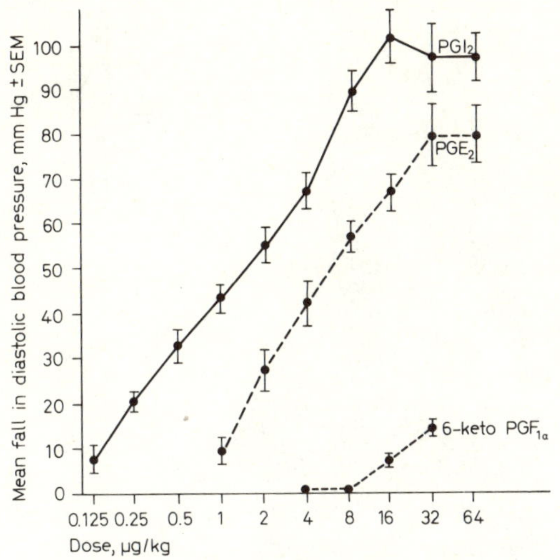

Fig. 1. Vasodepressor responses in rats to PGE_2, PGI_2 or 6-oxo-$PGF_{1\alpha}$. Mean falls (± SEM) in blood pressure plotted against the dose of each agent administered intravenously. Each point is the mean of 4–6 observations. From *Armstrong et al.* (8).

contraction followed by a longer lasting relaxation (13). Treatment of this tissue with 15-HPAA transforms this response into a contraction (fig. 2). PGH_2 produces in the bovine coronary artery either relaxation or sharp contraction followed by a long-lasting relaxation. The relaxation due to PG endoperoxide or the precursor arachidonic acid can be inhibited by 15-HPAA (26) and in some cases the relaxation induced by the endoperoxide is transformed into a contraction (fig. 3).

We have concluded from these experiments that the intrinsic activity of PG endoperoxides is a contraction which only becomes evident after preventing transformation of the endoperoxide into PGI_2. Additional evidence for this proposition comes from our work with porcine coronary artery strips where PGI_2 is a weaker constrictor than PGH_2 (27).

Dilatation of the coronary circulation induced by arachidonic acid in the isolated perfused heart of the guinea-pig is prevented by 15-HPAA treatment, accompanied by inhibition of the release of PGI_2 (measured by bioassay) or its stable end-product 6-oxo-$PGF_{1\alpha}$ (measured by t.l.c.) (95). These findings demonstrate that the metabolite of arachidonic acid present in vascular walls is PGI_2

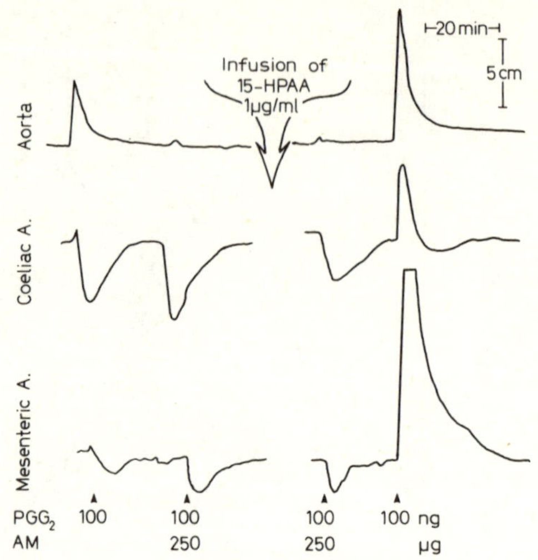

Fig. 2. The effect of 15-hydroperoxy arachidonic acid (15-HPAA) at 1 μg/ml on the responses to PGG_2 and PGI_2 on strips of rabbit aorta, rabbit coeliac and mesenteric arteries. PGG_2 (100 ng) contracted the rabbit aorta and caused a small contraction followed by a longer-lasting relaxation of the coeliac and mesenteric arteries. PGI_2, formed by incubating PGG_2 (100 ng) with 250 μg pig aortic microsomes (AM) did not contract rabbit aorta but relaxed the coeliac and mesenteric arteries, being more potent than the parent endoperoxide. A 5-min infusion of 15-HPAA (1 μg/ml) caused contraction of all tissues (not shown). An hour later PGG_2 (100 ng) gave a much greater contraction of rabbit aorta and now also contracted coeliac and mesenteric arteries. The activity of PGI_2 was not significantly altered after the 15-HPAA infusion. From *Bunting et al.* (13).

which could, therefore, play the regulatory role previously ascribed to PGE_2. Indeed, in the ductus arteriosus, where *Coceani and Olley* (16) proposed PGE_2 to be responsible for maintaining the fetal circulation, there are already two reports showing 6-oxo-$PGF_{1\alpha}$ to be the main metabolite (87, 89).

Several factors helped to mislead investigators studying arachidonic acid metabolism in vascular walls. First of all, PGI_2 is unstable and cannot be detected by acid chromatography. As pointed out by *Fried and Barton* (36), the claimed isolation of an isomer of PGI_2 in 1970 by *Pace-Asciak and Wolfe* (88) is incompatible with the known chemical properties. Secondly, the chromatographic mobility of the stable end-product 6-oxo-$PGF_{1\alpha}$ is very similar to PGE_2 or $PGF_{2\alpha}$ using most of the solvent systems (18). Thirdly, the commonly used bioassay tissues (rat stomach strip, chick rectum and rat colon) do not easily

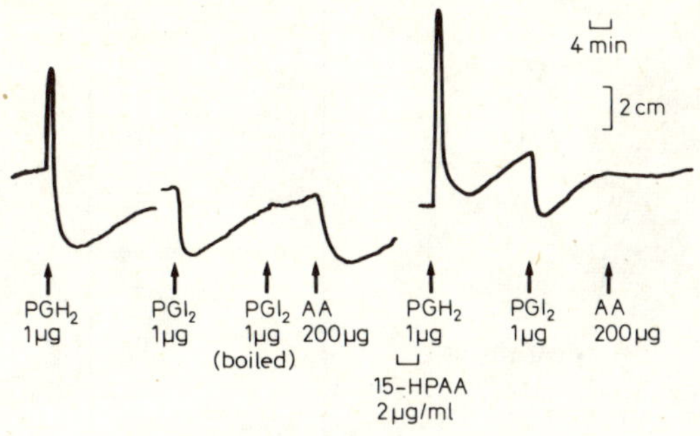

Fig. 3. Auxotonic recording from a superfused coronary artery strip. The effects of PGH_2 and sodium arachidonate (AA) are compared before and after treatment with 15-hydroperoxy arachidonic acid (15-HPAA) for 30 min as indicated by the horizontal bar. Boiling the microsomal suspension containing PGI_2 for 30 sec destroyed its activity. Note the change in effects of PGH_2 and AA after 15-HPAA treatment. From *Dusting et al.* (26).

differentiate between PGE_2 and 6-oxo-$PGF_{1\alpha}$ even though 6-oxo-$PGF_{1\alpha}$ is rather less active. This has led to the development of new bioassay tissues such as the bovine coronary artery which responds differently to PGE_2 and PGI_2 (73). Fourthly, the cross-reactivity of PG antibodies with 6-oxo-$PGF_{1\alpha}$ is not known but it is possible that such cross-reactivity has led to misinterpretations. In this context, it will be interesting to see whether 6-oxo-$PGF_{1\alpha}$ reacts with antibodies for PGA_2.

Finally, most of the previous work on arachidonic acid metabolism has been carried out in the presence of exogenous cofactors (such as glutathione or hydroquinone) and some of these substances may seriously disrupt the natural pathway of endoperoxide metabolism. Thus, in ram seminal vesicles where the formation of 6-oxo-$PGF_{1\alpha}$ had been previously described (14), *Cottee et al.* (18) have demonstrated that with low substrate concentrations and no cofactors, 6-oxo-$PGF_{1\alpha}$ is the main metabolite of arachidonic acid. Moreover, using a microsomal enzyme from vessel microsomes, *Salmon et al.* (92) have demonstrated that the presence of glutathione diverts the pathway from 6-oxo-$PGF_{1\alpha}$ to PGE_2.

Besides being a vasodilator, PGI_2 is the most potent inhibitor of platelet aggregation so far described (78). Its potency is 30–40 times that of PGE_1 and 10–20 times that of PGD_2 in human platelet-rich plasma against arachidonic acid-induced aggregation (fig. 4). It can also reverse the aggregation process (74). PGI_2 is also a potent inhibitor of aggregation induced by other agents including collagen, ADP and PG endoperoxides *in vitro* (56, 75), and *in vivo* it inhibits

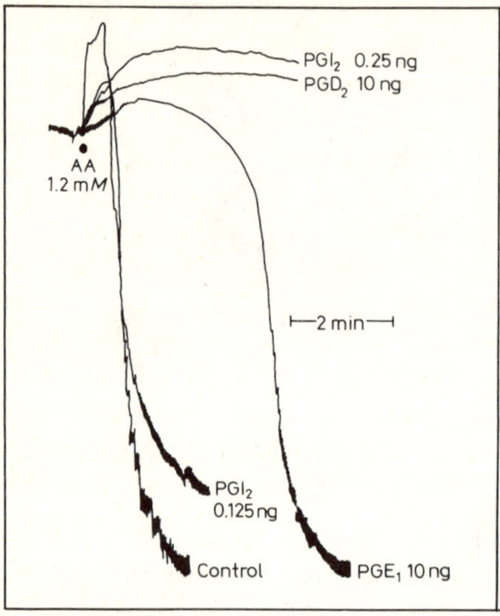

4

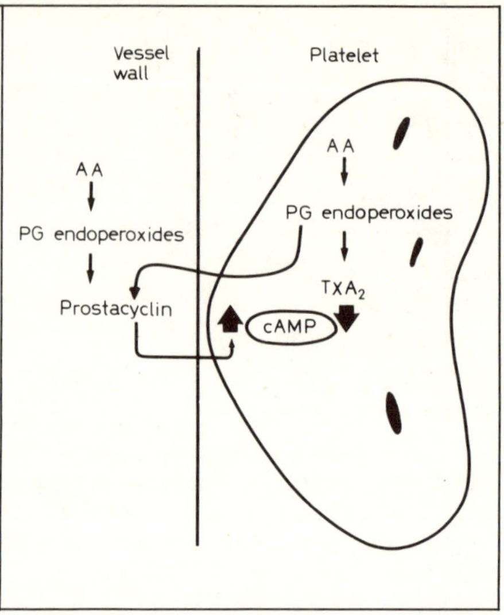

5

Fig. 4. Superimposed tracings showing inhibition of platelet aggregation by PGI_2, PGE_1 and PGD_2. Concentrations (per ml) of each PG (shown at the side of each tracing) were incubated for 1 min in platelet-rich plasma prior to the addition of arachidonic acid (1.2 m*M*). Note the difference in potency between the three PGs.

Fig. 5. Diagramatic representation of the cAMP control mechanism in platelets. Stimulation of platelets to aggregate leads to the formation of PG endoperoxide which can either be converted to TXA_2 which prevents stimulation of adenyl cyclase or escape and be utilized by PGI_2 synthetase. PGI_2 can also be synthesized by the vessel wall from arachidonic acid. The potent stimulatory effect of PGI_2 on the adenyl cyclase of platelets will increase cAMP and, as a consequence, inhibit platelet aggregation.

thrombus formation due to ADP in the microcirculation of the hamster cheek pouch (49). PGI_2 is synthesized by the vascular wall from endogenous precursors or from endoperoxides liberated by platelets during *in vitro* stirring (13, 43, 74). The enzyme responsible for PGI_2 formation is distributed throughout the vascular wall but is more concentrated in the intimal surface (76), decreasing progressively towards the adventitia. As do other less active PGs, such as PGE_1 and PGD_2, PGI_2 inhibits platelet aggregation by stimulating adenyl cyclase in the platelets leading to an increase in cyclic AMP (cAMP) levels (39, 99). In this respect, PGI_2 is 10 times more active than PGD_2 and 30 times more active than PGE_1. The stable metabolite 6-oxo-$PGF_{1\alpha}$ which is almost devoid of any anti-aggregatory activity is about 1,000 times less active than PGI_2 in increasing cAMP levels in platelets.

The PG endoperoxides and thromboxane A_2, the main metabolites of arachidonic acid in platelets, have an inhibitory activity on platelet adenylcyclase (70, 71). Because of this, we and others (39, 99) have proposed a regulatory mechanism on platelet cAMP which depends on a balance between thromboxane A_2 and PGI_2 activity and have postulated that platelet aggregability *in vivo* is controlled via a PGI_2 feedback mechanism (fig. 5). We demonstrated that incubation of platelet-rich plasma with vessel microsomes (75) or with fresh arterial tissue (13, 76) led to an inhibition of platelet aggregation which depended on the amount of tissue added and the time of incubation. This property of the microsomes or the fresh arterial tissue was abolished by previous treatment with 15-HPAA, the inhibitor of PGI_2 formation.

Recently, *Best et al.* (10) have demonstrated that incubation of platelet-rich plasma with fresh arterial tissue leads to an increase in cAMP which parallels the inhibition of aggregation. This effect of the arterial tissue is abolished by treatment of the tissue with tranylcypromine, another less active inhibitor of PGI_2 formation (43). These experiments, therefore, support our proposition of an homeostatic mechanism regulating platelet aggregability *in vivo* via a PGI_2 mechanism.

Whether a change of cAMP in vascular smooth muscle accompanies or even mediates the vasodilator effect of PGI_2 remains to be investigated.

3. Conclusion

The discovery of PGI_2 as the main metabolite of arachidonic acid in the vascular wall opens new perspectives for the understanding of the participation of PGs in vasodilatation and autoregulatory mechanisms of the cardiovascular system. Moreover, PGI_2 generation may explain many of the interactions postulated between platelets and the vascular wall. The formation of 6-oxo-$PGF_{1\alpha}$ has been demonstrated in other tissues such as the rat stomach fundus (86), the rat uterus (29), ram seminal vesicles (14, 18) and guinea-pig lung (21). The presence of this pathway in these tissues, especially when it is predominant, suggests that PGI_2 or 6-oxo-$PGF_{1\alpha}$ play other roles besides the ones we have defined here.

References

1 *Ahern, D.G. and Downing, D.T.:* Inhibition of prostaglandin biosynthesis by eicosa-5,8,11,14-tetraynoic acid. Biochim. biophys. Acta *210:* 456–461 (1970).

2 *Aiken, J.W.:* Inhibitors of prostaglandin synthesis specifically antagonize bradykinin and angiotensin-induced relaxations of the isolated coeliac artery from rabbit. Pharmacologist *16:* 295 (1974).

3 *Aiken, J.W.:* Effects of prostaglandin synthesis inhibitors on angiotensin tachyphylaxis in the isolated coeliac and mesenteric arteries of the rabbit. Pol. J. Pharmac. Pharm. *26:* 217–227 (1974).

4 *Aiken, J.W. and Vane, J.R.:* Intrarenal prostaglandin release attenuates the renal vasoconstrictor activity of angiotensin. J. Pharmac. exp. Ther. *184:* 678–687 (1973).

5 *Alexander, R.W.; Kent, K.M.; Pisano, J.J.; Keiser, H.R., and Cooper, T.:* Regulation of canine coronary blood flow by endogenously synthesized prostaglandins (abstr.). Circulation *7/8:* suppl. IV, p. 107 (1973).

6 *Änggård, E. and Bergström, S.:* Biological effects of an unsaturated trihydroxy acid ($PGF_{2\alpha}$) from normal swine lung. Acta physiol. scand. *58:* 1–12 (1963).

7 *Armstrong, J.M.; Chapple, D.; Dusting, G.J.; Hughes, R.; Moncada, S., and Vane, J.R.:* Cardiovascular actions of prostacyclin in chloralose anaesthetized dogs. Br. J. Pharmacol. *61:* 136P (1977).

8 *Armstrong, J.M.; Lattimer, N.; Moncada, S., and Vane, J.R.:* Comparison of the vasodepressor effects of prostacyclin and 6-oxo-prostaglandin $F_{1\alpha}$ with those of prostaglandin E_2 in rats and rabbits. Br. J. Pharmacol. *62:* 125–130 (1978).

9 *Bergström, S.; Carlson, L.A., and Weeks, J.R.:* The prostaglandins: a family of biologically active lipids. Pharmac. Rev. *20:* 1–48 (1968).

10 *Best, L.C.; Martin, T.J.; Russell, R.G.G., and Preston, F.E.:* Prostacyclin increases cyclic AMP levels and adenylate cyclase activity in platelets. Nature, Lond. *267:* 850–851 (1977).

11 *Block, A.J.; Feinberg, H.; Herbaczynska-Cedro, K., and Vane, J.R.:* Anoxic-induced release of prostaglandins in rabbit isolated hearts. Circulation Res. *36:* 34–42 (1975).

12 *Block, A.J.; Poole, S., and Vane, J.R.:* Modification of basal release of prostaglandins from rabbit isolated hearts. Prostaglandins *7:* 473–486 (1974).

13 *Bunting, S.; Gryglewski, R.J.; Moncada, S., and Vane, J.R.:* Arterial walls generate from prostaglandin endoperoxides a substance which relaxes strips of mesenteric and coeliac arteries and inhibits platelet aggregation. Prostaglandins *12:* 817–913 (1976).

14 *Chang, W.-C.; Murota, S.-I.; Matsuo, M., and Tsurufuji, S.:* A new prostaglandin transformed from arachidonic acid in carrageenin-induced granuloma. Biochem. biophys. Res. Commun. *72:* 1259–1264 (1976).

15 *Christ, E.J. and Dorp, D.A. van:* Comparative aspects of prostaglandins; Adv. Biosci., vol. 9, pp. 35–38 (Pergamon Press, Oxford 1972).

16 *Coceani, F. and Olley, P.M.:* The response of the ductus arteriosus to prostaglandins. Can. J. Physiol. Pharmacol. *51:* 220–225 (1973).

17 *Cohen, M.; Sztokalov, J., and Hinsch, E.:* The antihypertensive action of arachidonic acid in the spontaneously hypertensive rat and its antagonism by anti-inflammatory agents. Life Sci. *13:* 317–325 (1973).

18 *Cottee, F.; Flower, R.J.; Moncada, S.; Salmon, J.A., and Vane, J.R.:* Synthesis of 6-keto $PGF_{1\alpha}$ by ram seminal vesicle microsomes. Prostaglandins *14:* 413–423 (1977).

19 *Davies, H.A. and Horton, E.W.:* Output of prostaglandins from the rabbit kidney, its increase on renal nerve stimulation and its inhibition by indomethacin. Br. J. Pharmacol. *46:* 658–675 (1972).

20 *Davies, B.N.; Horton, E.W., and Withrington, P.G.:* The occurrence of prostaglandin E_2 in splenic venous blood of the dog following splenic nerve stimulation. Br. J. Pharmacol. Chemother. *32:* 127–135 (1968).

21 *Dawson, W.; Boot, J.R.; Cockerill, A.F.; Mallen, D.N.B., and Osborne, D.J.:* Release of novel prostaglandins and thromboxane after immunological challenge of guinea pig lung. Nature, Lond. *262:* 699–702 (1976).

22 *Deckere, E.A.M. de; Nugteren, D.H., and Ten Hoor, F.:* Prostacyclin is the major prostaglandin released from the isolated perfused rabbit and rat heart. Nature, Lond. *268:* 160–163 (1977).

23 *Downing, D.T.; Ahern, D.G., and Bachta, M.:* Enzyme inhibition by acetylenic compounds. Biochem. Biophys. Res. Commun. *40:* 218–223 (1970).

24 *Ducharme, D.W.; Weeks, J.R., and Montgomery, R.G.:* Studies on the mechanism of the hypertensive effects of prostaglandin $F_{2\alpha}$. J. Pharmac. exp. Ther. *160:* 1–10 (1968).

25 *Dunham, E.W. and Zimmerman, B.G.:* Release of prostaglandin-like material from dog kidney during nerve stimulation. Am. J. Physiol. *219:* 1279–1285 (1970).

26 *Dusting, G.J.; Moncada, S., and Vane, J.R.:* Prostacyclin (PGX) is the endogenous metabolite of arachidonic acid which relaxes coronary arteries. Prostaglandins *13:* 3–15 (1977).

27 *Dusting, G.J.; Moncada, S., and Vane, J.R.:* Prostacyclin (PGI_2) is a weak contractor of coronary arteries of the pig. Eur. J. Pharmacol. *45:* 301–304 (1977).

28 *Ellis, E.F.; Oelz, O.; Roberts, L.J.; Payne, A.L.; Sweetman, B.J.; Nies, A.S., and Oates, J.A.:* Coronary arterial smooth muscle contraction by a substance released from platelets: evidence that it is thromboxane A_2. Science, N.Y. *193:* 1135–1137 (1976).

29 *Fenwick, L.; Jones, R.L.; Naylor, B.; Poyser, N.L., and Wilson, N.H.:* Production of prostaglandins by the pseudopregnant rat uterus, *in vitro,* and the effect of tamoxifen with the identification of 6-keto prostaglandin $F_{1\alpha}$ as a major product. Br. J. Pharmacol. *59:* 191–199 (1977).

30 *Ferreira, S.H.:* Prostaglandins, aspirin-like drugs and analgesia. Nature new Biol. *240:* 200–203 (1972).

31 *Ferreira, S.H.; Moncada, S., and Vane, J.R.:* Indomethacin and aspirin abolish prostaglandin release from the spleen. Nature new Biol. *231:* 237–239 (1971).

32 *Ferreira, S.H.; Moncada, S., and Vane, J.R.:* Prostaglandins and the mechanism of analgesia produced by aspirin-like drugs. Br. J. Pharmacol. *49:* 86–97 (1973).

33 *Ferreira, S.H.; Moncada, S., and Vane, J.R.:* Some effects of inhibiting endogenous prostaglandin formation on the responses of the cat spleen. Br. J. Pharmacol. *47:* 48–58 (1973).

34 *Ferreira, S.H. and Vane, J.R.:* Prostaglandins: their disappearance from and release into the circulation. Nature, Lond. *216:* 868–873 (1967).

35 *Fitzpatrick, T.M.; Johnson, M.; Kot, P.A.; Ramwell, P.W., and Rose, J.C.:* Vasoconstrictor response to arachidonic acid in the isolated hind limb of the dog. Br. J. Pharmacol. *59:* 269–273 (1977).

36 *Fried, J. and Barton, J.:* Synthesis of 13,14-dehydroprostacyclin methyl ester: a potent inhibitor of platelet aggregation. Proc. natn. Acad. Sci. USA *74:* 2199–2203 (1977).

37 *Gilmore, N.; Vane, J.R., and Wyllie, J.H.:* Prostaglandins released by the spleen. Nature, Lond. *218:* 1135–1140 (1968).

38 *Gimbrone, M.A., jr. and Alexander, R.W.:* Angiotensin II stimulation of prostaglandin production in cultured human vascular endothelium. Science, N.Y. *189:* 219–220 (1975).

39 *Gorman, R.R.; Bunting, S., and Miller, O.V.:* Modulation of human platelet adenylate cyclase by prostacyclin (PGX). Prostaglandins *13:* 377–388 (1977).

40 *Greenberg, R.:* The effects of indomethacin and eicosa-5,8,11,14-tetraynoic acid on the response of the rabbit portal vein to electrical stimulation. Br. J. Pharmacol. *52:* 61–68 (1974).

41 *Greenberg, R.A. and Sparks, H.V.:* Prostaglandins and consecutive vascular segments of the canine hind limb. Am. J. Physiol. *216:* 567–571 (1969).

42 *Grodzinska, L.; Panczenko, B., and Gryglewski, R.J.:* Release of prostaglandin-like material from perfused mesenteric blood vessels of rabbits. J. Pharm. Pharmac. *28:* 40–43 (1976).

43 *Gryglewski, R.; Bunting, S.; Moncada, S.; Flower, R.J., and Vane, J.R.:* Arterial walls are protected against deposition of platelet thrombi by a substance (prostaglandin X) which they make from prostaglandin endoperoxides. Prostaglandins *12:* 685–713 (1976).

44 *Gryglewski, R.J. and Korbut, R.:* Prostaglandin feedback mechanism limits vasoconstrictor action of norepinephrine in perfused rabbit ear. Experientia *31:* 89–90 (1975).

45 *Hadazy, P. and Nador, T.:* Effects of indomethacin and PGE_1 on the vasoconstrictor responses of the rabbit ear artery to nerve stimulation. Prostaglandins *11:* 241–250 (1976).

46 *Hamberg, M. and Samuelsson, B.:* Detection and isolation of an endoperoxide intermediate in prostaglandin biosynthesis. Proc. natn. Acad. Sci. USA *70:* 899–903 (1973).

47 *Hedqvist, P.:* Autonomic neurotransmission; in *Ramwell* The prostaglandins, pp. 101–131 (Plenum Publishing, New York 1973).

48 *Herbaczynska-Cedro, K. and Vane, J.R.:* Prostaglandins as mediators of reactive hyperaemia in the kidney. Nature, Lond. *247:* 492 (1974).

49 *Higgs, G.A.; Moncada, S., and Vane, J.R.:* Prostacyclin (PGI_2) inhibits the formation of platelet thrombi induced by adenosine diphosphate (ADP) *in vivo.* Br. J. Pharmacol. *61:* 137P (1977).

50 *Hintze, T.H. and Kaley, G.:* Prostaglandins and the control of blood flow in the canine myocardium. Circulation Res. *40:* 313–320 (1977).

51 *Hollander, W.; Kramsch, D.M.; Franzblau, C.; Paddock, J., and Colombo, M.A.:* Suppression of atheromatous fibrous plaque formation by anti-proliferative and anti-inflammatory drugs. Circulation Res. *34:* 131–141 (1974).

52 *Horton, E.W. and Main, I.H.M.:* A comparison of the action of prostaglandin $F_{2\alpha}$ and E_1 on smooth muscle. Br. J. Pharmacol. *24:* 470–476 (1965).

53 *Ichikawa, S. and Yamada, J.:* Biological actions of free and albumin bound arachidonic acid. Am. J. Physiol. *203:* 681–684 (1962).

54 *Jackson, H.R. and Hall, R.C.:* The effect of aspirin on the response of the rabbit ear artery. Eur. J. Pharmacol. *21:* 107–110 (1973).

55 *Jacques, R.:* Arachidonic acid, an unsaturated fatty acid which produces slow contraction of smooth muscle and causes pain. Pharmacological and biochemical characteristics of its mode of action. Helv. physiol. pharmac. Acta *17:* 255–267 (1959).

56 *Johnson, R.A.; Morton, D.R.; Kinner, J.H.; Gorman, R.R.; McGuire, J.R.; Sun, F.F.; Whittaker, N.; Bunting, S.; Salmon, J.A.; Moncada, S., and Vane, J.R.:* The chemical characterization of prostaglandin X (prostacyclin). Prostaglandins *12:* 915–928 (1976).

57 *Kaley, G.:* The role of prostaglandins in vascular homeostasis. Fed. Proc. Fed. Am. Socs exp. Biol. *35:* 2358–2392 (1976).

58 *Kulkarni, P.S.; Roberts, R., and Needleman, P.:* Paradoxical endogenous synthesis of a coronary dilating substance from arachidonate. Prostaglandins *12:* 337–350 (1976).

59 *Larsson, C. and Änggård, E.:* Arachidonic acid lowers and indomethacin increases the blood pressure of the rabbit. J. Pharm. Pharmac. *25:* 653–655 (1973).

60 *Lonigro, A.J.; Itskovitz, H.D.; Crowshaw, K., and McGiff, J.C.:* Dependency of renal blood flow on prostaglandin synthesis in the dog. Circulation Res. *32:* 712–713 (1973).

61 *Malik, K.U. and McGiff, J.C.:* Modulation by prostaglandin E_1 of adrenergic transmission in isolated perfused rabbit and rat mesenteric arteries. Fed. Proc. Fed. Am. Socs exp. Biol. *35:* 763 (1975).

62 *Malik, K.U. and McGiff, J.C.:* Modulation by prostaglandins of adrenergic transmission in the isolated perfused rabbit and rat kidney. Circulation Res. *36:* 599–609 (1975).

63 *Malik, K.U. and McGiff, J.C.:* Cardiovascular actions of prostaglandins; in *Karim* Recent advances in prostaglandin research, vol. 3 (MTP, Lancaster 1976).

64 *McGiff, J.C.:* Prostaglandins as regulators of blood pressure. Hosp. Pract. *10:* 101–112 (1975).

65 *McGiff, J.C.; Crowshaw, K.; Terragno, N.A., and Lonigro, A.J.:* Release of a prostaglandin-like substance into renal venous blood in response to angiotensin II. Circulation Res. *26/27:* suppl. I, pp. 121–130 (1970).

66 *McGiff, J.C.; Crowshaw, K.; Terragno, N.A., and Lonigro, A.J.:* Renal prostaglandins: possible regulators of the renal actions of pressor agents. Nature, Lond. *227:* 1255–1257 (1970).

67 *McGiff, J.C.; Terragno, N.A.; Malik, K.U., and Lonigro, A.J.:* Release of a prostaglandin E-like substance from canine kidney by bradykinin. Circulation Res. *31:* 36–43 (1972).

68 *Messina, E.J.; Weiner, R., and Kaley, G.:* Inhibition of bradykinin vasodilatation and potentiation of norepinephrine and angiotensin vasoconstriction by inhibitors of prostaglandin synthesis in rat skeletal muscle. Circulation Res. *37:* 430–437 (1975).

69 *Messina, E.J.; Weiner, R., and Kaley, G.:* Prostaglandins and local circulatory control. Fed. Proc. Fed. Am. Socs exp. Biol. *35:* 2367–2375 (1976).

70 *Miller, O.V. and Gorman, R.R.:* Modulation of platelet cyclic nucleotide content by PGE_1 and the prostaglandin endoperoxide PGG_2. J. cyclic Nucl. Res. *2:* 79–87 (1976).

71 *Miller, O.V.; Johnson, R.A., and Gorman, R.R.:* Inhibition of PGE_1 – stimulated by cAMP accumulation in human platelets by thromboxane A_2. Prostaglandins *13:* 599–607 (1977).

72 *Moncada, S.:* Prostaglandin endoperoxides and thromboxanes: formation and effects. Arch. Pharmakol. *297:* 81–84 (1977).

73 *Moncada, S.; Ferreira, S.H., and Vane, J.R.:* Bioassay of prostaglandins and biologically active substances derived from arachidonic acid; in *Frölich* Advances in prostaglandin and thromboxane research, vol. 3 (Raven Press, Hewlett, in press).

74 *Moncada, S.; Gryglewski, R.J.; Bunting, S., and Vane, J.R.:* A lipid peroxide inhibits the enzyme in blood vessel microsomes that generates from prostaglandin endoperoxides the substance (prostaglandin X) which prevents platelet aggregation. Prostaglandins *12:* 715–733 (1976).

75 *Moncada, S.; Gryglewski, R.J.; Bunting, S., and Vane, J.R.:* An enzyme isolated from arteries transforms prostaglandin endoperoxides to an unstable substance that inhibits platelet aggregation. Nature, Lond. *263:* 663–665 (1976).

76 *Moncada, S.; Herman, A.G.; Higgs, E.A., and Vane, J.R.:* Differential formation of prostacyclin (PGX or PGI_2) by layers of the arterial wall. An explanation for the antithrombotic properties of vascular endothelium. Thromb. Res. *11:* 323–344 (1977).

77 *Moncada, S.; Higgs, E.A., and Vane, J.R.:* Human arterial and venous tissue generate prostacyclin (prostaglandin X), a potent inhibitor of platelet aggregation. Lancet *i:* 18–20 (1977).

78 *Moncada, S. and Vane, J.R.:* The discovery of prostacyclin (PGX) a fresh insight into arachidonic acid metabolism; in *Karsh and Fried* Biochemical aspects of prostaglandins and thromboxanes, pp. 155–177 (Academic Press, London/New York 1977).

79 *Mullane, K.M.; Armstrong, J.M., and McGiff, J.C.:* Potentiation by prostaglandins (PGs) of renal vascular sensitivity to pressor hormones in normotensive (NR) and genetically hypertensive (GHR) rats of the New Zealand strain; in *Samuelsson and*

Paoletti Advances in prostaglandin and thromboxane research, vol. 2, p. 954 (Raven Press, Hewlett 1976).

80 *Nakano, J.:* Cardiovascular actions; in *Ramwell* The prostaglandins, vol. 1, pp. 239–316 (Plenum Publishing, New York 1973).

81 *Nakano, J. and McCurdy, J.R.:* Cardiovascular effects of prostaglandin E_1. J. Pharmac. exp. Ther. *156:* 538–547 (1967).

82 *Needleman, P.:* The synthesis and function of prostaglandins in the heart. Fed. Proc. Fed. Am. Socs exp. Biol. *35:* 2376–2381 (1976).

83 *Needleman, P.; Kulkarni, P.S., and Raz, A.:* Coronary tone modulation: formation and actions of prostaglandins, endoperoxides and thromboxanes. Science, N.Y. *195:* 409–412 (1977).

84 *Needleman, P.; Marshall, G.R., and Johnson, E.M.:* Determinants and modification of adrenergic and vascular resistance in the kidney. Am. J. Physiol. *227:* 665–669 (1974).

85 *Nugteren, D.H. and Hazelhof, E.:* Isolation and properties of intermediates in prostaglandin biosynthesis. Biochim. biophys. Acta *326:* 448–461 (1973).

86 *Pace-Asciak, C.:* Isolation, structure and biosynthesis of 6-keto prostaglandin $F_{1\alpha}$ in the rat stomach. J. Am. chem. Soc. *98:* 2348–2349 (1976).

87 *Pace-Asciak, C. and Rangaraj, G.:* The 6-keto prostaglandin $F_{1\alpha}$ pathway in the lamb ductus arteriosus. Biochim. biophys. Acta *486:* 583–585 (1977).

88 *Pace-Asciak, C. and Wolfe, L.S.:* A novel prostaglandin derivative formed from arachidonic acid by rat stomach homogenates. Biochemistry *10:* 3657–3664 (1971).

89 *Powell, W.S. and Solomon, S.:* Formation of 6-oxo prostaglandin $F_{1\alpha}$ by arteries of the fetal calf. Biochem. biophys. Res. Commun. *75:* 815–822 (1977).

90 *Raz, A.; Isakson, P.C.; Minkes, M.S., and Needleman, P.:* Characterization of a novel metabolic pathway of arachidonate in coronary arteries which generates a potent endogenous coronary vasodilator. J. biol. Chem. *252:* 1123–1126 (1977).

91 *Rose, J.C.; Johnson, M.; Ramwell, P.W., and Kot, P.A.:* Effects of arachidonic acid on systemic arterial pressure, myocardial contractility and platelets in the dog. Proc. Soc. exp. Biol. Med. *147:* 652–655 (1974).

92 *Salmon, J.A.; Smith, D.R.; Flower, R.J.; Moncada, S., and Vane, J.R.:* Some characteristics of the prostacyclin-synthesizing enzyme in porcine aorta. Biochim. biophys. Acta (in press).

93 *Schrör, K.; Moncada, S., and Vane, J.R.:* Bradykinin – a powerful prostacyclin (PGX) releasing agent in the heart. Proc. Joint Italian-German Meet. in Pharmacology, Venice 1977 (in press).

94 *Schrör, K.; Moncada, S.; Ubatuba, F.B., and Vane, J.R.:* Formation of prostacyclin (PGX) causes decrease in the coronary vascular resistance during application of arachidonic acid in the rabbit isolated heart. Arch. Pharmakol. *297:* suppl. II, R31 (1977).

95 *Schrör, K.; Moncada, S.; Ubatuba, F.B., and Vane, J.R.:* Transformation of arachidonic acid and prostaglandin endoperoxides by the guinea-pig heart. Formation of RCS and prostacyclin. Eur. J. Pharmacol. (in press).

96 *Smith, J.B. and Willis, A.L.:* Aspirin selectively inhibits prostaglandin production in human platelets. Nature new Biol. *231:* 235–237 (1971).

97 *Staszewska-Barczak, J. and Vane, J.R.:* The role of prostaglandins in the local control of circulation. Clin. exp. Pharmacol. Physiol. *2:* suppl., pp. 71–78 (1975).

98 *Svensson, J. and Hamberg, M.:* Thromboxane A_2 and prostaglandin H_2: potent stimulators of the swine coronary artery. Prostaglandins *12:* 943–950 (1976).

99 *Tateson, J.E.; Moncada, S., and Vane, J.R.:* Effects of prostacyclin (PGX) on cyclic AMP concentrations in human platelets. Prostaglandins *13:* 389–397 (1977).

100 *Terragno, D.A.; Crowshaw, K.; Terragno, N.A., and McGiff, J.C.:* Prostaglandin synthesis by bovine mesenteric arteries and veins. Circulation Res. *36/37:* 76–80 (1975).
101 *Terragno, N.A.; Lonigro, A.J.; Malik, K.U., and McGiff, J.C.:* Bradykinin-induced vasodilatation and the release of prostaglandins. Circulation *44:* suppl. II, pp. 11–118 (1971).
102 *Terragno, N.A.; Lonigro, A.J.; Malik, K.U., and McGiff, J.C.:* The relationship of the renal vasodilator action of bradykinin to the release of a prostaglandin E-like substance. Experientia *28:* 437–439 (1972).
103 *Terragno, N.A.; Terragno, D.A.; Pacholczyk, D., and McGiff, J.C.:* Prostaglandins and the regulation of uterine blood flow in pregnancy. Nature, Lond. *249:* 57–58 (1974).
104 *Tuvemo, T. and Wide, L.:* Prostaglandin release from the human umbilical artery *in vitro.* Prostaglandins *4:* 689–694 (1973).
105 *Vane, J.R.:* Inhibition of prostaglandin synthesis as a mechanism of action for aspirin-like drugs. Nature, Lond. *231:* 232–235 (1971).
106 *Weiner, R. and Kaley, G.:* Influence of prostaglandin E_1 on the terminal vascular bed. Am. J. Physiol. *217:* 563–566 (1969).
107 *Zimmerman, B.G.; Ryan, M.J.; Gomer, S., and Kraft, E.:* Effect of the prostaglandin synthesis inhibitors indomethacin and eicosa-5,8,11,14-tetraynoic acid on adrenergic responses in the dog cutaneous vasculature. J. Pharmac. exp. Ther. *187:* 315–323 (1973).

S. Moncada, MD, Wellcome Research Laboratories, Langley Court, *Beckenham, Kent BR3 3BS* (England)

Mechanisms of Vasodilatation. Satellite Symp., 27th Int. Congr. Physiol. Sci., Wilrijk 1977, pp. 122–128 (Karger, Basel 1978)

Prostaglandin I_2 Synthesis and Vascular Effects in Isolated Coronary Artery[1]

Philip Needleman, Amiram Raz, Prasad S. Kulkarni, Angela Wyche and Sue E. Denny

Department of Pharmacology, Washington University Medical School, St. Louis, Mo.

The demonstration of endogenous synthesis of prostaglandins (PGs) in blood vessels has important implications with regard to the local vascular resistance. Such intrinsic biosynthesis of vasoactive substances can be produced by hormonal stimulation, neuronal stimulation, mechanical disturbances, and changes in oxygen tension. The focus of our work reported here involves the characterization of some of the chemical and biological properties of a new arachidonate product, now termed PGI_2. A schematic presentation of the arachidonate metabolic pathway is presented in figure 1. Since there is no storage form of prostaglandins or its precursor, activation requires lipase stimulation to release arachidonic acid from phospholipids. As will be described in blood vessels and heart, the major metabolic pathway is from arachidonate → endoperoxides → PGI_2 → 6-keto-$PGF_{1\alpha}$. This contrast with platelets where the sequence is arachidonate → endoperoxide → thromboxane A_2 → thromboxane B_2. Also shown in figure 1 are the known inhibitors (indicated with an x) of the various synthetic enzymes.

1. Contribution of Intrinsic Vascular PGs to Coronary Tone

Isolated spiral strips of beef and human coronary artery were constricted by PGE_2 and $PGF_{2\alpha}$ (ED_{50} – 500 ng/ml) but were relaxed by PGE_1 (6, 8). Inhibition of PG-cyclo-oxygenase by aspirin, indomethacin, or eicosa-5,8,11,14-tetraynoic acid (ETYA) led to a rapid and sustained increase in coronary tone which suggested continuous synthesis of endogenous vasodilator PG (6, 8).

[1] This work was supported by NIH Grants SCOR-HL17646, HE-14397, and HL-20787.

Acid-lipid extraction of media equilibrated with bovine coronary artery strips for 2 h yielded a PG-like substance which contracted rat stomach strips and had thin-layer chromatographic mobility comparable to PGE_1 or PGE_2 – solvent system A-1 (3, 6). Based on the responses to exogenous PGE_1 (dilator), PGE_2 and $PGF_{2\alpha}$ (constrictor) and to the increase in coronary muscle tension produced by cyclo-oxygenase inhibitors, the conclusion was previously drawn that intrinsic formation of PGE_1 exerts a relaxant effect on the inherent tone of the coronary arteries (6).

Alteration in the rate of PG biosynthesis within the blood vessel wall appears to be a determining factor regulating compensatory changes in the tone of bovine coronary arteries in response to reduced oxygen tension. The magnitude of the changes in spontaneous coronary tone were directly proportional, and the PG release inversely proportional to the oxygen tension of the bathing medium (7). The intrinsic rate of PG biosynthesis is increased under conditions of reduced oxygen tension, but is reduced at very low PO_2 (7). The reduced rate of PG production in anoxia is consistent with the requirement for molecular oxygen for the cyclo-oxygenase reaction. Coronary strips pretreated with indomethacin, ETYA, or aspirin responded to decreases in the bath oxygen tension with attenuated relaxation (7).

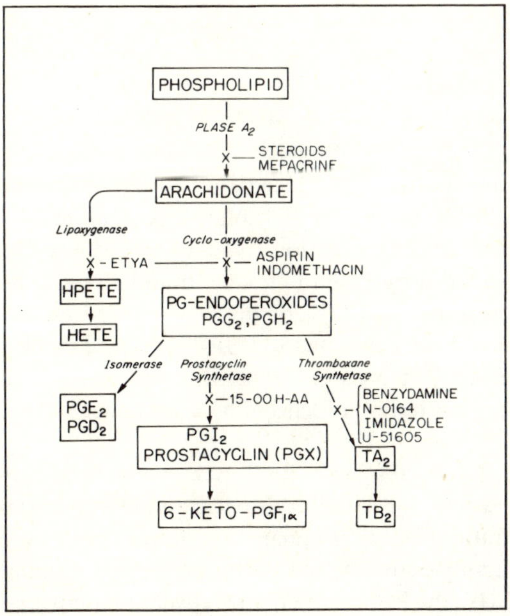

Fig. 1. Schematic representation of the arachidonate metabolic pathway.

2. Paradoxical Endogenous Synthesis of a Labile Coronary Dilating Substance from Arachidonate

Surprisingly, although the primary PGs PGE_2 and $PGF_{2\alpha}$ are constrictor (6, 8), administration of their precursor, arachidonic acid, caused relaxation of human as well as bovine and canine coronary arteries (8). This observation suggested that arachidonate was not converted by the coronary arteries to PGE_2 or $PGF_{2\alpha}$. However, the arachidonate induced coronary relaxation was inhibited by pretreatment with PG cyclo-oxygenase inhibitors and indomethacin addition to strips previously relaxed by arachidonate caused contraction (6, 8).

Since PGE_1, which relaxes the bovine and human coronary artery strips, cannot be formed from arachidonic acid, the coronary dilation due to arachidonate must be independent of PGE_1 formation. Linolenic and oleic acid also caused relaxation of the bovine coronary blood vessels both before and after indomethacin (8), thereby eliminating a direct nonspecific fatty acid effect as the cause of the arachidonate relaxation.

The intrinsic coronary relaxing substance produced from arachidonate appeared to be unstable, since addition of indomethacin after arachidonate resulted in coronary contraction (8). If the substance was stable after synthesis, then a cyclo-oxygenase inhibitor would not have been expected to exert an effect.

3. Effect of PG Endoperoxides on Bovine Coronary Arteries

A candidate for a relaxing substance produced from arachidonate may be the unstable endoperoxides PGG_2 or PGH_2. In isolated pig coronary artery strips, PGH_2 produces an initial small contraction followed by a slow increasing contraction similar to that produced by PGE_2 (2). Quantitation by gas chromatography-mass spectrometry revealed that 93% of the PGH_2 added to the bath with the pig coronary artery had been converted to PGE_2 (2). The data obtained with endoperoxides in pig coronary strips seem to conflict with the observation that bovine and human coronary arteries are relaxed by a substance synthesized from arachidonic acid (8).

PGH_2 was administered across a superfusion cascade which contained a bovine and porcine coronary artery and a rabbit thoracic aorta strip. Surprisingly, PGH_2 produced a concentration-dependent relaxation of the bovine coronary artery strip but contracted the pig coronary and rabbit aorta (12). Comparable results were obtained with PGH_3 (the PG-intermediate between eicosapentaenoic acid and PGE_3). In contrast, PGH_1 (the PG intermediate between dihomoγ-linolenate and PGE_1) which lacks the 5–6 double bond caused a dose-dependent contraction of bovine coronary artery strips as well as the porcine coronary and rabbit aorta (12). Thus, PGH_2 and PGH_3 caused a dose-dependent relaxation of bovine coronary artery strips, whereas PGH_1 produced coronary contraction. On the other hand, PGH_1, PGH_2, and PGH_3 constricted the pig coronary artery (12), in agreement with the report that PGH_2 contracted pig coronary artery (2). The primary PGs exert effects directionally opposite to their endoperoxide precursors. Thus, PGE_2 and PGE_3 contracted, and PGE_1 relaxed bovine coronary arteries (12). Although bovine coronaries do not possess thromboxane synthetase activity, they were profoundly constricted by exogenous thromboxane A_2 (2, 12).

4. Chemical Characterization of a Novel Metabolic Pathway of Arachidonate in Coronary Arteries which Generates a Potent Endogenous Coronary Vasodilator

The coronary relaxation induced by arachidonate and PGH_2 suggested that these substances may be converted to a PG other than PGE_2. Therefore, we studied the metabolism of [^{14}C] arachidonate by bovine coronary artery strips. Incubation of the [1-^{14}C] arachidonate labelled bovine coronary arteries for 4–6 h resulted in a gradual release of a PGE-like substance (0.4 ± 0.1 μg/g tissue/h) into the medium (as determined by bioassay) (15). Acid-lipid extraction of the medium followed by thin-layer chromatography showed the presence of a major radioactive product which co-migrated with standard PGE_2 in chromatography solvent system 'C' (chloroform:methanol:acetic acid:water 90:8:1:0.8) (15).

When the extract of the medium bathing a labelled coronary artery strip was chromatographed in a different solvent system (benzene:dioxane:acetic acid [60:30:3]), the 'PGE' peak separated into two peaks, one with chromatographic mobility identical to PGE_2 and the other with chromatographic mobility similar to $PGF_{2\alpha}$ (15). The lack of a radioactive peak, co-migrating with $PGF_{2\alpha}$ in chromatography system C, suggested that coronary arteries produce a substance different from PGE_2 or $PGF_{2\alpha}$. Alkali treatment (0.5 *N* KOH in methanol) of the total 'PGE' peak isolated from system C resulted in only partial conversion to PGB_2, whereas authentic PGE_2 was completely (>95%) converted. Sodium borohydride reduction of the 'PGE_2' peak from the coronary also gave only partial conversion to $PGF_{2\alpha}$ + $PGF_{2\beta}$ (15). The chemical and chromatographic properties of the unknown substance are similar to those of the major PG produced by the rat fundus homogenate which has been identified as 6-keto-$PGF_{1\alpha}$ (1, 13, 14).

5. Biological Characterization of the Unique PG Coronary Vasodilator

We found that addition of PGH_2 caused an immediate contraction of the rabbit thoracic aorta strip and a slightly delayed relaxation of the dog (or bovine) coronary artery (11). The same amount of PGH_2 was then incubated (2 min; room temperature) with bovine coronary artery microsomes and then tested. The PGH_2 activity (aorta contraction) had disappeared but an immediate profound coronary relaxation was produced. The relaxing activity disappeared if a 10-min delay was permitted before testing. The radioactive product obtained from long-term (15 min) incubations of ^{14}C-PGH_2 with blood vessel microsomes, was a mixture of 6-keto-$PGF_{1\alpha}$ and PGE_2. Thus, a labile intermediate between PGH_2 and 6-keto-$PGF_{1\alpha}$ seems to be responsible for the coronary relaxation induced by either arachidonic acid or by PGG_2 and PGH_2. The structure of the enzymatically produced labile intermediate between the PG endoperoxide and 6-keto-$PGF_{1\alpha}$ has been identified by *Johnson et al.* (12) and given the final name prostacyclin or PGI_2. The enzymatic conversion PGH_2 to PGI_2 by blood vessel microsomes was inhibited by 15-hydroperoxy arachidonic acid (15) and similarly pretreatment of bovine coronary strips blocked the conversion of PGH_2 into PGI_2 and thereby blocked the PGH_2-induced coronary relaxation.

We noted that incubation of PGH_1 with bovine coronary artery microsomes did not result in disappearance of the PGH_1-induced aorta contraction nor in alteration of the contractile effect on the coronary assay tissue (11). Thus, the conversion of endoperoxides to the labile relaxing substance apparently requires the presence of a 5–6 double bond. This bond is present in PGH_2 and PGH_3 (which dilate the bovine coronary) but is absent in PGH_1.

In independent investigations, it has been demonstrated that blood vessel microsomes from numerous sources (including aorta, mesenteric, coeliac, and coronary arteries) when incubated with PGH_2 produce the labile substance PGI_2 which proved to be a potent inhibitor of platelet aggregation and relaxes some vascular smooth muscles (4, 9, 10). The stable inactive metabolite of this pathway was also found to be 6-keto-$PGF_{1\alpha}$. These compounds were originally isolated and chemically characterized by *Pace-Asciak* (8, 9). The labile substance produced by coronary arteries had no effect on rabbit aorta but was an extremely potent inhibitor of platelet aggregation; these properties are very similar to those of prostacyclin (9).

The interaction between the arachidonate metabolites synthesized in blood vessels and platelets is of considerable patho-physiological significance. The two synthetic pathways appear to be physiologically antagonistic and the net response may reflect the algebraic sum of the two systems. Thus, stimulated platelets (which lack prostacyclin synthetase) release thromboxane A_2 which induces aggregation and vasoconstriction, whereas blood vessels (which lack thromboxane synthetase) synthesize a vasodilator that inhibits aggregation. Both substances are labile and therefore would be much more important in local vascular rather than systemic responses. Coronary tone, and possibly spasms, in ischemic zones of the myocardium may be influenced markedly by interplay between thromboxanes from platelets on the one hand and vasodilator endoperoxide products (i.e. PGI_2) synthesized endogenously on the other.

Summary

Arachidonic acid (AA) relaxes, whereas PGE_2 and $PGF_{2\alpha}$ contract, bovine and human coronary artery strips. The endogenous phospholipids of coronary artery strips were labelled with $[^{14}C]$ AA. The primary metabolite of AA released by the coronary strips was a novel prostaglandin, and some PGE_2. The novel PG was resistant to alkaline hydrolysis and sodium borohydride reduction. The mobility of this metabolite in multiple chromatography systems was similar to that of 6-keto-$PGF_{1\alpha}$. This metabolite demonstrated essentially no biologic activity. The precursor of 6-keto-$PGF_{1\alpha}$, PGH_2, relaxes bovine coronary artery strips, but the endoperoxide itself is rapidly degraded. A potent coronary relaxant is produced by incubating PGH_2 with microsomes from bovine coronary arteries. The coronary relaxant degraded to the stable end-product 6-keto-$PGF_{1\alpha}$. The biologically active component of this novel pathway of arachidonate metabolism appears to be an intermediate between PGH_2 and the final 6-keto-$PGF_{1\alpha}$-like-product. This intermediate, PGI_2, proves to a potent coronary vasodilator and a probable endogenous determinant of local vascular tone.

References

1 *Kalsner, S.:* Endogenous prostaglandin release contributes directly to coronary artery tone. Can. J. Physiol. Pharmacol. *53:* 560–565 (1975).

2 *Kulkarni, P.S.; Roberts, R., and Needleman, P.:* Paradoxical endogenous synthesis of a coronary dilating substance from arachidonate. Prostaglandins *12:* 337–353 (1976).

3 *Green, K. and Samuelsson, B.:* Thin layer chromatography of the prostaglandins. J. Lipid Res. *5:* 117–120 (1964).

4 *Kalsner, S.:* Intrinsic prostaglandin release. A mediator of anoxia-induced relaxation in an isolated coronary artery preparation. Blood Vessels *13:* 155–166 (1976).

5 *Ellis, E.F.; Oelz, O.; Roberts, L.J.; Payne, N.A.; Sweetman, B.J.; Nies, A.S., and Oates, J.A.:* Coronary arterial smooth muscle contraction by a substance released from platelets: evidence that it is thromboxane A_2. Science *193:* 1135 (1976).

6 *Needleman, P.; Kulkarni, P.S., and Raz, A.:* Coronary tone modulation: formation and actions of prostaglandins, endoperoxides and thromboxanes. Science *195:* 409–412 (1977).

7 *Raz, A.; Isakson, P.C.; Minkes, M.S., and Needleman, P.:* Characterization of a novel metabolic pathway of arachidonate in coronary arteries which generates a potent endogenous coronary vasodilator. J. biol. Chem. *252:* 1125–1126 (1977).

8 *Pace-Asciak, C.:* Isolation, structure and biosynthesis of 6-keto prostaglandin $F_{1\alpha}$ in the rat stomach. J. Am. chem. Soc. *98:* 2348–2349 (1976).

9 *Pace-Asciak, C. and Wolfe, L.S.:* A novel prostaglandin derivative formed from the arachidonic acid by rat stomach homogenates. Biochemistry, N.Y. *10:* 3657–3664 (1971).

10 *Dusting, G.J.; Moncada, S., and Vane, J.R.:* Prostacyclin (PGX) is the endogenous metabolite responsible for relaxation of coronary arteries induced by arachidonic acid. Prostaglandins *13:* 3–16 (1977).

11 *Needleman, P.; Raz, A.; Kulkarni, P.S.; Pure, E.; Wyche, A.; Denny, S.E., and Isakson, P.C.:* Biological and chemical characterization of a unique endogenous vasodilator prostaglandin produced in isolated coronary artery and in intact perfused heart. Intersciences Symp. on Prostaglandins, Santa Monica 1976.

12 *Johnson, R.A.; Morton, D.R.; Kinner, J.H.; Gorman, R.R.; McGuire, J.C.; Sun, F.F.; Whittaker, N.; Bunting, S.; Salmon, J.A.; Moncada, S., and Vane, J.R.:* The chemical structure of prostaglandin X (prostacyclin). Prostaglandins *12:* 915–928 (1976).

13 *Moncada, S.; Gryglewski, R.; Bunting, S., and Vane, J.R.:* An enzyme isolated from arteries transforms prostaglandin endoperoxides to an unstable substance that inhibits platelet aggregation. Nature, Lond. *263:* 663–665 (1976).

14 *Gryglewski, R.J.; Bunting, S.; Moncada, S.; Flower, R.J., and Vane, J.R.:* Arterial walls are protected against deposition of platelet thrombi by a substance (prostaglandin X) which they make from prostaglandin endoperoxides. Prostaglandins *12:* 685–713 (1976).

15 *Moncada, S.; Gryglewski, R.J.; Bunting, S., and Vane, J.R.:* A lipid peroxide inhibits the enzyme in blood vessel microsomes that generates from prostaglandin endoperoxides the substance which prevents platelet aggregation. Prostaglandins *12:* 715–738 (1976).

P. Needleman, MD, Department of Pharmacology, Washington University Medical School, *St. Louis, MO 63110* (USA)

Mechanisms of Vasodilatation. Satellite Symp., 27th Int. Congr. Physiol. Sci., Wilrijk 1977, pp. 129–136 (Karger, Basel 1978)

Heterogeneity in the Relaxation of Vascular Smooth Muscle

Noboru Toda

Department of Pharmacology, Shiga University of Medical Sciences, Seta, Ohtsu

Quantitative comparisons of a variety of vessels have demonstrated important differences in responsiveness to chemical and physical stimulation. Such heterogenous behavior has been observed among vascular smooth muscles taken from different regions of the same animal, from different species, from the same species at different ages or from intact and pathologic animals. This chapter focusses on the quantitative comparison of vascular relaxation induced *in vitro* by agents with different mechanisms of action.

1. Interventions which Release Endogenous Substances from the Vascular Wall

During contractions caused by prostaglandin (PG) $F_{2\alpha}$, the addition of *nicotine* causes: (1) a transient relaxation in helical strips of canine cerebral and coronary arteries; (2) a contraction followed by a relaxation of femoral arteries, and (3) a marked contraction of mesenteric and renal arteries (28–30). Hexamethonium and cocaine suppress the relaxation seen in cerebral arteries but not that observed in coronary arteries, suggesting that the latter relaxation is due to a non-specific, direct action on the smooth muscle cells. Relaxations of cerebral arteries are not influenced by sotalol, while those seen in femoral arteries are significantly attenuated, which indicates that in the latter vessels the inhibitory effect of nicotine is related to β-adrenergic stimulation. Figure 1 shows typical responses of a dog basilar artery to nicotine, isoproterenol and K^+ in the presence and absence of blocking agents. Hexamethonium suppresses only the response to nicotine, sotalol only the response to isoproterenol, and ouabain

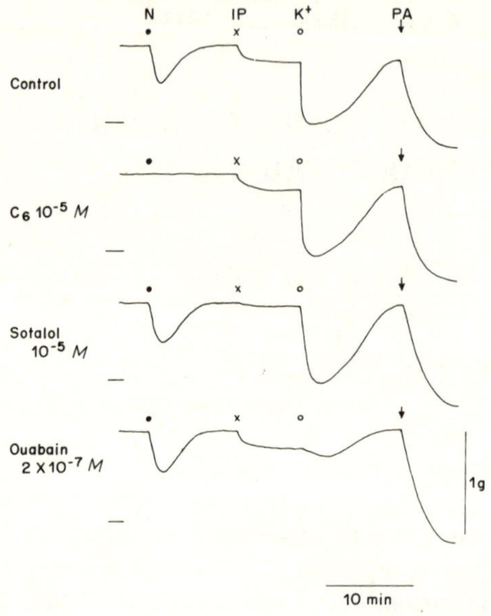

Fig. 1. Typical responses of a dog basilar artery to 10^{-4} *M* nicotine (N), 10^{-6} *M* isoproterenol (IP), 5 m*M* K^+ and 10^{-4} *M* papaverine (PA). Horizontal lines just left of the tracings represent the level prior to the addition of $PGF_{2\alpha}$.

reverses the response to K^+ but does not alter the other responses. Nicotine-induced relaxations of dog cerebral arteries are not influenced by a number of other antagonists, including atropine, metiamide, aminophylline, tetrodotoxin and aspirin; therefore, this relaxation is truly nicotinic in nature and is not due to activation of β-adrenergic, muscarinic, histaminergic (H_2), purinergic or PG-mediated mechanisms. It does not involve activation of the electrogenic Na^+ pump. It is possible that, in the cerebral artery, nicotine causes the release of unknown substance(s) from the blood vessel. In a recent study of middle cerebral arteries isolated from dogs in which an artificial subarachnoid hemorrhage had been induced (38), arteries from the bleeding side failed to respond to nicotine but exhibited relaxations in response to adenosine similar to those seen in arteries from the control side.

In contrast to nicotine, *tyramine* elicits a dose-related contraction of cerebral arteries from dogs as well as from other species (7, 40).

Relaxations similar to those induced by nicotine have been demonstrated in isolated cat middle cerebral arteries in response to *transmural electrical stimulation* (17). Using a variety of blocking agents, including guanethidine, phenoxybenzamine, atropine and tetrodotoxin, *Lee et al.* (17) concluded that the release of unknown substance(s) from the arteries may be involved in the stimulation-induced relaxation. In our trials with dog cerebral arteries contracted with $PGF_{2\alpha}$ or with serotonin, transmural electrical stimulation fails to produce such relaxations (34).

2. Potassium Ion

The addition of small amounts of *potassium ions* (K^+; 0.5–5 m*M*) causes relaxation of isolated arteries whether previously exposed to bathing media containing normal K^+ (26, 31) or deprived of the ion (3, 10, 16, 19). In the dog, the relaxation is greater in cerebral than in peripheral arteries (31). Treatment with ouabain abolishes the K^+-induced relaxation (4, 26) but not the relaxation induced by nicotine and isoproterenol (fig. 1). Substitution of Li^+ for Na^+ in the bathing media also abolishes the K^+-induced inhibition, which is inversely related to the external concentration of K^+ and directly related to the temperature of the bathing fluid. The content of Na^+ is greater in cerebral than in peripheral arteries, although extracellular space (measured with ^{14}C-inulin) is not significantly different in these arteries. These results suggest that the addition of K^+ elicits an activation of the electrogenic Na^+ pump; this results in hyperpolarization of the smooth muscle cells which in turn decreases cellular Ca^{++} (2, 13, 26, 31). The Na^+ pump may be less active in control solution in dog cerebral arteries than in peripheral vessels. The involvement of the Na^+ pump in the control of vascular tone has been postulated by *Hendrickx and Casteels* (15) and *Siegel et al.* (25) from electrophysiological studies using microelectrodes.

It has been shown that the vasodilator response to K^+ is attenuated in renal hypertensive rats and dogs (20, 21). These observations led to the suggestion that increased vascular resistance in hypertensive dogs may be related to a depressed activity of the ouabain-sensitive Na^+, K^+ pump in blood vessels (22).

3. Agents which Activate Specific Drug Receptors

Isoproterenol produces a dose-related relaxation of isolated arteries and veins, whether spontaneously active or made to contract with vasoconstrictor agents. The relaxation is markedly greater in dog coronary arteries than in other arteries (27). Only slight relaxation is seen in dog and human cerebral arteries.

According to *Park et al.* (23), the β-adrenergic receptor function in the aorta and pulmonary artery of newborn rabbits is not fully developed but matures during the first postnatal month. *Cohen and Berkowitz* (5) and *Fleisch et al.* (9) have demonstrated that aortic β-receptor activity decreases with increasing age in rabbits and rats. *Hayashi et al.* (14) have also observed an age-dependent decrease in the response of rabbit basilar arteries to isoproterenol. The latter drug increases the cyclic AMP level in the aorta from young rats but not in the aorta from old animals (8).

Decreased sensitivity to isoproterenol, as well as to adenosine and to nitroglycerine, has been demonstrated in aortic strips from hypertensive rats (6). In hypertensive animals, the vascular adenylate cyclase is less sensitive to hormones (including isoproterenol), and the vascular phosphodiesterase activity is decreased. The possible involvement of a decreased ratio of cyclic AMP:cyclic GMP in the increase in peripheral vascular tone, as well as in vascular smooth muscle proliferation underlying chronic hypertension, has been reviewed by *Amer* (1).

The presence of specific *dopamine* receptors in vascular smooth muscles has been postulated from *in situ* experiments reported by *Goldberg* (11) and from experiments with isolated vessels done by *Toda and Goldberg* (12, 36, 37). In dog arteries, with the exception of large femoral arteries, treated for 60 min with 10^{-5} *M* phenoxybenzamine, dopamine causes a dose-related relaxation and this response is greater in mesenteric and coronary than in cerebral arteries (32). The relaxation is not antagonized by propranolol in concentrations sufficient to suppress the response to isoproterenol. Atropine and aminophylline are also ineffective (35). Recently, *Hatano* and I could specifically antagonize dopamine-induced relaxations of dog coronary arteries using droperidol (unpublished data). This first demonstration of a specific antagonism against the dilator action of dopamine in isolated vessels strongly supports the hypothesis that specific dopamine receptors are present in vascular smooth muscle.

Acetylcholine elicits relaxation in most isolated small vessels from different species, but contraction in strips of rabbit arteries, including coronary and pulmonary. In the dog, the degree of relaxation differs in arteries from different regions, the potency being in the order of, coronary > mesenteric > cerebral (27). The inhibitory effect of acetylcholine is antagonized by atropine in a competitive manner. In dog cerebral arteries, hexamethonium reduces the relaxation to high concentrations of acetylcholine, atropine that to low concentrations, and the concomitant application of these antagonists abolishes the response. It appears that high concentrations of acetylcholine have the same effect on dog cerebral arteries as nicotine.

Adenosine-induced relaxations are approximately the same in different arteries from dogs (27). Aminophylline shifts the dose-response curve of adenosine to the right, and the inhibition is completely reversed by repeated washing (35).

4. Polypeptides

Bradykinin is a vasoactive nonapeptide formed from bradykininogen by the catalysis of kallikrein. Marked differences in the response to bradykinin are seen in a variety of dog arteries (33) and veins (39). Coronary, mesenteric, and renal arteries relax in response to the peptide, while cerebral, carotid, and femoral arteries contract. The bradykinin-induced relaxation is significantly attenuated by aspirin and indomethacin in concentrations which do not alter the response to adenosine. The effect of bradykinin is not influenced by atropine, propranolol, aminophylline, and metiamide. Release of a PGE-like substance by bradykinin has been reported in the isolated, perfused heart (18) and the involvement of PGs could explain the vasodilator action of bradykinin.

Angiotensin II induces a marked contraction of isolated dog mesenteric and femoral arteries as is seen in rabbit and rat aorta, but only a slight contraction of cerebral arteries. In some cerebral arteries, contracted with $PGF_{2\alpha}$, this peptide causes a transient relaxation. Angiotensin II antagonists, such as Sar^1, $Ileu^8$-angiotensin II and Sar^1, Ala^8-angiotensin II, abolish this relaxation but not the relaxation induced by K^+.

5. Non-Specific Vasodilators

Relaxations of isolated vessels by *papaverine, nitroglycerine,* and *sodium nitrate* ($NaNO_2$) are not specifically antagonized by known pharmacological blocking agents. These agonists cause a similar relaxation of dog cerebral, coronary, and mesenteric arteries (27). Maximal relaxation is obtained with $10^{-4}\ M$ papaverine. Nitroglycerine and $NaNO_2$ produce greater relaxations in large coronary arterial strips than in preparations from smaller coronary vessels (24).

6. Summary and Conclusion

The mean values of the maximal relaxations of isolated dog arteries obtained with various agents, as compared to those caused by $10^{-4}\ M$ papaverine, are summarized in table I. The effect of nicotine, bradykinin, and histamine in dog cerebral arteries is opposite to that observed in peripheral arteries. The degree of relaxation induced by isoproterenol, dopamine, acetylcholine and K^+ is different in dog cerebral and peripheral arteries, in aortas from young and old animals, and in arteries from intact and pathologic animals. Similar relaxations are observed in response to adenosine, PGE_1 and $NaNO_2$ in cerebral and peripheral arteries of the dog. The heterogeneity in the response of vascular

Table I. Relaxations of dog cerebral and peripheral arterial strips induced by a variety of agents

	Cerebral	Coronary	Mesenteric	Renal	Femoral
Papaverine (10^{-4} *M*)	1	1	1	1	1
Isoproterenol[a] (10^{-5} *M*)	0.09	0.71	0.39		
Acetylcholine[a] (10^{-4}–10^{-3} *M*)	0.11	0.32	0.23[b]	0.20[b]	
Dopamine[b] (5×10^{-4} *M*)	0.30	0.62	0.68	0.55	0.49
Nicotine (10^{-4} *M*)	0.25	↑	↑	↑	↑ and ↓
Histamine (2×10^{-4} *M*)	↑	0.76	0.81		
Bradykinin[a] (5×10^{-7} *M*)	↑	0.24	0.30	0.25	↑
Prostaglandin E_1 (10^{-5} *M*)	0.58	0.76	0.80	0.82	
Adenosine[a] (2×10^{-3} *M*)	0.71	0.80	0.59		
$NaNO_2$[a] (3×10^{-3} *M*)	0.71	0.89	0.77		
K^+ (5 m*M*)	0.66	0.19	0.21	0.36	0.30

Relaxations relative to those induced by 10^{-4} *M* papaverine are presented.
Preparations were contracted with $PGF_{2\alpha}$ except [a] which was with K^+.
[b] Treated with phenoxybenzamine. Upward arrows represent contractions.

smooth muscles to vasoactive agents is due to: (1) differences in the quantity or sensitivity of receptors in the case of isoproterenol, dopamine, and acetylcholine; (2) differences in the population ratio of two types of receptors through which opposite responses are induced (for instance, α- and β-adrenergic, α-adrenergic and dopaminergic, and histaminergic H_1 and H_2); (3) differences in the release of endogenous substances, such as norepinephrine, acetylcholine, PGs, ATP, K^+, polypeptides, or unknown substances; (4) differences in the inactivation of the drugs, and (5) differences in the activity of the membrane ATPase (electrogenic Na^+ pump), such as is seen with the response to small amounts of K^+.

References

1 *Amer, M.S.:* Cyclic nucleotides in disease: on the biochemical etiology of hypertension. Life Sci. *17:* 1021–1038 (1975).

2 *Anderson, D.K.:* Cell potential and the sodium-potassium pump in vascular smooth muscle. Fed. Proc. Fed. Am. Socs exp. Biol. *35:* 1294–1297 (1976).

3 *Biamino, G. and Wessel, H.-J.:* Potassium induced relaxation of vascular smooth muscle: a possible mechanism of exercise hyperemia. Pflügers Arch. *343:* 95–106 (1973).

4 *Chen, W.T.; Brace, R.A.; Scott, J.B.; Anderson, D.K., and Haddy, F.J.:* The mechanism of the vasodilator action of potassium. Proc. Soc. exp. Biol. Med. *140:* 820–824 (1972).

5 *Cohen, M.L. and Berkowitz, B.A.:* Age-related change in vascular responsiveness to cyclic nucleotides and contractile agonists. J. Pharmac. exp. Ther. *191:* 147–155 (1974).
6 *Cohen, M.L. and Berkowitz, B.A.:* Decreased vascular relaxation in hypertension. J. Pharmac. exp. Ther. *196:* 396–406 (1976).
7 *Edvinsson, L.; Aubineau, P.; Owman, C.; Sercombe, R., and Seylaz, J.:* Sympathetic innervation of cerebral arteries: prejunctional supersensitivity to norepinephrine after sympathectomy or cocaine treatment. Stroke *6:* 525–530 (1975).
8 *Ericsson, E. and Lundholm, L.:* Adrenergic β-receptor activity and cyclic AMP metabolism in vascular smooth muscle; variations with age. Mechanisms Ageing Development *4:* 1–6 (1975).
9 *Fleisch, J.H.; Maling, H.M., and Brodie, B.B.:* Beta-receptor activity in aorta: relations with age and species. Circulation Res. *26:* 151–162 (1970).
10 *Gellai, M. and Detar, R.:* Evidence in support of hypoxia but against high potassium and hyperosmolarity as possible mediators of sustained vasodilatation in rabbit cardiac and skeletal muscle. Circulation Res. *35:* 681–691 (1974).
11 *Goldberg, L.I.:* Cardiovascular and renal actions of dopamine: potential clinical applications. Pharmacol. Rev. *24:* 1–29 (1972).
12 *Goldberg, L.I. and Toda, N.:* Dopamine induced relaxation of isolated canine renal, mesenteric and femoral arteries contracted with prostaglandin $F_{2\alpha}$. Circulation Res. *36/37:* suppl., pp. I-97–I-102 (1975).
13 *Haddy, F.J.:* Potassium and blood vessels. Life Sci. *16:* 1489–1498 (1975).
14 *Hayashi, S.; Miyazaki, M., and Toda, N.:* Age-dependent changes in the response of isolated rabbit cerebral arteries to vasoactive agents. Jap. J. Pharmacol. *27:* suppl. 108P (1977).
15 *Hendrickx, H. and Casteels, R.:* Electrogenic sodium pump in arterial smooth muscle cells. Pflügers Arch. *346:* 299–306 (1974).
16 *Konold, P.; Gebert, G., and Brecht, K.:* The effect of potassium on the tone of isolated arteries. Pflügers Arch. ges. Physiol. *301:* 285–291 (1968).
17 *Lee, T.J.-F.; Su, C., and Bevan, J.A.:* Nonsympathetic dilator innervation of cat cerebral arteries. Experientia *31:* 1424–1426 (1975).
18 *Needleman, P.; Key, S.L.; Denny, S.E.; Isakson, P.C., and Marshall, G.R.:* Mechanism and modification of bradykinin-induced coronary vasodilatation. Proc. natn. Acad. Sci. USA *72:* 2060–2063 (1975).
19 *Norton, J.M. and Detar, R.:* Potassium and isolated coronary vascular smooth muscle. Am. J. Physiol. *222:* 474–479 (1972).
20 *Overbeck, H.W.:* Vascular responses to cations, osmolality, and angiotensin in renal hypertensive dogs. Am. J. Physiol. *223:* 1358–1364 (1972).
21 *Overbeck, H.W. and Clark, W.J.:* Vasodilator responses to K^+ in genetic hypertensive and in renal hypertensive rats. J. Lab. clin. Med. *86:* 973–983 (1975).
22 *Overbeck, H.W.; Pamnani, M.B.; Akera, T.; Brody, T.M., and Haddy, F.J.:* Depressed function of an ouabain-sensitive sodium-potassium pump in blood vessels from renal hypertensive dogs. Circulation Res. *38:* suppl., pp. II-48–II-52 (1976).
23 *Park, M.K.; Diehl, A.M., and Sunderson, J.M.:* Maturation of beta-adrenergic receptor activity of rabbit aorta and pulmonary artery. Life Sci. *19:* 321–328 (1976).
24 *Schnaar, R.L. and Sparks, H.V.:* Response of large and small coronary arteries to nitroglycerine, $NaNO_2$, and adenosine. AmJ. Physiol. *223:* 223–228 (1972).
25 *Siegel, G.; Jäger, R.; Nolte, J.; Bertsche, O.; Roedel, H., and Schröter, R.:* Ionic concentrations and membrane potential in cerebral and extra-cerebral arteries; in *Cervos-Navarro* Pathology of cerebral microcirculation, pp. 96–120 (de Gruyter, Berlin 1974).

26 *Toda, N.:* Responsiveness to potassium and calcium ions of isolated cerebral arteries. Am. J. Physiol. *227:* 1206–1211 (1974).

27 *Toda, N.:* The action of vasodilating drugs on isolated basilar, coronary and mesenteric arteries of the dog. J. Pharmac. exp. Ther. *191:* 139–146 (1974).

28 *Toda, N.:* Responses of isolated canine cerebral and peripheral arteries to nicotine. Proc. West. Pharmacol. Soc. *18:* 31–33 (1975).

29 *Toda, N.:* Nicotine-induced relaxation in isolated canine cerebral arteries. J. Pharmac. exp. Ther. *193:* 376–384 (1975).

30 *Toda, N.:* Regional differences in the response to nicotine in isolated canine arteries. Eur. J. Pharmacol. *35:* 151–160 (1976).

31 *Toda, N.:* Potassium-induced relaxation in isolated cerebral arteries contracted with prostaglandin $F_{2\alpha}$. Pflügers Arch. *364:* 235–242 (1976).

32 *Toda, N.:* Influence of dopamine and noradrenaline on isolated cerebral arteries of the dog. Br. J. Pharmacol. *58:* 121–126 (1976).

33 *Toda, N.:* Actions of bradykinin on isolated cerebral and peripheral arteries. Am. J. Physiol. *232:* H267–H274 (1977).

34 *Toda, N. and Fujita, Y.:* Responsiveness of isolated cerebral and peripheral arteries to serotonin, norepinephrine, and transmural electrical stimulation. Circulation Res. *33:* 98–104 (1973).

35 *Toda, N.; Hojo, M.; Sakae, K., and Usui, H.:* Comparison of the relaxing effect of dopamine with that of adenosine, isoproterenol and acetylcholine in isolated canine coronary arteries. Blood Vessels *12:* 290–301 (1975).

36 *Toda, N. and Goldberg, L.I.:* Dopamine-induced relaxation of isolated arterial strips. J. Pharm. Pharmacol. *25:* 587–589 (1973).

37 *Toda, N. and Goldberg, L.I.:* Effects of dopamine on isolated canine coronary arteries. Cardiovasc. Res. *9:* 384–389 (1975).

38 *Toda, N.; Ozaki, T., and Ohta, T.:* Cerebrovascular sensitivity to vasoconstricting agents induced by subarachnoid hemorrhage and vasospasm in dogs. J. Neurosurg. *46:* 296–303 (1977).

39 *Tsuru, H.; Ishikawa, N., and Shigei, T.:* Responsiveness of isolated dog veins to bradykinin and other bioactive peptides: distribution of sensitivity to bradykinin and possible correlation with genesis of the venous system. Blood Vessels *13:* 238–248 (1976).

40 *Urquilla, P.R.; Marco, E.J.; Balfagon, G., and Lluch, S.:* Adrenergic mechanisms in cerebral blood vessels: effect of tyramine on the isolated middle cerebral artery of the goat. Stroke *5:* 447–452 (1974).

N. Toda, MD, Department of Pharmacology, Shiga University of Medical Sciences, *Seta, Ohtsu 520-21* (Japan)

Mechanisms of Vasodilatation. Satellite Symp., 27th Int. Congr. Physiol. Sci., Wilrijk 1977, pp. 137–143 (Karger, Basel 1978)

Vasodilatation or Inhibition of Peripheral Vasoconstriction?

Jan M. van Nueten

Department of Pharmacology, Janssen Pharmaceutica Research Laboratories, Beerse

Vasodilatation, in contrast with passive distension, refers to an increase in blood vessel diameter due to relaxation of vascular smooth muscle cells. This implies that a certain degree of vascular tone must be present before vasodilatation can be induced. Hence, since most blood vessels suitable for *in vitro* experimentation are not myogenically active, one is forced in most cases to induce tone by giving vasoconstrictor agonists in order to demonstrate the relaxant effect of pharmacological agents thought to be vasodilators. As a consequence a certain controversy has risen, which to many pharmacologists working on isolated blood vessels is a matter of semantics, as to whether one should refer to vasodilator drugs or to inhibitors of vasoconstriction. The former would refer in particular to those agents which cause vasodilatation in the intact organism even in the absence of external vasoconstrictor stimulation, while the latter imply that preconstriction of the vessels under study is necessary before their inhibitory effect can be detected. The purpose of the present discussion is to examine whether such a distinction is helpful.

From all the data available so far, and independently of the origin of the blood vessels studied and of the presence or absence of myogenic tone, it is obvious that the basic contractile process is the same for all vascular smooth muscles, and ultimately depends upon the availability of activator Ca^{++} (4, 5, 20–22, 26). So far, no drug seems available which directly inhibits the interaction between the contractile proteins in the vascular smooth muscle cells, and all known vasodilator substances ultimately act by decreasing the available amount of activator Ca^{++}. There are, however, many ways by which such a decrease can be achieved.

The tone of any given blood vessel is the resultant of: (1) its intrinsic myogenic tone; (2) the amounts of vasoconstrictor agonists in the vicinity of the cell membrane, and (3) the presence of endogenous inhibitory substances. At

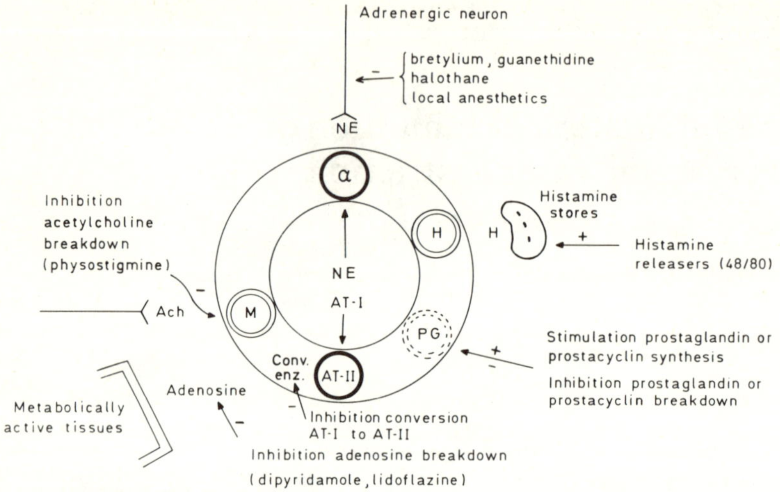

Fig. 1. Schematic representation of the ways to cause indirect vasodilatation. Ach = acetylcholine; NE = norepinephrine; AT = angiotensin; H = histamine; α = α-adrenergic; M = musculotropic; PG = prostaglandin-like substances; conv. enz. = converting enzyme. Black circles = receptors whose stimulation causes vasoconstriction of the smooth muscle; white circles = receptors whose stimulation causes vasodilatation. Modified after *Shepherd and Vanhoutte* (19).

first, vasodilatation may be produced in an 'indirect' way (fig. 1), that is by either decreasing the amounts of vasoconstrictor agonists in the vicinity of the smooth muscle cells, or by making more endogenous inhibitory substances available. Theoretically, both types of substances can originate from the blood perfusing the vessel, can be generated in the vessel wall itself or in its immediate vicinity, or can be liberated by autonomic nerve endings. Of the *vasoconstrictor agonists,* the physiologically more important ones are catecholamines (circulating or liberated at the adrenergic nerve endings) and angiotensin II. The liberation of norepinephrine by the adrenergic nerve endings can be markedly reduced by a number of naturally occurring substances such as acetylcholine, histamine, adenine nucleotides and 5-hydroxytryptamine; in addition, several therapeutic agents, such as bretylium tosylate, guanethidine, halothane, and local anesthetic agents cause vasodilatation by inhibition of peripheral adrenergic neurotransmission (3, 19, 22, 26).

Likewise, inhibition of the converting enzyme for angiotensin II may result in lower levels of the octapeptide and thus may have a vasodilator effect (16). Of

Table I. Mechanism of causing vasodilatation by direct interferences with vascular smooth muscle cells

Inhibition of drug receptor interaction
Alteration in membrane characteristics
Hyperpolarization
Local anesthetic effect
Chemical structure
Ca^{++} permeability
Alteration in intracellular processes
ATP regeneration
Removal of activator Ca^{++}

the *endogenous inhibitory substances,* of which the local level may be modulated by therapeutic agents, the more important ones are adenine nucleotides, prostaglandin-like substances and histamine. Theoretically, substances which augment the local synthesis of prostacyclin and certain prostaglandins could act as vasodilators (6, 23); likewise, the histamine release caused by compounds such as 48/80 explains the vasodilatation they cause (18). On the other hand, substances which inhibit the breakdown of endogenous inhibitors may depress vascular reactivity. This is the case, for example, for physostigmine in certain blood vessels innervated by cholinergic vasodilator nerves (25), or for substances such as dipyridamole and lidoflazine which delay the degradation of adenosine metabolism by decreasing its uptake in metabolic cells (13, 24).

A number of therapeutic agents relax blood vessels by their 'direct' effect on the vascular smooth muscle cells, and this can occur at different levels (table I). At first, a drug can prevent the binding of a vasoconstrictor agonist to its receptors on the cell membrane; the obvious example is the vasodilator effect of α-adrenolytic drugs (9). Other membrane characteristics can be affected. Among those the more important ones are: (1) membrane hyperpolarization, as seen most typically with β-adrenergic agonists (22); (2) local anesthetic effect (1, 15, 22); (3) alteration in the chemical structure of the membrane, for example, the transition from disulfide to sulfhydril groups (8, 17); (4) alteration of the membrane permeability for Ca^{++} (7, 10, 11, 27). Intracellular events also may be modified by drugs. Two examples are the interference with the regeneration of ATP, as seen with metabolic inhibitors, or the activation of the removal of activator Ca^{++}, which probably explains why drugs such as theophylline and papaverine cause vasodilatation (2, 14).

Of all these possible ways to cause relaxation of vascular smooth muscle cells, procedures which decrease the release of vasoconstrictor agonists or inhibit their binding to the membrane receptors can be considered to be true 'inhibitors of vasoconstriction'. All the other drugs theoretically will cause relaxation

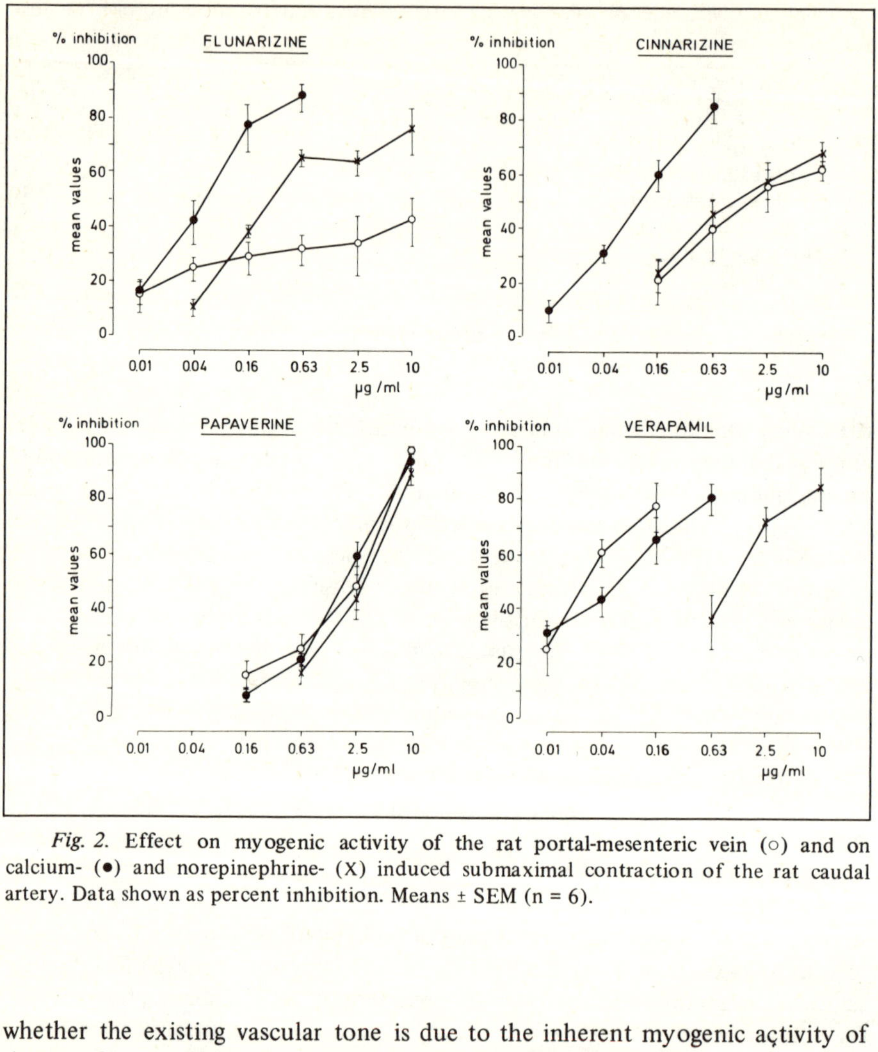

Fig. 2. Effect on myogenic activity of the rat portal-mesenteric vein (○) and on calcium- (●) and norepinephrine- (X) induced submaximal contraction of the rat caudal artery. Data shown as percent inhibition. Means ± SEM (n = 6).

whether the existing vascular tone is due to the inherent myogenic activity of the vascular smooth muscle cells, or to their response to vasoconstrictor agonists. This does not hold for all drugs said to decrease the membrane permeability for Ca^{++}; in particular, compounds such as flunarizine, a difluorated derivative of cinnarizine[1], dissociate between the increase in Ca influx caused by vasoconstrictor agonists and that due to the myogenic alteration in membrane potential (28).

[1] (E)-1-[*bis*(4-fluorophenyl)methyl]-4-(3-phenyl-2-propenyl)piperazine dihydrochloride.

To establish this dissociation, experiments were performed on spontaneously active portal-mesenteric veins of the rat, and on caudal arteries of the same species which lack myogenic activity (29). Flunarizine, and the classical Ca-inhibitors cinnarizine and verapamil were compared with the non-specific inhibitor papaverine. The effect of increasing concentrations of these agents was investigated on: (1) myogenic activity in the portal-mesenteric vein; (2) response of the caudal artery to norepinephrine, and (3) response of the depolarized caudal artery to increase in extracellular Ca^{++} concentration. The results are summarized in figure 2. Papaverine caused a similar depression in the three experimental conditions tested. Verapamil inhibited myogenic activity and the response to increasing Ca^{++}-concentrations significantly more than that to exogenous norepinephrine. Cinnarizine and flunarizine affected the response to Ca^{++} more than that to norepinephrine. However, unlike cinnarizine, flunarizine had little effect on the myogenic activity. Thus, it appears that flunarizine can indeed specifically inhibit the increased availability of activator Ca^{++} due to exogenous triggering of the vascular smooth muscle cells, but does not interfere with their potential to produce myogenic tone. This pharmacological behavior is in marked contrast with that obtained with verapamil and D-600 (12).

The question is whether such distinction has therapeutical implications. Obviously, vasodilator substances will be of use only in those conditions where there is an increased presence of vasoconstrictor agonists in the vicinity of the vascular smooth muscle cells. Substances such as flunarizine will cause vasodilatation specifically by rendering the smooth muscle cells unresponsive to these vasoconstrictor agonists, without affecting their myogenic tone; thus, they will not interfere with the normal metabolic control of tissue perfusion. In that regard, the distinction between vasodilators and drugs such as flunarizine which interfere only with vasoconstrictor stimuli may be of fundamental importance.

References

1 *Altura, B.M. and Altura, B.T.:* Effects of local anesthetics, antihistamines, and glucocorticoids on peripheral blood flow and vascular smooth muscle. Anesthesiology *41:* 197–214 (1974).

2 *Berti, F.; Fumagalli, R.; Folco, G.C.; Omini, C., and Bernareggi, V.:* Role of cyclic 3',5'-AMP on contraction and relaxation of perfused rat caudal artery. Pharmacol. Res. Commun. *6:* 519–527 (1974).

3 *Bevan, J.A. and Su, C.:* Sympathetic mechanisms in blood vessels: nerve and muscle relationships. Annu. Rev. Pharmacol. *13:* 269–285 (1973).

4 *Bohr, D.F.:* Electrolytes and smooth muscle contraction. Pharmac. Rev. *16:* 85–111 (1964).

5 *Bohr, D.F.:* Vascular smooth muscle updated. Circulation Res. *32:* 665–672 (1973).

6 *Bunting, S.; Gryglewski, R.; Moncada, S., and Vane, J.R.:* Arterial walls generate from

prostaglandin endoperoxides a substance (prostaglandin X) which relaxes strips of mesenteric and coeliac arteries and inhibits platelet aggregation. Prostaglandins *12:* 897–913 (1976).

7 *Fleckenstein, A.; Nakayama, K.; Fleckenstein-Grün, G., and Byon, Y.K.:* Interaction of vasoactive ions and drugs with Ca-dependent excitation-contraction coupling of vascular smooth muscle; in *Carafoli, Clementi, Dabrikowski and Margreth* Calcium transport in contraction and secretion, pp. 555–566 (North-Holland, Amsterdam 1975).

8 *Fleish, J.H.; Krzan, M.C., and Titus, E.:* Pharmacologic receptor activity of rabbit aorta. Effect of dithiothreitol and *n*-ethylmaleimide. Circulation Res. *33:* 284–290 (1973).

9 *Furchgott, R.F.:* The classification of adrenoceptors (adrenergic receptors). An evaluation from the standpoint of receptor theory; in *Blaschko and Muscholl* Catecholamines, pp. 283–335 (Springer, Berlin 1972).

10 *Godfraind, T. and Kaba, A.:* The role of calcium in the action of drugs on vascular smooth muscle. Archs int. Pharmacodyn. Thér. *196:* suppl., pp. 35–49 (1972).

11 *Golenhofen, K.:* P- and T-systems for calcium activation in vascular smooth muscle; in *Wolf and Werthessen* The smooth muscle of the artery, pp. 46–59 (Plenum Publishing, New York 1975).

12 *Golenhofen, K. and Weston, A.H.:* Differentiation of calcium activation systems in vascular smooth muscle; in *Betz* Ionic actions on vascular smooth muscle, pp. 21–25 (Springer, Berlin 1976).

13 *Kübler, W. und Bretschneider, H.J.:* Kompetitive Hemmung der katalysierten Adenosindiffusion als Mechanismus der coronarerweiternden Wirkung eines Pyrimidopyrimidin-Derivates. Pflügers Arch. ges. Physiol. *280:* 141–147 (1964).

14 *Kukovetz, W.R.; Pöch, G.; Wurm, A.; Holzmann, S., and Paietta, E.:* Effect of phosphodiesterase-inhibition on smooth muscle tone; in *Betz* Ionic actions on vascular smooth muscle, pp. 124–131 (Springer, Berlin 1976).

15 *Muldoon, S.M.; Verbeuren, T.J., and Vanhoutte, P.M.:* Effects of amide-linked local anesthetics on adrenergic neuroeffector junction in cutaneous veins of dog. J. Pharmac. exp. Ther. *196:* 723–736 (1976).

16 *Needleman, P.; Doublas, J.R.; Jakschik, B.A.; Blumberg, A.L.; Isakson, P.C., and Marshall, G.R.:* Angiotensin antagonists as pharmacological tools. Fed. Proc. Fed. Am. Socs exp. Biol. *35:* 2488–2493 (1976).

17 *Needleman, P.; Jakschik, B., and Johnson, E.M.:* Sulfhydryl requirement for relaxation of vascular smooth muscle. J. Pharmac. exp. Ther. *187:* 324–331 (1973).

18 *Paton, W.D.M.:* Compound 48/80: a potent histamine liberator. Br. J. Pharmacol. *6:* 499–508 (1951).

19 *Shepherd, J.T. and Vanhoutte, P.M.:* Veins and their control, pp. 1–269 (Saunders, Philadelphia 1975).

20 *Shepherd, J.T. and Vanhoutte, P.M.:* Skeletal-muscle blood flow: neurogenic determinants; in *Zelis* The peripheral circulations, pp. 3–35 (Grune & Stratton, New York 1975).

21 *Somlyo, A.P. and Somlyo, A.V.:* Vascular smooth muscle. I. Normal structure, pathology, biochemistry, and biophysics. Pharmacol. Rev. *20:* 197–272 (1968).

22 *Somlyo, A.P. and Somlyo, A.V.:* Vascular smooth muscle. II. Pharmacology of normal and hypertensive vessels. Pharmacol. Rev. *22:* 251–353 (1970).

23 *Staszewska-Barczak, J. and Vane, J.R.:* The role of prostaglandins in the local control of circulation. Clin. exp. Pharmacol. Physiol. *2:* suppl., pp. 71–78 (1975).

24 *Belle, H. van:* The disappearance of adenosine in blood. Effect of lidoflazine and other drugs. Eur. J. Pharmacol. *11:* 241–248 (1970).

25 *Hee, R. van and Vanhoutte, P.M.:* Cholinergic inhibition of adrenergic neurotransmission in dog's gastric arteries. Archs int. Pharmacodyn. Thér. *220:* 348 (1976).

26 *Vanhoutte, P.M.:* Heterogeneity in vascular smooth muscle; in *Kaley and Altura* Microcirculation, vol. 2 (Academic Press, New York, in press, 1977).

27 *Nueten, J.M. van:* Comparative bioassay of vasoactive drugs using isolated perfused rabbit arteries. Eur. J. Pharmacol. *6:* 286–293 (1969).

28 *Nueten, J.M. van and Janssen, P.A.J.:* Comparative study of the effects of flunarizine and cinnarizine on smooth muscles and cardiac tissues. Archs Int. Pharmacodyn. Thér. *204:* 37–55 (1973).

29 *Wade, D.N. and Beilin, L.J.:* Vascular resistance in the perfused isolated rat tail. Br. J. Pharmacol. *38:* 20–36 (1970).

J.M. van Nueten, Department of Pharmacology, Janssen Pharmaceutica Research Laboratories, *B–2340 Beerse* (Belgium)

Mechanisms of Vasodilatation. Satellite Symp., 27th Int. Congr. Physiol. Sci.,
Wilrijk 1977, pp. 144–151 (Karger, Basel 1978)

Inhibitors of Calcium Influx[1]

Théophile Godfraind and Nicole Morel[2]
Laboratoire de Pharmacodynamie Générale et de Pharmacologie, UCL, Bruxelles

The dependency upon the external calcium concentration of contractile responses of vascular smooth muscle to catecholamines strongly suggest that the latter increase Ca^{++} influx. The quantitative analysis of this dependency uses different methodologies which include: (1) measurement of contractile responses; (2) estimates of transmembrane calcium fluxes, and (3) characterization of the calcium channels. The same methodological approaches obviously also can be used when investigating the mode of action of pharmacological agents thought to inhibit the Ca^{++} influx.

1. Contractile Responses

The common procedure to assess the ability of a pharmacological compound to inhibit Ca^{++} influx, is to depolarize the studied preparation in Ca^{++}-free solution, and then to gradually increase the concentration of the ion in the bathing solution; this procedure then can be repeated in presence of increasing concentrations of the putative Ca^{++} influx antagonist (fig. 1). This experimental approach has been used to identify a number of pharmacological compounds as so-called 'calcium antagonists'; they are tentatively listed in table I. They share the property to decouple the drug-receptor interaction from the contractile response, presumably by interfering with the Ca^{++} metabolism in smooth muscle. In addition, they have, in most cases, other potent pharmacological actions. Of the drugs available so far, cinnarizine and flunarizine appear to be the most selective blockers of Ca^{++} influx. This specificity is documented best for cinnarizine. Figure 1 illustrates the calcium dose-response curves of rabbit

[1] This work has been supported by FRSM (Grant No. 20121).
[2] *N. Morel* is Boursier de l'IRSIA.

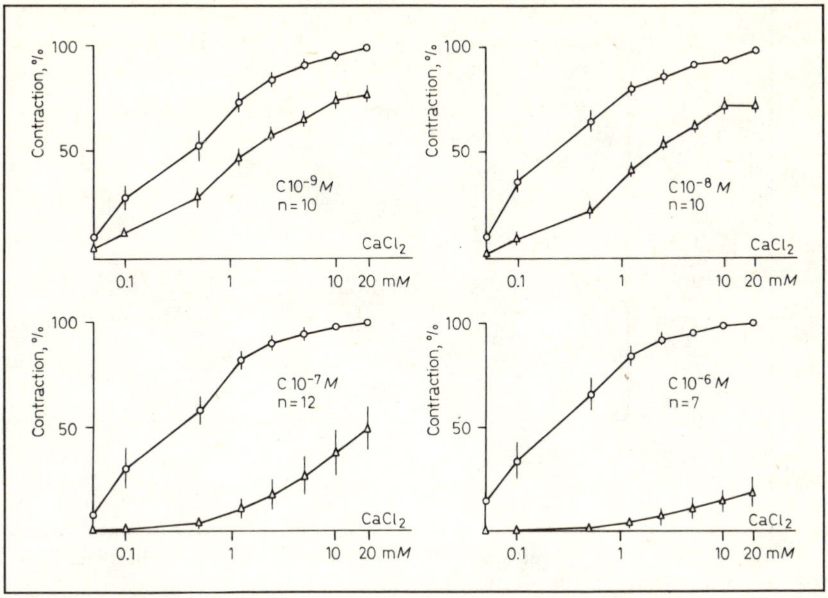

Fig. 1. The action of cinnarizine (C) on the contraction evoked by calcium in depolarized arteries. Dose-effect curves of $CaCl_2$ on depolarized rabbit mesenteric arteries in absence of cinnarizine (○) and 90 min after its addition to the medium (△). From *Godfraind and Kaba* (8), with permission of the publishers.

Table I. Classification of drugs[1] with antagonistic properties to the function of calcium in smooth muscles (receptor-response decoupling agents)

1. With action on the central nervous system	chlorpromazine (4)
2. With local anesthetic properties	procaine (9)
	tetracaine (3)
3. With curarizing properties	neomycine (13)
4. With cardiodepressive action	verapamil (4)
	prenylamine (4)
	D600 (4)
	nifedipine (4)
	diltiazem (4)
	perhexiline (4)
5. With an action on phosphodiesterase	papaverine (9)
6. With antianginal action	lidoflazine (9)
7. With an action on the metabolism of monovalent cations	suloctidil (6)
8. Selective blockers of Ca influx	cinnarizine (9)
	flunarizine (12)

[1] Only clinically used drugs are listed here.

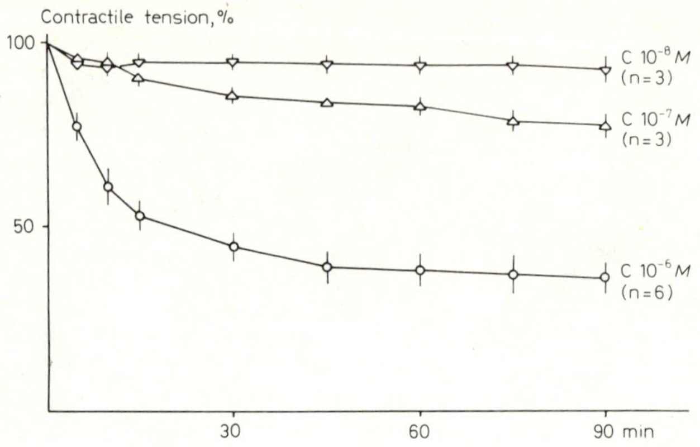

Fig. 2. Relaxation by cinnarizine (C) of contractile rabbit mesenteric arteries. From *Godfraind and Kaba* (8).

mesenteric arteries, previously equilibrated in a depolarizing solution (high K^+) without Ca^{++}. The preparations are very sensitive to increases in Ca^{++} concentrations, and 50 μM of the ion already trigger a contraction. The response to Ca^{++} is attenuated by incubation with cinnarizine, in concentrations as low as 10^{-9} *M* (fig. 1). If cinnarizine is added during a Ca^{++}-evoked contraction, a dose-dependent relaxation is noted (fig. 2). In the same preparations, cinnarizine reduces the contraction evoked by adrenaline in the presence of Ca^{++}, but not the residual response obtained with the catecholamine in Ca^{++}-free depolarizing solution; the latter residual response is thought to be due to the release of intracellular bound Ca^{++}. Similar results are obtained in rat aorta. They favor the interpretation that cinnarizine acts at the smooth muscle cell membrane, to decrease the availability of Ca ion, by reducing its influx (8, 9).

2. Transmembrane Calcium Fluxes

Several attempts have been made to estimate the changes in cytoplasmic Ca^{++} occurring during excitation of smooth muscle. Lanthanum replaces Ca^{++} on superficial binding sites and does not penetrate the cell (13). On the basis that it might block transmembrane Ca^{++} fluxes, it has been proposed that the Ca^{++} content of a muscle washed in La^{+++} solution is an estimate of cellular Ca^{++} (1).

Table II. ^{45}Ca content (mmole/kg wet weight) of rat aorta incubated for 20 min in ^{45}Ca physiological solution and washed either in physiological solution or in 50 m*M* La solution (6)

Time in non-radioactive solution, min	^{45}Ca content of aorta washed in non-radioactive solutions	
	physiological solution (4)	50 m*M* La solution (4)
2	0.49 ± 0.01	0.25 ± 0.03
5	0.17 ± 0.01	0.10 ± 0.01
10	0.08 ± 0.005	0.08 ± 0.009
15	0.064 ± 0.004	0.069 ± 0.006
30	0.033 ± 0.001	0.043 ± 0.004
60	0.017 ± 0.001	0.028 ± 0.002

In La solution, ^{45}Ca efflux was faster than in physiological solution during the first 5 min of washout; thereafter, it was lower but not completely absent. ^{45}Ca content of controls before washing = 2.27 ± 0.045 (6).

The evidence that La^{+++} blocks both Ca^{++} binding to superficial sites and Ca^{++} influx is provided by the observation that the calcium diffusion space in the presence of lanthanum is similar to the inulin diffusion space (2). However, La^{+++} does not completely block Ca^{++} efflux, as shown in several smooth muscles (table II) (2, 5).

In recent experiments, Ca^{++} fluxes across the smooth muscle cell membrane were estimated by measuring the Ca^{++} turnover in the La^{+++}-resistant Ca^{++} fraction (5). This fraction corresponds to the amount of Ca^{++} which is not displaced when the tissue is washed in 50 m*M* $LaCl_3$. In the latter solution, most of the Ca^{++} displacement occurs during the first 5 min; Ca^{++} efflux from intracellular stores appears to be delayed, which allows to estimate the La^{+++}-resistant Ca^{++} fraction. Noradrenaline increases the rate of uptake of ^{45}Ca into this Ca^{++} fraction: in the resting rat aorta, this uptake equals 6 µmole ^{45}Ca kg^{-1} min^{-1}, and in presence of noradrenaline 10^{-5} *M*, it increases to 36 µmole ^{45}Ca kg^{-1} min^{-1}. There is no net gain in tissue Ca^{++}, and Ca^{++} efflux increases in a similar manner. The increase in Ca^{++} uptake is dose-dependent; a response of 50% of the maximum is produced by 2×10^{-8} *M* noradrenaline which is the ED_{50} for the contractile response to the catecholamine. In the presence of phentolamine, the dose-response curves for noradrenaline on ^{45}Ca uptake were displaced in a manner suggesting a competitive antagonism (fig. 3). The pA_2 for phentolamine was 7.8. In view of the similarity of this pA_2 value with that

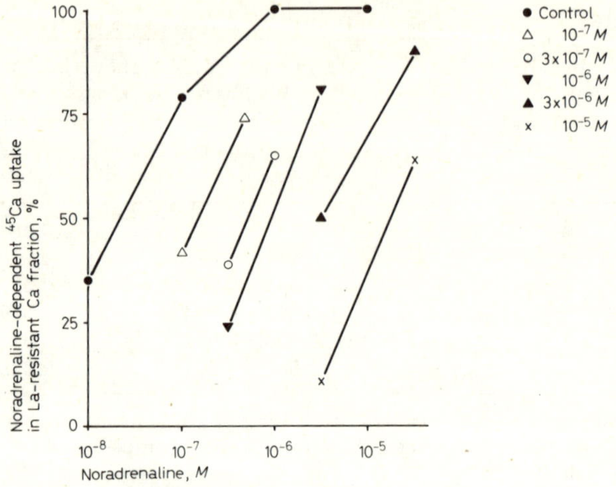

Fig. 3. Noradrenaline dose-effect curve on the ^{45}Ca influx in the presence and absence of phentolamine. Modified from *Godfraind* (5).

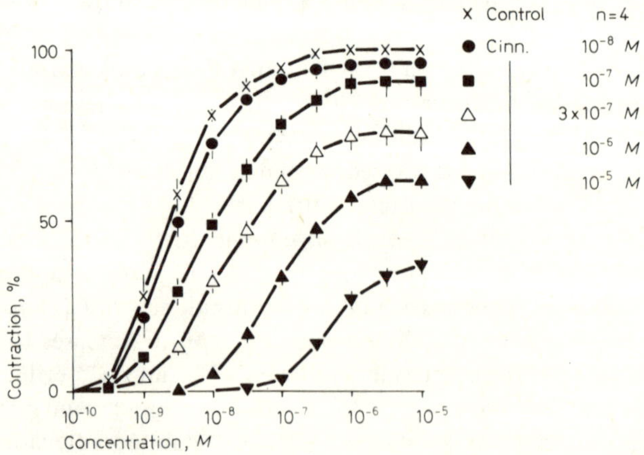

Fig. 4. The action of cinnarizine on the contractile response of rat aorta to noradrenaline (*Godfraind*, unpublished).

obtained for the antagonistic effects of phentolamine on the contractile responses to noradrenaline, it seems that the activation of α-adrenergic receptors is responsible both for the contraction and for the increased rate of Ca^{++} exchange in the La^{+++}-resistant Ca^{++} fraction. The rate of exchange from this fraction appears to be the biologically important component of Ca^{++} exchange across smooth muscle cell membranes.

Cinnarizine is a non-competitive antagonist for noradrenaline in the rat aorta. When the preparation is incubated with the drug, the maximum response to the catecholamines is depressed, whereas the ED_{50} is unchanged (fig. 4). As in the mesenteric arteries, cinnarizine inhibits the component of the contraction sensitive to extracellular calcium. The noradrenaline sensitive Ca^{++} influx is depressed by cinnarizine in the same range of concentrations: in both cases, the concentration of cinnarizine producing a 50% inhibition equals $7.9 \times 10^{-7} M$. The noradrenaline-sensitive Ca^{++} efflux is barely affected by cinnarizine. This indicates that blockade of the contraction can be accounted for by blockade of Ca^{++} influx.

3. Calcium Channels

We have studied the uptake of ^{3}H-cinnarizine by the isolated rat aorta (11). For ^{3}H-cinnarizine concentrations between $10^{-8} M$ and $10^{-6} M$, the uptake by the tissue reaches a steady state level after 90 min of incubation. For higher concentrations no equilibrium is reached. Therefore, the latter concentrations cannot be taken into account for further analysis. The cellular uptake of ^{3}H-cinnarizine at equilibrium shows a linear and a saturable component; it fits the equation:

$$U = a\,C_m + \frac{b\,C_m{}^n}{K' + C_m{}^n}, \qquad (1)$$

where U is the equilibrium tissue content; C_m the ^{3}H-cinnarizine concentration in the medium; a the proportionality constant for the linear uptake; b the capacity constant for the saturable binding sites; and K' is equal to $(C_m)_{50}^n$, $(C_m)_{50}$ being the concentration needed for half saturation and n the Hill coefficient. Values for these parameters are reported in table III where they are compared with those for the ouabain uptake; the latter has been shown in other tissues to be the sum of non-saturable and saturable binding (7, 10).

The saturable ^{3}H-cinnarizine uptake corresponds to the range of cinnarizine concentrations which produce inhibition of the contraction evoked by potassium depolarization (fig. 1). When the uptake of ^{3}H-cinnarizine is studied in the

Table III. Parameters describing the uptake of cinnarizine and ouabain by the rat aorta (11)

	Cinnarizine	Ouabain
a Clearance of the linear uptake, 1/kg wet weight	100	0.43
b Capacity of the saturable binding (μmol/kg wet weight)	16	17
Concentration required for half saturation, *M*	1.2×10^{-7}	2×10^{-4}
n Hill coefficient	1.62	1.0

presence of another Ca antagonist, D600 (10^{-4} *M*), the cinnarizine-saturable binding is absent. It is likely that the excess of D600 competes with cinnarizine for the binding to calcium channels. This then means that the measurement of the saturable binding of ^{3}H-cinnarizine provides a reasonable means of estimating the number of Ca^{++} channels. The observation that the Hill coefficient for the saturable uptake is higher than 1 indicates that each channel might bind more than 1 cinnarizine molecule and that this binding exhibits a positive cooperativity. Assuming a 1:2 stoichiometry, the upper limit for calcium channels density should be half the density of Na pumping sites as estimated for the ouabain uptake.

We have also studied the binding of the ^{3}H-cinnarizine to subcellular components of rat aorta. After homogenization of the tissues and differential centrifugation, a nuclear, a mitochondrial, and a microsomal fraction are obtained. The ^{3}H-cinnarizine uptake is then determined in the following medium: protein 75 μg/ml, KCl 100 m*M*, EGTA 1 m*M*, Tris-maleate 20 m*M* (pH 7.4). The uptake of ^{3}H-cinnarizine by the nuclear and the mitochondrial fractions is linear for concentrations ranging from 2.5×10^{-8} *M* up to 5×10^{-7} *M*. By contrast, the binding to the microsomal fraction showed a saturable component in the same range of concentrations. The concentration producing the half saturation equals 8×10^{-8} *M*, a value close to the one found in the whole tissue. These observations indicate that in the whole tissue, the saturable binding of cinnarizine occurs at the level of the plasma membrane.

4. Conclusion

The experimental data summarized show that vascular relaxation can be achieved by an inhibition of calcium influx. Cinnarizine is acting specifically on this process by blockade of calcium channels activated during vascular smooth muscle excitation.

References

1 *Breemen, C. van; Farinas, B.R.; Casteels, R.; Gerba, D.; Wuytack, F., and Deth, R.:* Factors controlling cytoplasmic Ca^{2+} concentration. Phil. Trans. R. Soc. *265:* 57–71 (1973).

2 *Burton, J. and Godfraind, T.:* Sodium-calcium sites in smooth muscle and their accessibility to lanthanum. J. Physiol., Lond. *241:* 287–298 (1974).

3 *Feinstein, M.B. and Paimre, M.:* Pharmacological action of local anesthetics on excitation-contraction coupling in striated and smooth muscle. Fed. Proc. Fed. Am. Socs exp. Biol. *28:* 1643–1648 (1969).

4 *Fleckenstein, A.:* Specific pharmacology of calcium in myocardium, cardiac pacemakers, and vascular smooth muscle. Annu. Rev. Pharmac. Tox. *17:* 149–166 (1977).

5 *Godfraind, T.:* Calcium exchange in vascular smooth muscle, action of noradrenaline and lanthanum. J. Physiol., Lond. *260:* 21–35 (1976).

6 *Godfraind, T.:* Actions of suloctidil and of the Ca antagonist cinnarizine on the rat aorta. Archs int. Pharmacodyn. Thér. *221:* 342–343 (1976).

7 *Godfraind, T. and Ghysel-Burton, J.:* Binding sites related to ouabain-induced stimulation or inhibition of the sodium pump. Nature, Lond. *265:* 165–166 (1977).

8 *Godfraind, T. and Kaba, A.:* Blockade or reversal of contraction induced by calcium and adrenaline in depolarized arterial smooth muscle. Br. J. Pharmacol. *36:* 549–560 (1969).

9 *Godfraind, T. and Kaba, A.:* The role of calcium in the action of drugs on vascular smooth muscle. Archs int. Pharmacodyn. Thér. *196:* suppl., pp. 35–49 (1972).

10 *Godfraind, T. and Lesne, M.:* The uptake of cardiac glycosides in relation to their actions in isolated cardiac muscle. Br. J. Pharmacol. *46:* 488–497 (1972).

11 *Godfraind, T. and Morel, N.:* An upper limit to the number of calcium channels in smooth muscle as estimated with cinnarizine. Eur. J. Pharmacol. *41:* 245–246 (1977).

12 *Nueten, J.M. van and Janssens, P.A.J.:* Comparative study of the effects of flunarizine and cinnarizine on smooth muscles and cardiac tissues. Archs int. Pharmacodyn. Thér. *204:* 37–55 (1973).

13 *Weiss, G.G. and Goodman, F.R.:* Effects of lanthanum on contraction calcium distribution and ^{45}Ca movements in intestinal smooth muscle. J. Pharmac. exp. Ther. *169:* 46–55 (1969).

Prof. *T. Godfraind,* Université de Louvain, Faculté de Médecine, Laboratoire de Pharmacodynamie Générale et de Pharmacologie, Avenue Emmanuel-Mounier 73, UCL 7350, *B–1200 Bruxelles* (Belgium)

Mechanisms of Vasodilatation. Satellite Symp., 27th Int. Congr. Physiol. Sci., Wilrijk 1977, pp. 152–157 (Karger, Basel 1978)

Inhibitors of Cyclic AMP Breakdown

Jean-Claude Stoclet and Claire Lugnier

Laboratoire de Pharmacodynamie, Faculté de Pharmacie, Université Louis-Pasteur, Strasbourg

In the early seventies, different groups reported evidence supporting the hypothesis that adenosine 3′,5′-monophosphate (cyclic AMP) is the mediator of β-adrenergic vasodilatation (19). Different drugs exerting a relaxant effect on vascular smooth muscle were found to inhibit cyclic AMP breakdown *in vitro* and to potentiate catecholamine induced relaxation and accumulation of cyclic AMP (22, 26). Up until now and despite a variety of experimental approaches, it has not been possible to establish the role of cyclic AMP in smooth muscle function. It is beyond the scope of this article to discuss this subject which has recently been reviewed in detail (19).

The problem of the participation of cyclic AMP in the relaxing effects of drugs that inhibit cyclic AMP breakdown *in vitro* is still complicated by the fact that all these drugs, including cyclic AMP analogues, also act on other cellular mechanisms. This review presents a brief and non-exhaustive summary of some of the more significant advances in the pharmacology of phosphodiesterase inhibitors in vascular myocytes. The question of the influence of these drugs on guanosine 3′,5′-monophosphate (cyclic GMP) metabolism will also be discussed on account of the theory of the dualism of action of the two cyclic nucleotides.

1. Drugs that Vasodilate and Inhibit Cyclic AMP Breakdown

In acellular preparations, cyclic AMP breakdown is inhibited by a large variety of chemical compounds described in recent reviews (2, 6, 27). These compounds belong to different groups with no obvious structural relationships and display a wide spectrum of pharmacological properties. Vasodilatating or smooth muscle relaxing effects have been described for most of them, including cyclic AMP derivatives, theophylline and derivatives, papaverine and related isoquinolines, imidazolidinones, pyrazolopyridines and diazoxide, for instance.

In the case of methylxanthines, isoquinoline compounds, and imidazolidinones, a correlation has been found between potencies of inhibition of *in vitro* cyclic AMP hydrolysis and relaxing potencies on vascular smooth muscle strips (14, 16) or *in vivo* vasodilating and antihypertensive effects (21). This type of correlation supports the hypothesis that phosphodiesterase inhibition and the ensuing cyclic AMP accumulation mediate the relaxing effects of these drugs. However, this mechanism of action is not yet established. In several recent studies, a quantitative correlation between the ability of drugs to increase the cyclic AMP level and to produce relaxation could not be demonstrated (1, 7, 8, 12, 19). Furthermore, the mechanisms proposed to explain the postulated role of cyclic AMP in smooth muscle relaxation have not yet been established (19).

2. Drugs with Multiple Effects

It is obvious that different mechanisms may contribute to the action of drugs which cause or prevent vasodilatation. Verapamil and related 'Ca^{++} antagonists' (14) as well as phenothiazine derivatives (24) relax vascular smooth muscle at lower concentrations than those predicted from their potency as phosphodiesterase inhibitors. This is because their action is exerted by inhibition of Ca^{++} influx in the case of verapamil or blockade of α-receptors in the case of phenothiazines (24). Papaverine inhibits prostaglandin E_2-induced contraction (15), impairs phosphorylative oxydations and may also directly impair Ca^{++} availability (9). Other phosphodiesterase inhibitors showing anti-aggregant effects on platelets also inhibit prostaglandin effects (6).

Theophylline antagonizes the vasodilatating properties of adenosine in an apparently competitive manner (20) and also alters Ca^{++} binding to membranes (13). In the same line but with opposite consequences, it was recently suggested that smooth muscle relaxation induced by an exogenously applied dibutyryl analogue of cyclic AMP might result from an effect of this drug on a 'purinergic' receptor sensitive to many purine nucleotides and nucleosides (5). Thus, the relaxing effect of dibutyryl-cyclic AMP on smooth muscle strips cannot be considered as specific evidence for a relaxing role of intracellular cyclic AMP.

Finally, it should be mentioned that some phosphodiesterase inhibitors, such as dipyridamole and 1-methyl-3-isobutyl xanthine (MIX), inhibit the release of cyclic AMP from cells in culture (11). The importance of cyclic AMP extrusion from vascular smooth muscle cells *in situ* is unknown. Its inhibition may further complicate the interpretation of the effects of these drugs.

3. Cyclic Guanosine 3',5'-Monophosphate

Recently, *Schultz et al.* (23) observed that many vasodilators and/or smooth muscle relaxants including thiocyanate increase the cyclic GMP content of the vas deferens of the rat. This finding led the authors to suggest that cyclic GMP may play the role of a negative feedback signal: its increase under the influence of Ca ions activating contraction might in turn decrease Ca^{++} release or influx into the cytoplasm. The drugs that were shown to increase the cyclic GMP content include diazoxide, hydralazine, minoxidil, nitroglycerol, dipyridamole, lidoflazine and the 'Ca^{++} antagonists' verapamil, prenylamine, and cinnarizine. The mechanism of cyclic GMP accumulation is unknown at the present time, but it is of obvious interest to test the effect of these drugs on the different molecular forms of phosphodiesterase that have recently been isolated from vascular smooth muscles (see below). For instance, the potent antihypertensive agents diazoxide and hydralazine weakly inhibit cyclic AMP breakdown (4). A direct or indirect effect on cyclic GMP metabolism might explain their pharmacological properties. The finding that papaverine inhibits the hydrolysis of cyclic GMP (18) in addition to its well known action on cyclic AMP hydrolysis also raises the question of the participation of cyclic GMP in the relaxing effect of this drug (see below).

4. Selective Inhibition of Phosphodiesterases

Many pharmacological effects other than vasodilatation may result from phosphodiesterase inhibition. It is beyond the scope of this article to discuss these effects which include cardiac stimulation, inhibition of platelet aggregation, diuretic effects, bronchodilatation and antiasthmatic properties, psychotropic, anti-inflammatory, anti-neoplasic effects, etc. (2, 6). Obviously, the various drugs do not exert all these effects simultaneously. For instance, under normal circumstances, tranquilizing doses of benzodiazepines do not produce marked vasodilatation. Metabolic and pharmacokinetic factors may and probably do explain some of these differences. Also, some of the above-mentioned pharmacological properties may not be mediated by cyclic nucleotides. However, it was recently suggested that selective pharmacological effects may result from selective inhibition of different molecular forms of phosphodiesterase which have been separated from mammalian tissues where they are differently distributed (27, 29).

Phosphodiesterase activity from pig coronary arteries has been chromatographically resolved into two forms (28). One has a greater Vmax and a higher Km for cyclic AMP than for cyclic GMP. This form is sensitive to Ca^{++}-dependent activation by a thermostable protein. The second form, which seems similar to most particulate phosphodiesterase, is sometimes described as the 'low Km cyclic AMP' or 'negatively cooperative' phosphodiesterase. It is characterized by relative selectivity towards cyclic AMP as substrate, lack of sensitivity to the activator protein mentioned above, and non-linear kinetics.

Whatever the mechanism of this anomalous behaviour, it prevents the determination of the Ki value of inhibitors. As stated in a critical review by *Wells and Hardman* (29), these two forms are widely distributed in mammalian tissues and appear to exist in different proportions in various cell types. Since both forms are probably present in crude tissue extracts, it is only possible to determine which phosphodiesterase is affected by inhibitors and which type of inhibition is involved if they are separated. Papaverine and MIX have an inverse selectivity, the former being 3–4 times more active on the non activable form which uses cyclic AMP as preferred substrate than on the activable form (30). We have recently found that relaxing concentrations of papaverine indeed increase the cyclic AMP level in rat aorta strips without affecting the cyclic GMP level (10). Thus, it seems that selective inhibition of a phosphodiesterase form may induce selective accumulation of the corresponding cyclic nucleotide.

The existence of these different forms and of a Ca^{++}-dependent protein activator may also explain why many different chemical structures inhibit this complex enzyme system: there is a great diversity of possible mechanisms of interaction. For example, phenothiazine derivatives are reported to prevent activation by the protein activator (27).

Amer et al. (2, 3) have reported an increase in the hydrolysis of cyclic AMP by homogenates of arteries and hearts of hypertensive rats. An increase in the ratio of cyclic GMP to cyclic AMP levels occurs simultaneously. These observations, together with the finding that antihypertensive agents with different mechanisms of action all normalize the cyclic GMP/cyclic AMP ratio, led *Amer* to speculate that alterations in cyclic nucleotide metabolism are related to the development of hypertension. We have recently reported that an increase in cyclic AMP hydrolysis precedes the onset of hypertension while an increase in cyclic GMP hydrolysis follows the raise of blood pressure in spontaneously and mineralocorticoid hypertensive rats (17, 25). It would be important to learn whether these modifications bear on one or the other of the phosphodiesterase forms.

It should be mentioned that none of the molecular forms of phosphodiesterase has been purified to homogeneity and that their properties may be artifactually modified by the extraction procedure (see review 19).

5. Conclusion

The recent separation and characterization of different molecular forms of phosphodiesterase, which are more or less selectively inhibited by drugs, provide a basis for the search for compounds acting more specifically on one or the other of these phosphodiesterases and/or increasing selectively cyclic AMP or cyclic GMP levels in cells. There is no doubt that such compounds would be useful tools in elucidating the role of the phosphodiesterases in the regulation of cyclic nucleotide levels and in analyzing the function of these cyclic nucleotides. It is too early to judge whether the development of such compounds may provide useful therapeutic agents. Indeed, we do not know if the phosphodiesterases of the intact cell are identical to those which have been isolated. Up until now, attention has focussed on the possible (but as yet not established) role of cyclic AMP and cyclic GMP on the contractile properties of vascular smooth muscle. Cyclic GMP may also play a role in vascular smooth muscle proliferation that takes place in hypertensive arteries. One may speculate that the occurrence of other cyclic nucleotides and their participation in other functions may be discovered in the future. At the present time, the link between *in vitro* inhibition of phosphodiesterase and a pharmacological effect on vascular smooth muscle is not established.

References

1 *Allen, J.C.:* Ca^{++} binding properties of canine aortic microsomes: lack of effects of cAMP. Blood Vessels *14:* 91–104 (1977).

2 *Amer, M.S. and Kreighbaum, W.E.:* Cyclic nucleotide phosphodiesterases: properties, activators, inhibitors, structure activity. J. pharm. Sci. *64:* 1–36 (1975).

3 *Amer, M.S.; Perhach, J.L.; Gomol, A.W., and McKinney, G.R.:* Cyclic nucleotides and the mechanism of action of antihypertensive drugs. Pharmacologist *16:* 288 (1974).

4 *Andersson, R.:* Cyclic AMP as a mediator of the relaxing action of papaverine, nitroglycerine, diazoxide and hydralazine in intestinal and vascular smooth muscle. Acta pharmac. tox. *32:* 321–336 (1973).

5 *Bülbring, E. and Hardman, J.H.:* Effects on smooth muscle of nucleotides and dibutyryl analogues of cyclic nucleotides; in *Worcel and Vassort* Smooth muscle pharmacology and physiology, pp. 125–134 (INSERM, Paris 1976).

6 *Chasin, M. and Harris, D.N.:* Inhibitors and activators of cyclic nucleotide phosphodiesterase. Adv. cyclic Nucleotide Res., vol. 7, pp. 225–290 (Raven Press, Hewlett 1976).

7 *Collins, G.A. and Sutter, M.C.:* Quantitative aspects of cyclic AMP and relaxation in the rabbit anterior mesenteric portal vein. Can. J. Physiol. Pharmacol. *53:* 989–997 (1975).

8 *Daniel, E.E. and Crankshaw, J.:* Relation of cAMP to relaxation of pulmonary artery. Blood Vessels *11:* 295–311 (1974).

9 *Demesy-Waeldele, F. and Stoclet, J.C.:* Papaverine, cyclic AMP and the dependence of the rat aorta on extracellular calcium. Eur. J. Pharmacol. *31:* 185–194 (1975).

10 *Demesy-Waeldele, F. and Stoclet, J.C.:* Effect of papaverine on cyclic nucleotide levels in the isolated rat aorta. Eur. J. Pharmacol. *46:* 63–66 (1977).

11 *Vellis, J. De; Inglish, D., and Brooker, G.:* Paradoxical effects of actinomycin D, acetoxycycloheximide, and methylisobutylxanthine on cyclic AMP metabolism. Proc. natn. Acad. Sci. USA *33:* 507 (1974).

12 *Diamond, J. and Blisard, K.S.:* Effects of stimulant and relaxant drugs on tension and cyclic nucleotide levels in canine femoral artery. Mol. Pharmacol. *12:* 688–692 (1976).

13 *Kalsner, S.:* Mechanism of potentiation of contractor responses to catecholamines by methylxanthines in aortic strips. Br. J. Pharmacol. *43:* 379–388 (1971).

14 *Kukovetz, W.R.; Pöch, G.; Wurm, A.; Holzmann, S., and Paietta, E.:* Effect of phosphodiesterase inhibition on smooth muscle tone; in *Betz* Ionic actions on vascular smooth muscle, pp. 124–131 (Springer, Berlin 1976).

15 *Levy, J.V.:* Papaverine antagonism of prostaglandin E_2-induced contraction of rabbit aortic strips. Res. Commun. Chem. Pathol. Pharmacol. *5:* 297–310 (1973).

16 *Lugnier, C.; Bertrand, Y., and Stoclet, J.C.:* Cyclic nucleotide phosphodiesterase inhibition and vascular smooth muscle relaxation. Eur. J. Pharmacol. *19:* 134–136 (1972).

17 *Lugnier, C.; Michon, T., and Stoclet, J.C.:* Evolution of aortic and cardiac cyclic AMP phosphodiesterase during the onset of mineralocorticoid hypertension in the rat. Br. J. Pharmacol. *58:* 434P–435P (1976).

18 *Lugnier, C. and Stoclet, J.C.:* Inhibition by papaverine of cGMP and cAMP phosphodiesterases from rat heart. Biochem. Pharmacol. *23:* 3071–3074 (1974).

19 *Namm, D.H. and Leader, J.P.:* Occurrence and function of cyclic nucleotides in blood vessels. Blood Vessels *13:* 24–47 (1976).

20 *Olsson, R.A.; Davis, C.J.; Khouri, E.M., and Patterson, R.E.:* Evidence for adenosine receptor on the surface of dog coronary myocytes. Circulation Res. *39:* 93–98 (1976).

21 *Pettinger, W.; Sheppard, H.; Palkoski, Z., and Renyi, E.:* Angiotensin antagonism and antihypertensive activity of phosphodiesterase inhibiting agents. Life Sci. *12:* 49–62 (1973).

22 *Pöch, G. and Kukovetz, W.R.:* Studies on the possible role of cyclic AMP in drug induced coronary vasodilatation. Adv. cyclic Nucleotide Res., vol. 1, pp. 195–211 (Raven Press, Hewlett 1972).

23 *Schultz, K.D.; Schultz, K., and Schultz, G.:* Sodium nitroprusside and other smooth muscle-relaxants increase cyclic GMP levels in rat ductus deferens. Nature, Lond. *265:* 750–751 (1977).

24 *Stoclet, J.C.; Demesy, F., and Scheftel, J.M.:* Effects on vascular smooth muscle of 3′,5′-cyclic nucleotide phosphodiesterase inhibitors and antagonism with Ca^{2+} ions. J. Pharmacol. *3:* 554 (1972).

25 *Stoclet, J.C.; Michon, T. et Lugnier, C.:* Métabolisme des nucléotides 3′5′-cycliques dans l'aorte et dans le cœur durant l'installation de l'hypertension spontanée du rat Okamoto. J. Pharmacol. (in press).

26 *Triner, L.; Nahas, G.G.; Vulliemoz, Y.; Overweg, N.I.A.; Verodky, M.; Habif, D.V., and Ngai, S.H.:* Cyclic AMP and smooth muscle function. Ann. N.Y. Acad. Sci. *185:* 458–476 (1971).

27 *Weiss, B. and Hait, W.N.:* Selective cyclic nucleotide phsophodiesterase inhibitors as potential therapeutic agents. Annu. Rev. Pharmac. Tox. *17:* 441–448 (1977).

28 *Wells, J.N.; Baird, C.E.; Wu, Y.T., and Hardman, J.G.:* Cyclic nucleotide phosphodiesterase activities of pig coronary arteries. Biochim. biophys. Acta *384:* 430–442 (1975).

29 *Wells, J.N. and Hardman, J.G.:* Cyclic nucleotide phosphodiesterases. Adv. cyclic Nucleotide Res., vol. 8 (in press).

30 *Wells, J.N.; Wu, Y.J.; Baird, C.E., and Hardman, J.G.:* Phosphodiesterases form porcine coronary arteries: inhibition of separated forms by xanthines, papaverine, and cyclic nucleotides. Mol. Pharmacol. *11:* 775–783 (1975).

Dr. *J.C. Stoclet,* Laboratoire de Pharmacodynamie, Faculté de Pharmacie, Université Louis-Pasteur, *F–67083 Strasbourg Cédex* (France)

Mechanisms of Vasodilatation. Satellite Symp., 27th Int. Congr. Physiol. Sci., Wilrijk 1977, pp. 158–164 (Karger, Basel 1978)

Organic Nitrates, Sodium Nitroprusside, and Vasodilatation

V.A.W. Kreye [1]

IInd Physiological Institute, University of Heidelberg, Heidelberg

Sodium nitroprusside and the nitrates preferentially relax tonic smooth muscle (vascular, bronchial), whereas their relaxant effects on phasic smooth muscles (gastrointestinal, ureteral, uterine) are transient and incomplete (14, 16). We discuss additional evidence suggesting that nitroprusside and the nitrates dilate vascular smooth muscle by the same cellular mechanism.

1. Independence of Sodium Nitroprusside (NP) and Nitroglycerin (GTN) Action from Extracellular Calcium

In organ bath studies on arteries (8, 14), an apparent competitive antagonism between NP and Ca^{++} ions suggested an interference of NP with Ca^{++} influx. However, controversial results were obtained from ^{45}Ca flux studies with NP (2, 9, 14). In the absence of Ca^{++} and presence of La^{+++}, the noradrenaline-induced contraction of rabbit aorta, which depends on the release of intracellular Ca^{++} (3), is still depressed by NP (14). Similar results were obtained with canine saphenous veins (22). In rat aortic strips made to contract with K ions, we investigated the effect of variations of the rinsing fluid on the speed of relaxation; whereas omission of Ca^{++} from the rinsing fluid has little influence on absolute or relative speed of relaxation ($-dT/dt_{max}$ or $-dT/dt_{max}/\Delta t$, respectively), 10^{-4} *M* NP increases these parameters by more than 100% – an effect independent from the Ca^{++} concentration in the rinsing fluid (13). NP and GTN retain their vasodilator effect as opposed to verapamil and prenylamine, when they are applied to bovine coronary arteries during K^+ contractures, despite absence of Ca^{++} and presence of La^{+++} (7). Thus, NP and GTN relax vascular

[1] Supported by Deutsche Forschungsgemeinschaft within SFB 90 'Cardiovasculäres System'.

smooth muscle independently of extracellular Ca^{++}, and a possible interference with transmembrane Ca fluxes is not a necessary prerequisite to understand their vasodilator action.

2. Effect on Membrane Potential and Ionic Fluxes (^{36}Cl, ^{24}Na)

Häusler and Thorens (9) described a dose-dependent hyperpolarization of up to 12 mV caused by NP in the rabbit main pulmonary artery. Drugs may induce hyperpolarization of vascular smooth muscle by at least four mechanisms: (1) activation of an electrogenic Na pump; (2) reduction of Na^+ permeability; (3) reduction of Cl^- permeability – since the equilibrium potential E_{Cl} is much smaller than the resting membrane potential (E_m) in vascular smooth muscle (19, 23); and (4) augmentation of K^+ permeability.

In organ bath experiments on the rat aorta, we studied the effects of temperature changes and of ouabain on the action of NP, and of Cl^- replacement by isethionate on the action of NP, GTN, and isosorbide dinitrate. To study ^{24}Na efflux, pieces of rabbit aorta were loaded with ^{24}Na, and their radioactivity measured continuously during superfusion at 37 °C with physiological salt solution (PSS) containing 10^{-7} *M* noradrenaline. 10^{-5} *M* NP was added after 50 min and withdrawn after 90 min. To study Cl^- efflux, pieces of rabbit aorta were loaded with ^{36}Cl for 90 min after an initial relaxation period for 2 h in normal PSS. Under constant rotation, the loaded preparations were transferred through a series of vials with 5 ml PSS at 37 °C. 10^{-7} *M* noradrenaline was present from the onset, and either NP or GTN were added after 20 min. In a control series, relaxation was induced by omission of Ca^{++}. Finally, the tissues were dissolved and their radioactivity measured. Rates of efflux were calculated as $\Delta cpm/\Delta t$/total cpm at time t. In the study on ^{36}Cl efflux, cpm were added retrogradely and, by the method of least squares, the efflux equations $y = c \cdot e^{-kt}$ determined for the late linear phase of the ^{36}Cl efflux curve.

At 27 °C, the rat aorta is as sensitive to the vasodilator action of NP as at 37 °C. Incubation with 10^{-4} *M* ouabain for 1 h does not alter the responsiveness to NP. ^{24}Na efflux from rabbit aorta is not increased by 10^{-5} *M* NP (fig. 1). Thus, NP obviously does not hyperpolarize the cell by stimulation of an electrogenic Na pump. The Na^+ permeability is probably also not affected by NP, since, under steady state conditions, active Na^+ efflux should equal passive Na^+ influx (23), and a reduction of Na^+ permeability would result indirectly in a decreased Na^+ efflux; this did not occur in our experiments.

In accordance with *Wahlström* (24), 10^{-7} *M* noradrenaline increases ^{36}Cl efflux from vascular smooth muscle considerably ($k = 0.0713 \pm 0.0018$ min^{-1}; $\bar{x} \pm$ SE; n = 10) in comparison to controls (0.0399 ± 0.0031 min^{-1}). Both GTN and NP proved to be powerful inhibitors of the noradrenaline-induced increase

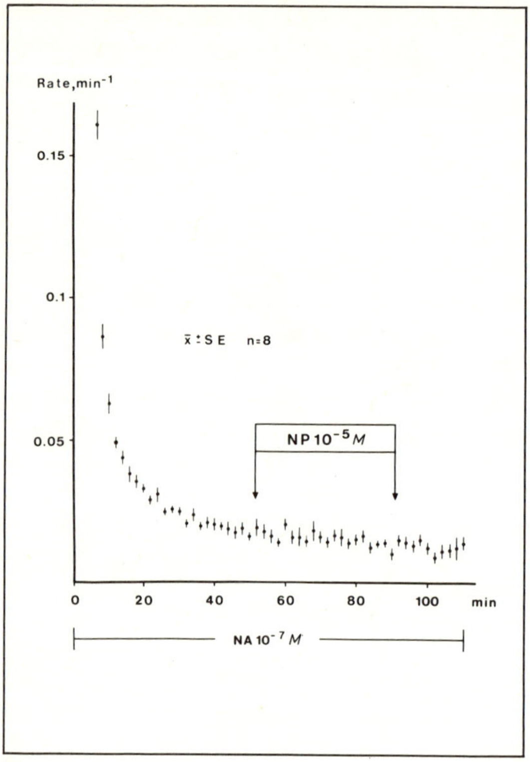

Fig. 1. ^{24}Na efflux from rabbit aorta in the presence of noradrenaline (NA), and effect of addition of sodium nitroprusside (NP) to the superfusate at min 50–90. Ordinate: Rate of efflux, Δcpm/Δt/total cpm at time t.

of ^{36}Cl efflux (fig. 2). The rate coefficients k of the efflux equation dropped from 0.0724 and 0.0719 min^{-1} with noradrenaline alone to 0.0452 and 0.0424 min^{-1} after addition of 10^{-5} *M* GTN (n = 5) or NP (n = 7), respectively. This *chloride antagonism* is a specific effect of the drugs, and not simply a consequence of the relaxation process. When vasodilatation is induced by withdrawal of Ca^{++} rather than by NP or GTN, no effect on ^{36}Cl efflux can be demonstrated.

NP is devoid of any α-adrenergic blocking activity (14), and no such an effect has been claimed to be involved in the action of GTN. Therefore, our findings may reflect an inhibitory influence of NP and GTN on the Cl^- permeability of vascular smooth muscle. Especially for NP, it is conceivable that this large anion may interfere with the kinetics of small anions at the level of the cell membrane.

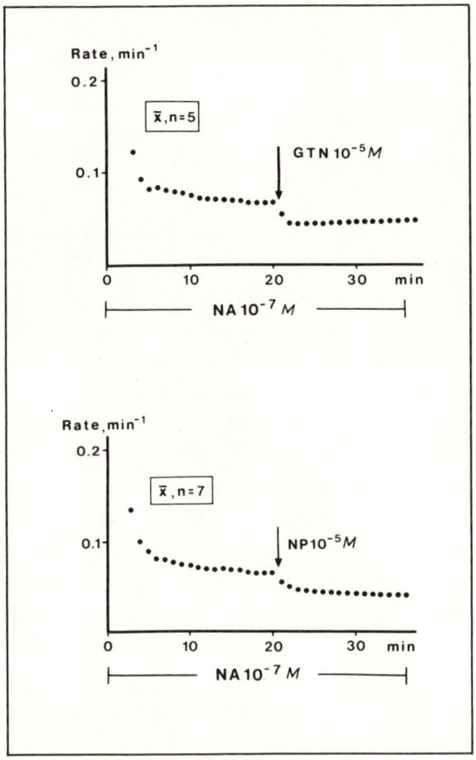

Fig. 2. ^{36}Cl efflux from rabbit aortae in the presence of noradrenaline (NA), and effect of nitroglycerin (GTN, above) and sodium nitroprusside (NP, below). Ordinates: Rate of efflux, Δcpm/Δt/total cpm at time t.

An estimate of E_{Cl} based on the zero intercepts of efflux equations, an ^{3}H-sorbitol space of 50–58%, and a dry weight/wet weight ratio of 0.22 resulted in values between −30 and −40 mV. Although these are higher than those reported for other types of vascular smooth muscle (10, 19, 23), they are considerably lower than the resting E_m of large arteries of the rabbit (20, 21). Hence, our results can explain the reported hyperpolarization produced by NP, and they suggest that GTN acts as a vasodilator in a similar manner, by causing hyperpolarization. Since we have not performed studies on K^+ efflux, the possibility of an increased K^+ permeability underlying the vasodilator effect of these drugs must be kept in mind.

In organ bath studies performed on the rat aorta, replacement of Cl^- by isethionate ions suppresses the inhibitory effect of NP, GTN and isosorbide dinitrate on K^+ contractures but not that on responses to noradrenaline.

Although we have no ready explanation for this finding, it is in favour of the assumption that both NP and the organic nitrates produce vasodilatation by the same cellular mechanism.

3. Effect of NP and GTN on Cellular Cyclic GMP Levels

NP (14) and GTN (5) do not affect cAMP levels in vascular tissues, whereas they are potent stimulants of guanylate cyclase. They increase cGMP levels in various smooth muscles, including rat vas deferens (18), rat myometrium (6), bovine trachea (11), canine femoral artery (5), rabbit carotid and main pulmonary artery (4), and hog spleen artery (1). The increases in cGMP levels are large – e.g. over 1,500% in canine femoral arteries with 2×10^{-4} *M* GTN (5), or over 4,000% in hog spleen artery with 10^{-4} *M* NP (1). 8-Bromo-cGMP, a lipophylic derivative of cGMP, relaxes bovine trachea (11), rat vas deferens, and rat aorta (17), suggesting a role of cGMP in NP- and GTN-induced vasodilatation. On the other hand, several non-relaxant drugs, such as carbachol, acetylcholine, or histamine, also increase cGMP levels in smooth muscle (5, 11) and, as regards NP, this drug does not relax rat vas deferens (14), although it increases cGMP considerably in this tissue (18). Thus, a causal relation between cellular cGMP and drug-induced vasodilatation has yet to be demonstrated.

4. Other Metabolic Effects

Although several biochemical effects of the organic nitrates have been established (15), it is not clear whether these are major determinants of their vasodilator action. As regards sodium nitroprusside, this drug releases CN^- on breakdown, but it does not reduce cellular ATP at vasodilator concentrations (14). Recently, we found that CN^- completely antagonizes the vasodilator effect of sodium nitroprusside. However, since other metabolic inhibitors do not have such an action, it seems unlikely that metabolic inhibition is involved in this process (12).

References

1 *Böhme, E.; Graf, H., and Schultz, G.:* Effects of sodium nitroprusside and other smooth muscle relaxants on cyclic GMP formation in smooth muscle and platelets. Adv. cyclic Nucleotide Res., vol. 9 (in press, 1978).

2 *Breemen, C. van:* Transmembrane calcium transport in vascular smooth muscle; in *Bevan, Burnstock, Johansson, Maxwell and Nedergard* Vascular neuroeffector mechanisms, pp. 67–79 (Karger, Basel 1976).

3 *Breemen, C. van; Farinas, B.R.; Gerba, P., and McNaughton, E.D.:* Excitation-contraction coupling in rabbit aorta studied by the lanthanum method for measuring cellular calcium influx. Circulation Res. *30:* 44–54 (1972).

4 *Burkard, W.P.:* Effects of sodium nitroprusside on contractile state and cyclic nucleotide levels in rabbit arteries. Arch. Pharmakol. *297:* suppl. II, R12 (1977).

5 *Diamond, J. and Blisard, K.S.:* Effects of stimulant and relaxant drugs on tension and cyclic nucleotide levels in canine femoral artery. Mol. Pharmacol. *12:* 688–692 (1976).

6 *Diamond, J. and Holmes, T.G.:* Effects of potassium chloride and smooth muscle relaxants on tension and cyclic nucleotide levels in rat myometrium. Can. J. Physiol. Pharmacol. *53:* 1099–1107 (1975).

7 *Fermum, R.; Klinner, U. und Meisel, P.:* Versuche zum Wirkungsmechanismus von Gefässspasmolytika. I. Wirkung von Nitroprussid-Natrium, Nitroglycerin, Prenylamin und Verapamil an der arretierten Kalium-induzierten Kontraktur isolierter Koronararterien. Acta biol. med. germ. *35:* 1347–1358 (1976).

8 *Häusler, G.:* The effects of sodium nitroprusside (SNP) on vascular smooth muscle. Experientia *31:* 729 (1975).

9 *Häusler, G. and Thorens, D.:* The pharmacology of vasoactive antihypertensives; in *Bevan, Burnstock, Johansson, Maxwell and Nedergard* Vascular neuroeffector mechanisms, pp. 232–241 (Karger, Basel 1976).

10 *Jones, A.W.:* Altered ion transport in large and small arteries from spontaneously hypertensive rats and the influence of calcium. Circulation Res. *34/35:* suppl. I, pp. I-117–I-122 (1974).

11 *Katsuki, S. and Murad, F.:* Regulation of adenosine cyclic 3′,5′-monophosphate levels and contractility in bovine tracheal smooth muscle. Mol. Pharmacol. *13:* 330–341 (1977).

12 *Kreye, V.A.W.:* Inhibition of nitroprusside-induced relaxation of isolated vascular smooth muscle by cyanide. Cause of clinical tachyphylaxis? Arch. Pharmakol. *297:* suppl. II, R30 (1977).

13 *Kreye, V.A.W. and Lüth, J.B.:* Effect of sodium nitroprusside, temperature, and calcium withdrawal on the relaxation speed of vascular smooth muscle; in *Betz* Ionic actions on vascular smooth muscle, pp. 145–149 (Springer, Berlin 1976).

14 *Kreye, V.A.W.; Baron, G.D.; Lüth, J.B., and Schmidt-Gayk, H.:* Mode of action of sodium nitroprusside on vascular smooth muscle. Arch. Pharmakol. *288:* 381–402 (1975).

15 *Needleman, P. and Johnson, E.M., jr.:* The pharmacological and biochemical interaction with sulfhydryls. Possible correlations with the mechanism for tolerance development, vasodilatation, and mitochondrial and enzyme reactions; in *Needleman* Organic nitrates. Handbook of experimental pharmacology, vol. 40, pp. 97–114 (Springer, Berlin 1975).

16 *Nickerson, M.:* Vasodilator drugs; in *Goodman and Gilman* The pharmacological basis of therapeutics, pp. 727–743 (Macmillan, New York 1975).

17 *Schultz, K.D.:* Effects of 8-Br-cGMP on the tonus of rat vas deferens and aorta. Arch. Pharmakol. *297:* suppl. II, R12 (1977).

18 *Schultz, K.D.; Schultz, K., and Schultz, G.:* Sodium nitroprusside and other smooth muscle-relaxants increase cyclic GMP levels in rat ductus deferens. Nature, Lond. *265:* 750–751 (1977).

19 *Siegel, G.; Roedel, H.; Nolte, J.; Hofer, H.W., and Bertsche, O.:* Ionic composition and ion exchange in vascular smooth muscle, in *Bülbring and Shuba* Physiology of smooth muscle, pp. 19–39 (Raven Press, Hewlett 1976).

20 *Somlyo, A.V.; Häusler, G., and Somlyo, A.P.:* Cyclic adenosine monophosphate:

potassium-dependent action on vascular smooth muscle membrane potential. Science *169:* 490–491 (1970).

21 *Speden, R.N.:* Excitation of vascular muscle; in *Bülbring, Brading, Jones and Tomita* Smooth muscle, pp. 558–588 (Arnold, London 1970).

22 *Verhaege, R.H. and Shepherd, J.T.:* Effect of nitroprusside on smooth muscle and adrenergic nerve terminals in isolated blood vessels. J. Pharmac. exp. Ther. *199:* 269–277 (1976).

23 *Wahlström, B.A.:* Ionic fluxes in the rat portal vein and the applicability of the Goldman equation in predicting the membrane potential from flux data. Acta physiol. scand. *89:* 436–448 (1973).

24 *Wahlström, B.A.:* A study on the action of noradrenaline on ionic content and sodium, potassium and chloride effluxes in the rat portal vein. Acta physiol. scand. *89:* 522–530 (1973).

Prof. Dr. med. *V.A.W. Kreye,* IInd Physiological Institute, University of Heidelberg, Im Neuenheimer Feld 326, *D–6900 Heidelberg* (FRG)

Mechanisms of Vasodilatation. Satellite Symp., 27th Int. Congr. Physiol. Sci., Wilrijk 1977, pp. 165–172 (Karger, Basel 1978)

Intravenous Anesthetic Agents and Vascular Smooth Muscle Function

Bella T. Altura and Burton M. Altura

Department of Physiology, SUNY Downstate Medical Center, Brooklyn, N.Y.

Controversy exists as to the effects of intravenous anesthetic agents on the peripheral vasculature (1, 8, 11, 12, 14–16, 20, 24, 27, 29). Some of this dilemma is probably a result of concentration and species differences, as well as differences in regional vasculature examined and methods used. The evidence available suggests that most of the intravenous anesthetics (e.g. barbiturates, ethanol, ketamine, steroids, etc.) produce different degrees of peripheral vasodilatation (11, 12, 14, 16, 18–20, 27, 29). Until recently, it was thought that this vasodilatation was due to the effects of these agents on the central and autonomic nervous systems (14, 16, 24, 29). However, recent *in vitro* experiments and *in vivo* microcirculatory studies suggest that concentrations of intravenous anesthetics used to induce surgical anesthesia can exert direct depressant effects on vascular smooth muscle.

1. Influence of Various General Anesthetic Agents on Baseline Tension and Spontaneous Mechanical Activity of Isolated Arteries and Veins

In vitro, many of the commonly used general anesthetic agents (e.g. halothane, diethyl ether, barbiturates, ethanol, ketamine, and steroids) cause a dose-dependent inhibition of the spontaneous mechanical activity in venous and arterial smooth muscles as well as lowering resting tension in arteries which are capable of developing basal tone (1, 3, 6–9, 11, 15). In all cases, the greater the concentration of the anesthetic agent, the more rapid the inhibition and the more the resting tension is lowered. Venous smooth muscle seems to be more sensitive to these inhibitory actions than arterial smooth muscle (1, 6, 8, 9, 11, 15). Most of these inhibitory actions are observed with concentrations of the anesthetics corresponding to those achieved during surgical anesthesia (6, 8, 9, 11, 14, 15, 19, 29). The relative order of potency to inhibit vascular myogenic tone is similar to that to induce anesthesia, e.g. thiopental > secobarbital ≅ pentobarbital > amobarbital > phenobarbital > barbital (8). If venous smooth muscle cells are more sensitive to these inhibitory actions than are arterial smooth muscle cells, this may help to explain why, in particular, barbiturates, alcohol, halothane, and ether compromise the control of the cardiovascular system (11, 12, 14, 16, 24, 27, 29), produce hypotension and venous pooling, and exacerbate the effects of circulatory shock (11, 14, 18, 20, 27).

2. Effects of Anesthetics on Contractile Responses Elicited by Neuro-Humoral Agents

In addition to their direct actions, anesthetics could cause vasodilatation by inhibition of the constrictor action of endogenous neuro-humoral substances that help maintain vascular tone (2–9, 11, 13, 15, 21, 30). Figures 1 and 2 illustrate that intravenous anesthetics (e.g. barbiturates, ketamine) markedly, and dose-dependently, attenuate the contractile responses of both arteries and

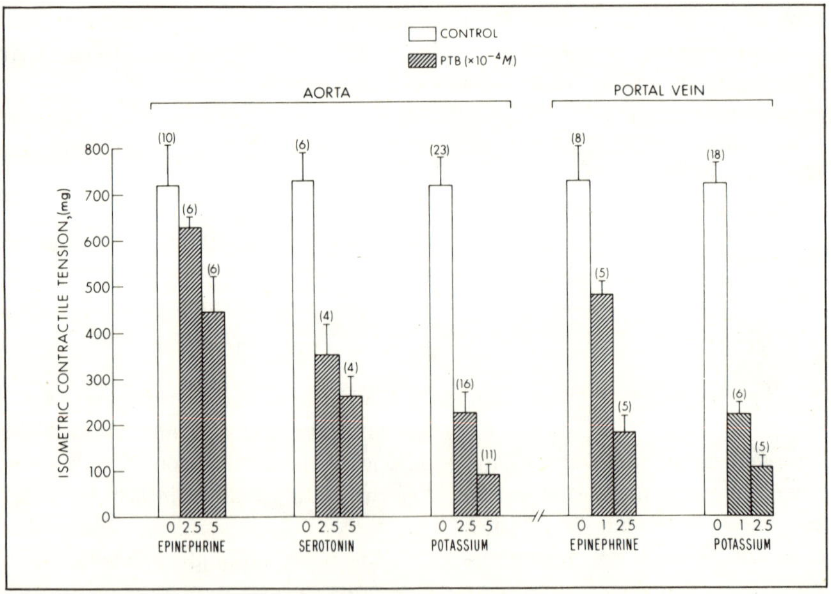

Fig. 1. Different sensitivity of equi-potent epinephrine, serotonin, and KCl-induced contractions on rat aortic strips and portal veins to inhibition by pentobarbital ($\times 10^{-4}$ *M*). Numbers in parentheses indicate different number of preparations utilized. Bars represent 1 SE. All experimental mean values (± SE) are significantly different from controls (without PTB) ($p < 0.01$). From *Altura and Altura* (9).

veins to catecholamines, serotonin, angiotensin, vasopressin, and potassium ions. Similar findings have been observed with halothane, ethanol, and steroids (2–5, 11, 15). None of the inhibitory effects induced by these anesthetics are modified or attenuated by known pharmacologic antagonists (1–4, 6, 8, 11, 13, 15). Other *in vitro* experiments indicate that arteries and veins made to contract with amines, angiotensin, vasopressin, KCl, and $BaCl_2$ relax in a dose-dependent manner upon the addition of anesthetic concentrations of barbiturates, halothane, ketamine, steroids, and ethanol (2–9, 15, 30).

3. Effects of Intravenous Anesthetics on Calcium Chloride-Induced Contractions of Vascular Muscle

The above data suggest that a variety of intravenous anesthetic agents act rather non-specifically on vascular smooth muscle to inhibit the effect of vasoconstrictor substances. Thus, barbiturates, ethanol, ketamine, and steroids may act directly to reduce the availability of Ca^{++} for the contractile process. If

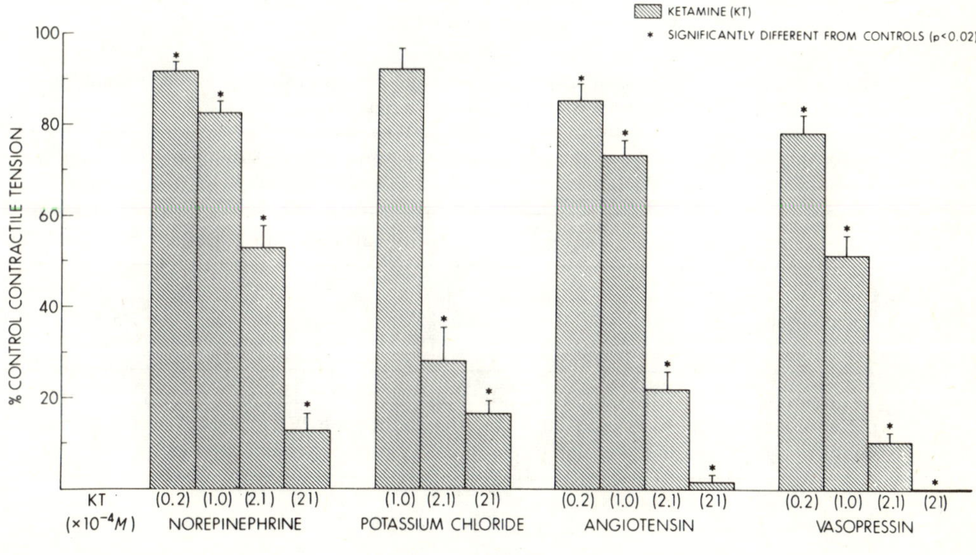

Fig. 2. Different sensitivity of equi-potent norepinephrine, KCl, angiotensin II, and vasopressin-induced contractions on rat aortic strips to inhibition by ketamine HCl (× 10^{-4} *M*). Experimental values with asterisks are significantly different from controls (without KT).

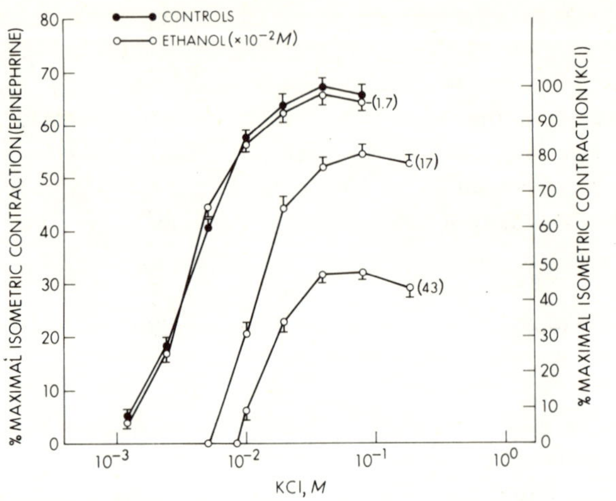

Fig. 3. Influence of ethanol (× 10^{-2} *M*) on KCl concentration-effect curves in rat aortic strips. Values are means ± 1 SE. From *Altura et al.* (6).

Table I. Relative sensitivities of $CaCl_2$-, epinephrine-, and potassium-induced contractions of rat aortic strips to the inhibitory action of barbiturates

Vasoactive stimulus	Concentration of barbiturate required to exert threshold inhibitory action (× 10^{-5} *M*)[1]				
	thiopental	secobarbital	pentobarbital	amobarbital	phenobarbital
$CaCl_2$	5	5	7.5	10	25
Epinephrine	10	10	25	25	100
Potassium	5	10	10	10	50

[1] The lowest concentration of barbiturate either needed to increase significantly the threshold concentration of vasoactive stimulant for contraction, ED_{50} and/or needed to reduce significantly the maximum tension of the agonist. From *Altura and Altura* (8).

so, one would anticipate that calcium-induced contractions of K^+-depolarized vascular muscles would be the most sensitive to the intravenous anesthetic inhibition, as is the case (table I) (2, 6, 8, 9, 15). Although other vasoactive agents, such as catecholamines and serotonin, may mobilize an intracellular and/or tightly bound membrane pool of Ca to cause contraction of vascular

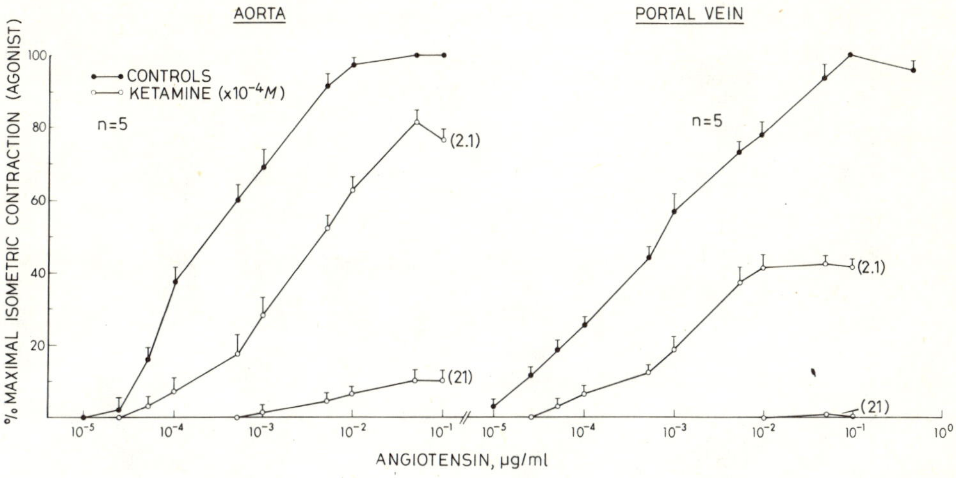

Fig. 4. Effects of ketamine ($\times$ 10^{-4} *M*) on angiotensin II dose-response curves in rat aortic strips and rat portal veins.

muscle (10, 28), anesthetics could interfere with the mobility of these intracellular Ca ions, since they penetrate both membranes and cells rapidly (26, 29). This would mean that intravenous anesthetics may act at the vascular muscle membrane as well as intracellularly. The observed shifts of the log dose-response curves for agonists (fig. 3, 4) (2, 6, 8, 9, 15), as well as for Ca^{++}, together with the reduction in maximum developed tension observed in the presence of the anesthetics, would be consistent with this hypothesis. *Price* (25) and *Nayler and Szeto* (23) have recently suggested that halothane and barbiturates may produce depression of cardiac muscle by interfering with Ca^{++} at several cellular sites.

4. Effects of Intravenous Anesthetics on the Microcirculation

Recent quantitative microcirculatory observations in the rat mesentery and the bat wing, using closed-circuit television microscopy, have revealed that most of the intravenous anesthetics so far tested dilate arterioles and, in some cases, muscular venules when administered in doses which induce anesthesia in these species (fig. 5) (1, 2, 15, 17–20). In addition, these intravenous anesthetics can inhibit, non-specifically, constrictions caused by humoral agonists and Ba ions (1, 2, 20, 22) (fig. 6). Overall, it is likely that concentrations of many different intravenous anesthetics used to induce surgical anesthesia can exert profound

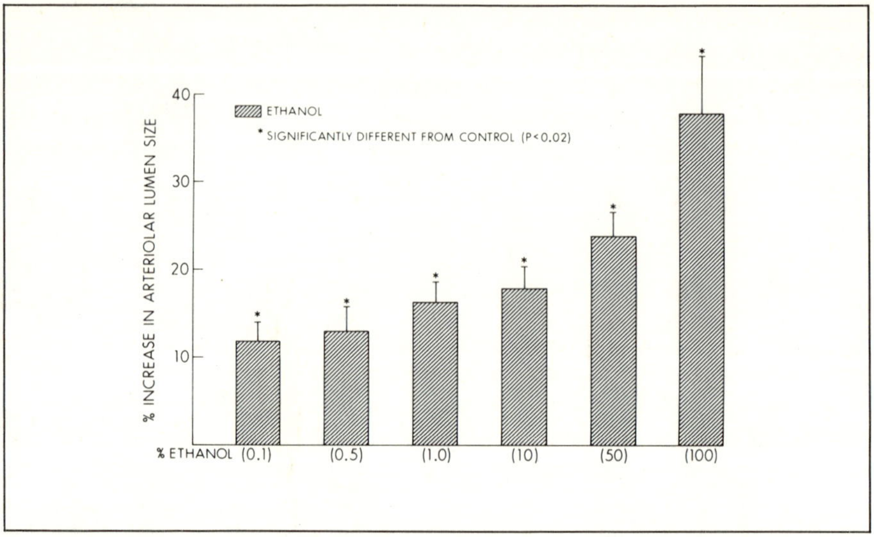

Fig. 5. Influence of different doses of topically applied ethanol (%) on rat mesenteric terminal arterioles (18–23 μm i.d.). All values in presence of ethanol are significantly different from paired control (before ethanol) ($p < 0.02$). Ethanol was applied for 2–5 min.

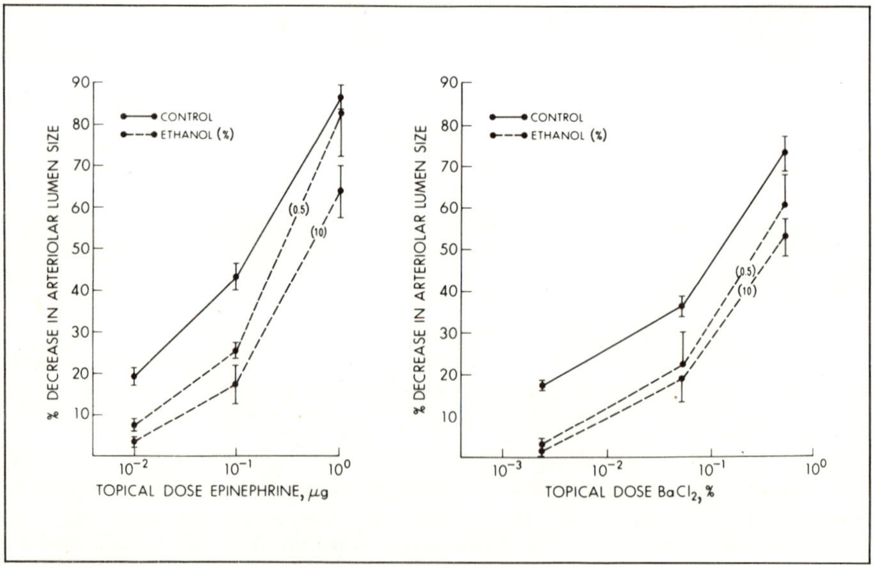

Fig. 6. Inhibition of epinephrine and barium chloride-induced rat mesenteric arteriolar constrictor responses by topical application of ethanol (%).

depressant effects on macro- as well as micro-vascular smooth muscles. Although these studies provide good evidence that a part, or all, of the anesthetic-induced vasodilatation is related to altered movement and/or translocation of Ca^{++}, one must also consider the possibilities that interferences with the function of the central nervous system and/or with autonomic reflexes, as well as metabolic effects of these drugs, may play important roles (14, 16, 24, 26, 28, 29).

5. Conclusion

Although administration of certain anesthetics (e.g. halothane, barbiturates, ethanol, ketamine, steroids) may not always result in peripheral vasodilatation, it is generally agreed that deep anesthesia, when induced by these agents, will result in dilatation and hypotension. Data obtained on isolated arteries and veins, as well as *in vivo* on microcirculatory vessels, suggest that the direct vascular actions of halothane and intravenous anesthetics may contribute to the decrease of vasoconstrictor tone associated with induction of anesthesia. At least two mechanisms appear to contribute to this vasodilator effect: (1) inhibition of the normal rhythm or vasomotion (i.e. spontaneous mechanical activity) of vascular smooth muscle; (2) depression of the contractile responses to endogenous neuro-humoral substances that play a role in maintaining vascular tone. These dilator actions may be related to interference with movement and/or translocation of Ca^{++}.

References

1 *Altura, B.M.:* Pharmacology of venular smooth muscle: new insights. Microvasc. Res. (in press).

2 *Altura, B.M. and Altura, B.T.:* Peripheral vascular actions of glucocorticoids and their relationship to protection in circulatory shock. J. Pharmac. exp. Ther. *190:* 300–315 (1974).

3 *Altura, B.M. and Altura, B.T.:* Vascular smooth muscle and prostaglandins. Fed. Proc. Fed. Am. Socs exp. Biol. *35:* 2360–2366 (1976).

4 *Altura, B.M. and Altura, B.T.:* Vascular smooth muscle and neurohypophyseal hormones. Fed. Proc. Fed. Am. Socs exp. Biol. *36:* 1853–1861 (1977).

5 *Altura, B.M. and Edgarian, H.:* Ethanol-prostaglandin interactions in contraction of vascular smooth muscle. Proc. Soc. exp. Biol. Med. *152:* 334–336 (1976).

6 *Altura, B.M.; Edgarian, H., and Altura, B.T.:* Differential effects of ethanol and mannitol on contraction of arterial smooth muscle. J. Pharmac. exp. Ther. *197:* 352–361 (1976).

7 *Altura, B.T. and Altura, B.M.:* Effects of pentobarbital sodium on isolated rat arterial and venous smooth muscle. Pharmacologist *16:* 303 (1974).

8 *Altura, B.T. and Altura, B.M.:* Barbiturates and aortic and venous smooth-muscle function. Anesthesiology *43:* 432–444 (1975).

9 *Altura, B.T. and Altura, B.M.:* Pentobarbital and contraction of vascular smooth muscle. Am. J. Physiol. *229:* 1635–1640 (1975).
10 *Bohr, D.F.:* Vascular smooth muscle updated. Circulation Res. *32:* 665–672 (1973).
11 *Clark, S.C. and MacCannell, K.L.:* Vascular responses to anesthetic agents. Can. Anaesth. Soc. J. *22:* 20–33 (1975).
12 *Conahan, T.J.:* New intravenous anesthetics. Surg. Clins N. Am. *55:* 851–859 (1975).
13 *Dalske, H.F.:* Effects of ethanol and its interaction with vasoactive agents in bovine vascular smooth muscle. Archs int. Pharmacodyn. Thér. *218:* 54–65 (1975).
14 *Dundee, J.W. and Wyant, G.M.:* Intravenous anesthesia (Churchill-Livingstone, Edinburgh 1974).
15 *Edgarian, H. and Altura, B.M.:* Ethanol and contraction of venous smooth muscle. Anesthesiology *44:* 311–317 (1976).
16 *Greisheimer, E.M.:* The circulatory effects of anesthetics; in *Hamilton and Dow* Handbook of physiology, vol. III, pp. 2485–2487 (Am. Physiological Society, Washington 1963).
17 *Harris, P.D.; Hodoval, L.F., and Longnecker, E.:* Quantitative analysis of microvascular diameters during pentobarbital and thiopental anesthesia in the bat. Anesthesiology *35:* 337–342 (1971).
18 *Hershey, S.G.; Zweifach, B.W., and Rovenstine, E.A.:* Effects of depth of anesthesia on behavior of peripheral vascular bed. Anesthesiology *14:* 245–254 (1953).
19 *Longnecker, D.E.; Miller, F.N., and Harris, P.D.:* Small artery and vein responses to ketamine HCl in the bat wing. Anesth. Analg. *53:* 64–68 (1974).
20 *Longnecker, D.E. and Harris, P.D.:* Anesthesia and microcirculation; in *Kaley and Altura* Microcirculation, vol. III (University Park Press, Baltimore, in press).
21 *Lundy, P.M.; Gowdey, C.W., and Colhoun, E.H.:* Further experimental evidence for a smooth muscle depressant effect of ketamine. Archs int. Pharmacodyn. Thér. *216:* 57–62 (1975).
22 *Miller, F.N. and Wiegman, D.L.:* Anesthesia induced alteration of small vessel responses to norepinephrine. Eur. J. Pharmacol. *44:* 331 (1977).
23 *Nayler, W.G. and Szeto, J.:* Effect of sodium pentobarbital on calcium in mammalian heart muscle. Am. J. Physiol. *222:* 339–344 (1972).
24 *Price, H.L.:* General anesthesia and circulatory homeostasis. Physiol. Rev. *40:* 187–218 (1960).
25 *Price, H.L.:* Calcium reverses myocardial depression caused by halothane. Anesthesiology *41:* 576–579 (1974).
26 *Seeman, P.:* The membrane actions of anesthetics and tranquilizers. Pharmacol. Rev. *24:* 583–655 (1972).
27 *Sharpless, S.K.:* Hypnotics and sedatives. I. The barbiturates; in *Goodman and Gilman* The pharmacological basis of therapeutics, pp. 98–120 (Macmillan, New York 1970).
28 *Shepherd, J.T. and Vanhoutte, P.M.:* Veins and their control (Saunders, Philadelphia 1975).
29 *Wylie, W.D. and Churchill-Davidson, H.C.:* A practice of anesthesia; 3rd ed. (Lloyd-Luke, London 1972).
30 *Yamanaka, I. and Dowdy, E.G.:* The effects of ketamine on spiral-cut strips of rabbit aorta. Anesthesiology *40:* 222–227 (1974).

Dr. *B.T. Altura*, Department of Physiology, Box 31, SUNY Downstate Medical Center, 450 Clarkson Avenue, *Brooklyn, NY 11203* (USA)

Mechanisms of Vasodilatation. Satellite Symp., 27th Int. Congr. Physiol. Sci., Wilrijk 1977, pp. 173–180 (Karger, Basel 1978)

Vascular Relaxation and Aging

B.A. Berkowitz

Department of Physiological Chemistry and Pharmacology, Roche Institute of Molecular Biology, Nutley, N.J.

1. In vitro *Studies*

In both pharmacology and therapeutics, it is probably the rule rather than the exception to expect differences in drug responses as a function of age. However, it was not until 1970 that *Fleisch et al.* (17) focused attention on an important difference between the vascular response of old and young animals. They utilized the isolated helically cut thoracic aortic strip preparation described by *Furchgott and Bhadrakom* (19) and found that the β-receptor agonist isoproterenol, relaxed rat and rabbit aorta obtained from young but not old animals. A summary of these and subsequent findings are shown in table I. The ability of the thoracic aortas to relax after treatment with isoproterenol was lost by 90 days in rats and by 2 years in rabbits. There was a degree of drug and organ specificity to this response since sodium nitrite, not a β-agonist, could relax almost completely the thoracic aorta of both old and young animals. Moreover, tracheal or stomach smooth muscle exhibited either no (17) or only a small decrease (1) in response to isoproterenol with aging. It was concluded that the thoracic aorta undergoes functional changes with aging with a loss of β-receptor activity. In agreement with these authors, *Ericsson and Lundholm* (14) and *Gulati et al.* (22) have reported that rats and rabbits respectively exhibit decreased relaxation of the aorta to isoproterenol with aging. Relaxation to isoproterenol of the aorta and pulmonary artery from rabbits younger than 10 days old was less than that observed in aortas from 1- and 4-month-old animals, yet the relaxation following sodium nitrite did not change with age (16). The relaxation of the pulmonary artery of rats (59–294 days old) and rabbits (53–788 days old) induced by isoproterenol also decreased with aging (31). Thus, with the exception of the tissue obtained from newborn rabbits, aortic and pulmonary artery relaxation produced by β-receptor agonists declines with age.

Table I. Evidence for a decreased vascular relaxation with aging

Relaxant agonist	Blood vessel	Species	References
Isoproterenol	aorta	rat, rabbit	(14, 16, 17)
Isoproterenol	pulmonary artery	rabbit	(16)
Cyclic AMP	aorta, mesenteric artery	rat	(5, 14)
Cyclic GMP	aorta	rat	(5)
Papaverine	aorta	rat	(5)
Nitroglycerin	aorta	rabbit	(16)
Dopamine	aorta	rat	(6, 28)

The species of animal from which the blood vessels are obtained can be an important variable in studying vascular relaxation and aging. For example, in the aorta obtained from cats, no β-agonist-induced relaxation was observed in either 3-month-old or 2-year-old animals (17). To measure relaxation of blood vessels *in vitro,* a certain degree of tone must be produced by a contractile agonist. For the rabbit, the choice of contractile agonist is important for the subsequent quantitation of relaxation. Using KCl to contract the rabbit aorta, 8–19 times more nitroglycerin was required for relaxation (ED_{50}) in 2-year-old animals than in tissues from 2-month-old animals. If histamine was the contractile agonist, only 4 times as much nitroglycerin was necessary to produce the equivalent relaxation response in the older rabbit aorta. Thus, although the greatest response in rabbits is a loss of relaxation to β-receptor agonists, there also appears to be some loss of relaxation in general with aging, possibly indicating a generalized deficiency in the relaxation mechanism of aortic tissue from the older rabbits. There is no explanation why the aortic response to nitroglycerin is somewhat more variable with aging in rabbits than rats.

There could be differences in the mechanism of aging in arteries and veins. Unlike the aorta, pulmonary and mesenteric artery, the relaxation of the portal vein obtained from rabbits or rats does not vary with age when exposed to either isoproterenol or nitroglycerin (16); it will be important to establish if relaxation of other veins is similar to the portal vein. Comparison of data obtained from the aorta or arterial preparations with that obtained from the portal vein was complicated by the fact that the arteries were helically cut or used as a ring of smooth muscle whereas the vein was cut longitudinally (25). Indeed, recent comparative studies with longitudinal portal vein preparations suggest that it is not a good model for other vascular beds (11).

Dopamine is now used in the treatment of hemodynamic imbalances in shock due to myocardial infarction or failure. Selective vascular dilatation is believed to contribute to the efficacy of this drug. It is unknown whether the efficacy of dopamine varies with age in man; in rats, dopamine-induced relaxation is most apparent in the aorta of animals less than 2 months old (6, 28). It was of interest that apomorphine, which usually is considered a dopamine-mimetic drug, cannot be characterized as such on the isolated rat aorta. Apomorphine relaxed the rat aorta but this relaxation did not decline between 5 and 10 weeks of age, whereas that of dopamine fell off markedly (6).

2. In situ *and* in vivo *Studies*

Systolic blood pressure, diastolic blood pressure, and total peripheral resistance increased with age. *Goldman* (21) has questioned why this occurs even in the absence of vascular disease and reviewed the evidence for the provocative hypothesis that the elevated peripheral resistance was to a large degree under physiological control and not simply a passive product of atherosclerosis and vessel rigidity. More recently, it was shown that acetylcholine was less effective in dilating the renal vasculature in older men (24). The sensitivity to the vascular effects of isoproterenol was reduced in hypertensive patients (27). However, in this study the patients averaged 6 years older than normotensive control subjects. Thus, it is difficult to conclude if the altered relaxation is secondary to the hypertension or also influenced by the differences in age.

Although further *in vivo* studies on vascular relaxation and aging in man are required, they will have to explain an unexpected finding. There is data challenging the existence of physiologically active β-receptors responsive to neuronally released norepinephrine. The approach taken to study the effects of this transmitter as it is released from nerves was that if the norepinephrine activated both α- and β-receptors in blood vessels, then the administration of β-blocking drugs should potentiate vasoconstriction. On the other hand, α-blocking drugs should improve vasodilatation. No *in vivo* studies in the young have been reported but in the adult dog and man these phenomena could be demonstrated only for injected norepinephrine but not for neuronally released norepinephrine (3, 12, 20).

3. Biochemical Mechanisms

Cyclic adenosine 3′,5′-monophosphate (cyclic AMP) has been proposed as the intracellular mediator for the relaxation of vascular smooth muscle produced by β-adrenergic receptor agonists – for review, see *Cohen and Berkowitz* (7). Cyclic AMP as well as cyclic guanosine monophosphate both relaxed aortic strips from young rats (3–5 weeks old) to a greater extent than those from older rats (9–13 weeks old). Similar results were obtained with phosphodiesterase inhibitors and papaverine which should elevate the concentration of endogenous cyclic nucleotides. Nitroglycerin, which does not apparently relax blood vessels via the cyclic nucleotide system, relaxed the aortas from young and old rats equally well (29). It can be concluded that the diminished aortic response to isoproterenol and β-receptor stimulants with aging may be explained not only by a decrease in β-receptor number or activity but also by a decrease in vascular sensitivity to the cyclic nucleotides.

Studies on the vascular adenylate cyclase from animals of different ages have suffered from the fact that homogenization of the blood vessels results in an apparent loss of sensitivity of the enzyme to β-agonists (10). Using intact arteries which could respond to β-agonists with an increase of cyclic AMP, we were unable to find any direct correlation between the loss of the relaxant ability of β-agonists and reduced ability of vascular relaxants to stimulate cyclic AMP. However, since the basal concentration of cyclic AMP in blood vessels does fall with age (10, 14), direct studies to quantitate the number and affinity of β-receptors in blood vessels as a function of age will be worthwhile.

Calcium storage and transport is of obvious importance in vascular relaxation. However, surprisingly little is known about vascular calcium disposition with aging (33). Serotonin and norepinephrine but not potassium appear to become more dependent on extracellular calcium to produce contraction of the aorta from 1-year-old compared to 1-month-old rats. In older rats, either less calcium is available or its storage is altered (9). Comparable studies on the role of calcium in relaxation as a function of age have not as yet been undertaken.

4. Relationship to Vascular Function and Disease

Are alterations in vascular relaxation with aging related to aberrant vascular function and do they occur prematurely in diseases of the blood vessels? Several interesting possibilities can be discussed. In normal maturing rats (8–10 weeks old), there is a correlation between the decline in β-agonist-induced relaxation of the aorta and mesenteric artery and the increase in blood pressure that occurs with maturity (17).

There is also a decrease in vascular relaxation that occurs with hypertension (8, 34), but this differs from that seen with aging. In aortic strip preparations from spontaneously hypertensive rats or renal hypertensive rats, all relaxant agonists (isoproterenol, adenosine, and nitroglycerin) are less effective, whereas in aging those agonists which work through the cyclic nucleotide system are those which are primarily less effectively. A fascinating aspect of these studies was the observation that antihypertensive doses of reserpine improved aortic relaxation (8).

Vascular relaxation is also decreased in the aorta obtained from diabetic animals (2) and isoproterenol is poorly effective in lowering blood pressure *in vivo* in diabetic rats. Since the pathology of diabetes is in many ways similar to aging (23, 36), the relationship between diabetes and aging on vascular function and chemistry merits continued study.

There is no question of the importance of structural changes in blood vessels with advancing age (33). The physical properties of vessels are altered with aging and include decreased distensibility, increased vessel wall stiffness and decreased ability to relax (4, 5). The fibrosis of the arterial walls which may mediate these functional defects of aging and disease are associated with increased amounts of collagen in the vascular walls. Studies on the control of collagen disposition and metabolism in blood vessels appear to be a promising new approach to vascular regulation (26, 30).

One of the most consistent observations in the literature is that the nitroglycerin-induced relaxation of blood vessels is more resistant to a decline with aging than that produced by β-receptor agonists (5, 16). Nitroglycerin is therapeutically the most effective of the anti-anginal drugs. There is no serious debate that vasodilatation is a crucial part of the mechanism of action of nitroglycerin. Thus, we already may be empirically taking advantage of the fact that the vascular relaxant ability of nitroglycerin continues in spite of advancing age.

5. Endocrine Control of Vascular Relaxation

It is now clear that thyroid hormones are important regulators of vascular relaxation (15, 18, 37). In the absence of thyroid hormone, the aorta is poorly responsive to isoproterenol's relaxant action (fig. 1). However, it responds well

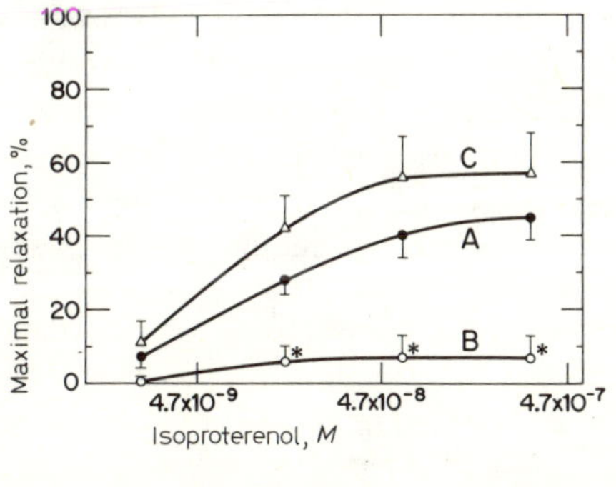

Fig. 1. Dose-response curves for isoproterenol on aortic strips removed from (A) control rats, (B) thyroidectomized rats, and (C) thyroxine-treated (3 μg/100 g/day) thyroidectomized rats. Thyroid was removed from 4- to 5-week-old rats and the animals studied 3 months later. These rats were spontaneously hypertensive Wistar rats. Similar results are obtained with normotensive Sprague-Dawley rats.

to nitroglycerin-induced relaxation (32). In the case of excess thyroid hormones, there is *in vitro* and *in vivo* evidence for increased sensitivity of vascular β-receptors to catecholamines (18, 37).

We were interested in determining if the decrease in aortic relaxation with aging could be reversed by endocrine manipulation. Thyroid hormones administered for 3 weeks to 2-year-old adult rats partially restored β-agonist-mediated relaxation. *Denckla* (13) has proposed that a substance of pituitary origin can block the effect of thyroid hormones on peripheral tissue, and this blockade is responsible for some of the pathological consequences of aging. Indeed, hypophysectomy plus administration of thyroid hormones almost completely reverses the loss of vascular relaxation with aging. Thus, deficits in vascular relaxation with aging can be reversed and may not be solely structural in nature but may also be of biochemical origin. It has recently been shown for the heart that thyroid hormones actually regulate the number of β-receptors (35) and this mechanism may participate in the improvement of vascular relaxation caused by administration of thyroid hormones.

6. Conclusion

There is convincing data for a decline in vascular relaxation with aging obtained from *in vitro* studies. Studies of the alterations in the relaxation of the aortic strip preparation provides a useful model for aging. Measurement of vascular relaxation and vasodilatation *in vivo* in young compared to old subjects remains to be done. It has been proposed that in those vessels where relaxation wanes with age this response is mediated by a diminished β-receptor or cyclic nucleotide activity. Recent evidence indicates that thyroid hormones and possibly other hormones of the pituitary modulate vascular relaxation. Administration of excess thyroid cause a rejuvenation of β-agonist-induced aortic relaxation which had disappeared in older animals. Clearly, age is a variable which must be closely controlled in future studies examining vascular relaxation and vasodilatation. A better understanding of vascular relaxation will offer new approaches to the goal of improving the impaired cardiovascular function that occurs with aging.

References

1 *Aberg, G.; Adler, G., and Ericsson, E.:* The effect of age on β-adrenoceptor activity in tracheal smooth muscle. Br. J. Pharmacol. *47:* 181–182 (1973).

2 *Berkowitz, B.A. and Cohen, M.L.:* Diminished vascular cyclic 3′5′-adenosine monophosphate function in diabetes. Abstr. 6th Int. Congr. Pharmacol., 1975, p. 228.

3 *Brick, I.; Hutchison, K.J., and Roddie:* A comparison of the effects of circulating

nonadrenaline and vasoconstrictor nerve stimulation on forearm blood vessels. J. Physiol., Lond. *189:* 27–28 (1967).

4 *Cliff, W.J.:* Blood vessels, pp. 106–107 (Cambridge University Press, London 1976).

5 *Cohen, M.L. and Berkowitz, B.A.:* Age-related changes in vascular responsiveness to cyclic nucleotides and contractile agonists. J. Pharmac. exp. Ther. *191:* 147–155 (1974).

6 *Cohen, M.L. and Berkowitz, B.A.:* Differences between the effects of dopamine and apomorphine on rat aortic strips. Eur. J. Pharmacol. *34:* 49–58 (1975).

7 *Cohen, M.L. and Berkowitz, B.A.:* Role of cyclic nucleotides in hypertension; in *Scriabine and Sweet* New antihypertensive drugs, p. 25 (Spectrum, New York 1976).

8 *Cohen, M.L. and Berkowitz, B.A.:* Decreased vascular relaxation in hypertension. J. Pharmac. exp. Ther. *196:* 396–406 (1976).

9 *Cohen, M.L. and Berkowitz, B.A.:* Vascular contraction: effect of age and extracellular calcium. Blood Vessels *13:* 139–154 (1976).

10 *Cohen, M.L.; Blume, A.S., and Berkowitz, B.A.:* Vascular adenylate cyclase: role of age and guanine nucleotide activation. Blood Vessels *14:* 25–42 (1977).

11 *Cohen, M.L. and Wiley, K.:* Specific enhancement of norepinephrine-induced contraction in rat veins after beta adrenergic antagonists. J. Pharmac. exp. Ther. *201:* 406–416 (1977).

12 *Daws, P.M. and Faulkner, D.C.:* The effects of propranolol on vascular responses to sympathetic nerve stimulation. Br. J. Pharmacol. *53:* 517–524 (1975).

13 *Denckla, W.D.:* Role of the pituitary and thyroid glands in the decline of minimal O_2 consumption with age. J. clin. Invest. *53:* 572–581 (1974).

14 *Ericsson, E. and Lundholm, L.:* Adrenergic β-receptor activity and cyclic AMP metabolism in vascular smooth muscle: variations with age. Mech. Aging Devmnt *4:* 1–6 (1975).

15 *Field, F.P.; Janis, R.A., and Triggle, P.J.:* Relationship between aortic reactivity and blood pressure of renal hypertensive hyperthyroid and hypothyroid rats. Can. J. Physiol. Pharmacol. *51:* 344–535 (1973).

16 *Fleisch, J.H. and Hooker, C.S.:* The relationship between age and relaxation of vascular smooth muscle in the rabbit and rat. Circulation Res. *38:* 243–249 (1976).

17 *Fleisch, J.H.; Maling, H.M., and Brodie, B.B.:* Beta-receptor activity in aorta. Circulation Res. *26:* 151–162 (1970).

18 *Freedberg, A.S. and Hamolsky, M.W.:* Effects of thyroid hormones on certain nonendocrine orgon systems; in *Greep and Astwood* Handbook of physiology, vol. III, pp. 435–468 (Am. Physiological Society, Washington 1974).

19 *Furchgott, R.F. and Bhadrakom, S.:* Reactions of strips of rabbit aorta to epinephrine, isopropylarternenol sodium nitrite and other drugs. J. Pharmac. exp. Ther. *108:* 129–143 (1953).

20 *Glick, G.; Epstein, S.E.; Wechaler, A.S., and Braunwald, E.:* Physiological differences between the effects of neuronally released and bloodbone norepinephrine on beta adrenergic receptors in the arterial bed of the dog. Circulation Res. *21:* 217–227 (1967).

21 *Goldman, R.:* Speculations of vascular changes with age. J. Am. Geriat. Soc. *18:* 765–779 (1970).

22 *Gulati, O.D.; Methew, B.P.; Parikh, H.M., and Krishmamurty, V.S.R.:* Beta adrenergic receptor of rabbit thoracic aorta in relation to age. Jap. J. Pharmac. *23:* 259–268 (1973).

23 *Hamlin, C.; Kohn, R.R., and Luschin, J.H.:* Apparent accelerated aging of human collagen in diabetes mellitus. Diabetes *24:* 902–904 (1977).

24 *Hollenberg, N.K.; Adams, D.F.; Solomon, H.S.; Rashid, A.; Abrams, H.L., and Merril, J.P.:* Senescence and the renal vasculature in normal man. Circulation Res. *34:* 309–316 (1974).

25 *Hughes, J. and Vane, J.R.:* An analysis of the responses of the isolated portal vein of the rabbit to electrical stimulation and to drugs. Br. J. Pharmac. Chemother. *30:* 46–60 (1967).

26 *Iwatsuki, K.; Cardinale, G.J.; Spector, S., and Udenfriend, S.:* Reduction of blood pressure and vascular collagen in hypertensive rats by β-aminopriopionitrile. Proc. natn. Acad. Sci. USA *74:* 360–362 (1977).

27 *London, G.M.; Safar, M.E.; Weiss, Y.A., and Milliez, D.L.:* Isoproterenol sensitivity and total body clearance of propranolol in hypertensive patients. J. clin. Pharmacol. *16:* 174–182 (1976).

28 *Murakami, W. and Shibata, S.:* Evidence for a dopamine sensitive receptor in the young rat aorta. Res. Commun. Chem. Path. Pharm. *13:* 349–352 (1976).

29 *Needleman, P.; Jakschik, B., and Johnson, E.M.:* Sulfhydryl requirement for relaxation of vascular smooth muscle. J. Pharmac. exp. Ther. *187:* 324–331 (1973).

30 *Ooshima, A.; Fuller, G.C.; Cardinale, G.J.; Spector, S., and Udenfriend, S.:* Increased collagen synthesis in blood vessels of hypertensive rats and its reversal by antihypertensive agents. Proc. natn. Acad. Sci. USA *71:* 3019–3023 (1974).

31 *Park, M.K.; Diehl, A.M., and Sunderson, J.M.:* Maturation of beta-adrenergic receptor activity of rabbit aorta and pulmonary artery. Life Sci. *19:* 321–328 (1976).

32 *Rioux, F. and Berkowitz, B.A.:* Role of the thyroid gland in the development and maintenance of spontaneous hypertension in rats. Circulation Res. *40:* 306–312 (1977).

33 *Somlyo, A.P. and Somlyo, A.V.:* Vascular smooth muscle. Pharmac. Rev. *20:* 197–272 (1970).

34 *Triner, L.; Vulliemoz, Y.; Verosky, M., and Manger, W.M.:* Cyclic adenosine monophosphate and vascular reactivity in spontaneously hypertensive rats. Biochem. Pharmacol. *24:* 743–745 (1975).

35 *Williams, L.T.; Lefkowitz, R.J.; Watanabe, A.M.; Hathaway, D.R., and Besch, H.P., jr.:* Thyroid hormone regulation of β-adrenergic receptor number. J. biol. Chem. *252:* 2787–2789 (1977).

36 *Whittingham, S.; Mathews, J.D.; Mackay, J.R.; Stocks, A.E.; Unger, B., and Martin, F.I.R.:* Diabetes mellitus, autoimmunity and aging. Lancet *i:* 763–766 (1971).

37 *Zsoter, T.; Rom, H., and Chappel, C.:* Effect of thyroid hormones on vascular response. J. Lab. clin. Med. *64:* 433–441 (1964).

B.A. Berkowitz, PhD, Department of Physiological Chemistry and Pharmacology, Roche Institute of Molecular Biology, *Nutley, NJ 07110* (USA)

Mechanisms of Vasodilatation. Satellite Symp., 27th Int. Congr. Physiol. Sci., Wilrijk 1977, pp. 181–186 (Karger, Basel 1978)

Vascular Relaxation in Hypertensive Rats[1]

S. Shibata and J.B. Cheng

Department of Pharmacology, School of Medicine, University of Hawaii, Honolulu

The vascular contractile responsiveness of hypertensive animals has received extensive attention (1, 10, 12). However, to our knowledge, only limited reports, with inconsistent results, have appeared regarding the ability to relax of vascular smooth muscle taken from hypertensive animals (3, 10, 14). Thus, it is still uncertain whether the response of vascular smooth muscle from hypertensive animals to various dilator agents is reduced, enhanced, or unaltered when compared to that of normotensive animals. Therefore, the response to vasodilator drugs of different parts of the aorta of the spontaneous hypertensive rat (SHR) was compared with that of renal hypertensive and normotensive Wistar rats (NWR), at different ages.

1. Isoproterenol and Acetylcholine

Thoracic and abdominal aortic strips from both SHR and NWR were made to contract with serotonin (10^{-5} *M*); the relaxation induced by isoproterenol and acetylcholine during such contractions decreased with increasing age (fig. 1, 2). Age-related changes in isoproterenol induced (5, 6) and dopamine induced (2, 9) relaxations have been reported. The present experiments demonstrate that increasing age is also associated with a decrease in the response to acetylcholine.

The isoproterenol- and acetylcholine-induced relaxation was less pronounced in the thoracic aorta from the SHR than in that from the NWR. In fact, at the age of 22–25 weeks, the SHR aorta, unlike the NWR aorta, showed no detectable relaxing response to either of these agents. *Spector et al.* (10) have reported that the thoracic aorta from the SHR is more responsive to isoproterenol than that from normotensive rats. In contrast, *Triner et al.* (14) and

[1] Partially supported by the Hawaii Heart Association.

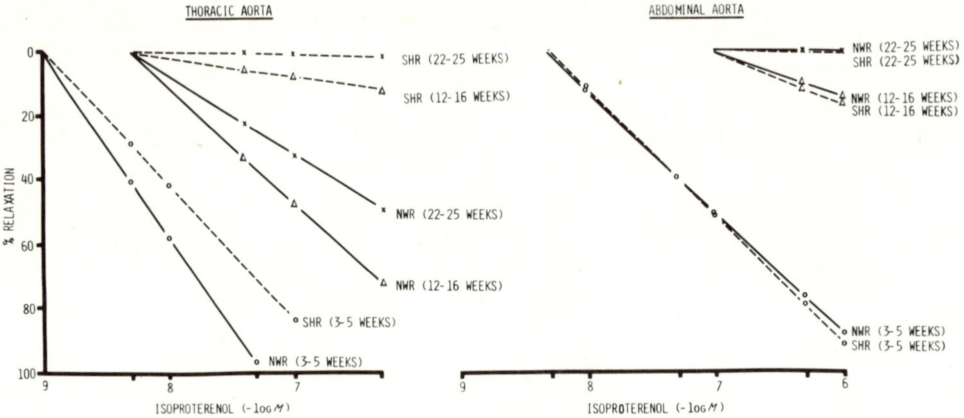

Fig. 1. Effects of isoproterenol on thoracic or abdominal aortae of SHR and NWR at the different ages. Each point represents at least 6 observations from the different animals. Each corresponding point is significantly different between SHR and NWR thoracic aortae ($p < 0.05$) but not in abdominal aortae. Systolic blood pressures (mm Hg): SHR: 105 ± 3 in 3–5 weeks, 195 ± 5 in 12–16 weeks, and 205 ± 8 in 22–25 weeks; NWR: 104 ± 5 in 3–5 weeks, 121 ± 3 in 12–16 weeks, and 135 ± 5 in 22–25 weeks (n = 25).

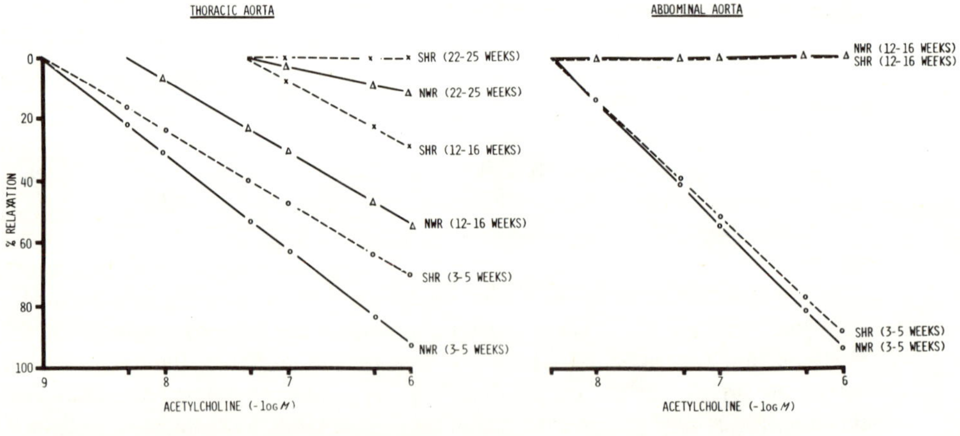

Fig. 2. Effects of acetylcholine on thoracic or abdominal aortae of SHR and NWR at the different ages. Each point represents at least 6 observations from the different animals. Each corresponding point is significantly different between SHR and NWR thoracic aortae ($p < 0.05$) but not in abdominal aortae.

Cohen and Berkowitz (3) found that isoproterenol-induced relaxation was less in the thoracic aorta from the SHR. Our results agree with the latter two reports. Furthermore, the present experiments demonstrate that such a defect in relaxation is observed in the SHR even in a prehypertensive state. Previously, our group (7) reported a diminished response to isoproterenol in both prehypertensive and established SHR atria. The thoracic aorta from the renal hypertensive rat (SBP = 186 ± 8 mm Hg, n = 8) showed a similar impairment of the relaxation to isoproterenol when compared to the aorta from sham-operated control rats. In contrast to the thoracic aorta, however, the isoproterenol-induced (SHR and renal hypertensive rats) or acetylcholine-induced (SHR) relaxation of the abdominal aorta was not different from that obtained in control animals.

The defect in the relaxing ability of the thoracic aorta from prehypertensive SHR suggests that its low reactivity to vasodilator neurohumoral substances might not be causally related to the high blood pressure. In addition, long-term reserpine treatment of prehypertensive or established SHR prevented or reduced, respectively, the increase in blood pressure, but did not influence the decrease in the relaxing response of the aorta. This also supports the idea that the defect in relaxation is not a secondary consequence of hypertension.

2. Nitroglycerin and Papaverine

There was no demonstrable difference in the relaxing response to nitroglycerin and papaverine in either the thoracic or the abdominal aortae from SHR compared with the NWR; this confirms our earlier observations (12) but not those of *Cohen and Berkowitz* (3), who described an attenuated response to nitroglycerin in the SHR aorta. Reserpine treatment does not affect the relaxations induced by nitroglycerin or papaverine in either NWR or SHR.

3. Mg^{++} and Mn^{++}

The thoracic aorta of the SHR relaxes less when exposed to solutions with high Mg^{++} or high Mn^{++} concentrations than does the same blood vessel of the normotensive rat (fig. 3, 4). In the NWR, but not in the SHR, the thoracic aorta shows an age-dependent decrease in the Mg^{++}-induced relaxation. Thus, the relaxing effect of Mg^{++} is even greater in older NWR (22–25 weeks) than in young SHR (3–5 weeks). On the other hand, in the abdominal aorta, an age-related decrease in the relaxing response to Mg^{++} was observed which did not differ in SHR and NWR. Thoracic aortae from young SHR (3–5 weeks) exhibited a reduced relaxing response to Mn^{++} even when compared with aortae

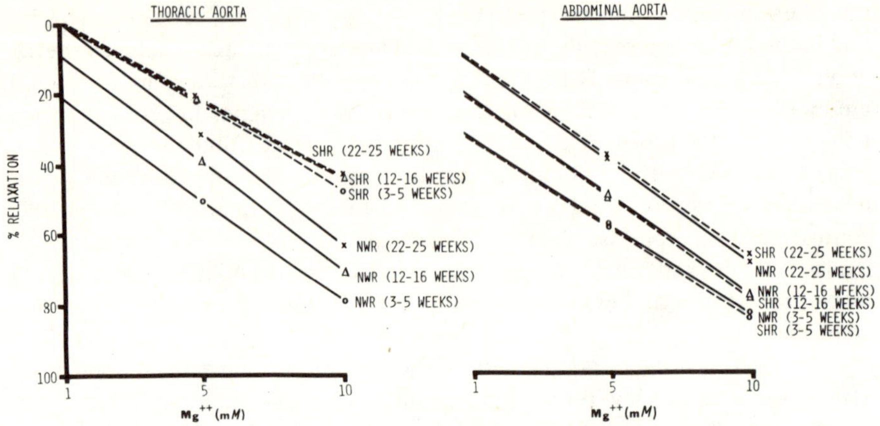

Fig. 3. Effects of Mg^{++} on thoracic or abdominal aortae of SHR and NWR at the different ages. Each point represents at least 6 observations from the different animals. Each corresponding point is significantly different between SHR and NWR thoracic aortae ($p < 0.05$) but not in abdominal aortae.

from old NWR (22–25 weeks). In the abdominal aorta, there was no difference in the Mn^{++}-induced relaxation between SHR and NWR.

The impairment of the cation-induced relaxations is much greater than that of responses induced by isoproterenol. This fact suggests that the difference in relaxation between the thoracic aortae from SHR and NWR could be associated with differences in Ca^{++} regulation by the vascular smooth muscle cells, since the mechanical changes induced by Mg^{++} and Mn^{++} are related to Ca^{++} translocation at the cellular and the subcellular levels (4, 11).

4. ^{45}Ca Efflux

Further suggestion of the involvement of a change in Ca^{++} homeostasis in the SHR comes from Ca^{++} efflux studies. The SHR thoracic aorta shows a smaller Ca^{++} efflux than that from the NWR; the Ca^{++} efflux of abdominal aorta was similar in SHR and NWR. *Shibata et al.* (13) reported that the ^{45}Ca uptake is significantly greater, but the ^{45}Ca influx significantly less in the aorta of the SHR than in that of the NWR; they suggested that there is a difference in translocation in Ca^{++} at the cellular level between SHR and control rats. In addition, several investigators have recently reported that ATP-dependent calcium uptake by the plasma membrane-enriched (16) or the microsomal (8, 15) fraction from the SHR aorta is significantly less than that from the NWR, and

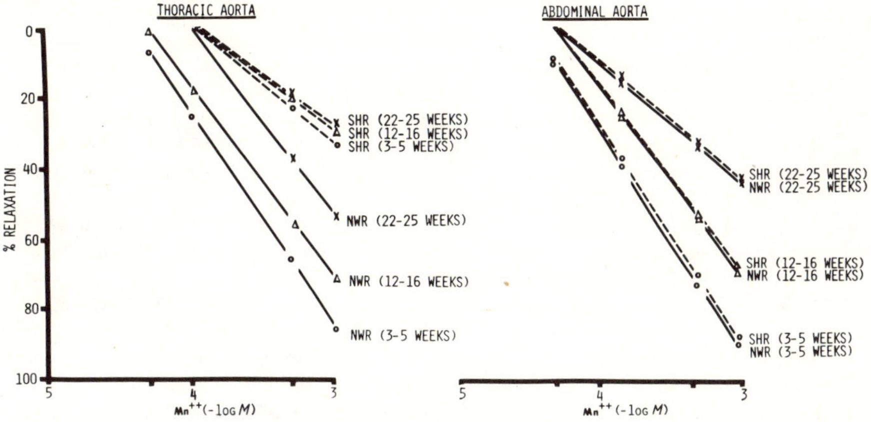

Fig. 4. Effects of Mn^{++} on thoracic or abdominal aortae of SHR and NWR at the different ages. Each point represents at least 6 observations from the different animals. Each corresponding point is significantly different between SHR and NWR thoracic aortae ($p < 0.05$) but not in abdominal aortae.

have postulated that a decreased rate of calcium transport occurs in the aorta of the SHR. These data suggest that in the thoracic aorta, the difference in the Ca^{++} binding properties and Ca^{++} permeability of the vascular smooth muscle cells from the SHR, compared with the NWR, might be causally related to the altered vascular reactivity to vasodilator substances.

5. Conclusion

In the thoracic aorta, the relaxations induced by isoproterenol, acetylcholine, Mg^{++} and Mn^{++} are less in the SHR (prehypertensive and hypertensive stages) than in the normotensive Wistar rat; in the abdominal aortae, no such differences are obvious. There is no difference in the relaxant effect of nitroglycerin and papaverine on the thoracic and abdominal aortae from these animals. The relaxation response to isoproterenol, but not to the other relaxants tested, is less in the aorta of renal hypertensive rats than in that of control animals. After long-term reserpine treatment, the thoracic aorta of the SHR still relaxes less on exposure to the test agents. The ^{45}Ca efflux is less in the thoracic aorta of the SHR; no difference is found between the abdominal aortae of SHR and NWR. These results suggest that the decreased relaxation of the SHR thoracic aorta may be attributed to a change in the musculature. However, for unknown reasons, this defect is not seen in the abdominal aorta.

References

1 *Clineschmidt, B.V.; Geller, R.G.; Govier, W.C., and Sjoerdsma, A.:* Reactivity to norepinephrine and nature of the alpha adrenergic receptor in vascular smooth muscle of a genetically hypertensive rat. Eur. J. Pharmacol. *10:* 45–50 (1970).

2 *Cohen, M.L. and Berkowitz, B.A.:* Differences between the effects of dopamine and apomorphine on rat aortic strips. Eur. J. Pharmacol. *34:* 49–58 (1975).

3 *Cohen, M.L. and Berkowitz, B.A.:* Decreased vascular relaxation in hypertension. J. Pharmac. exp. Ther. *196:* 396–409 (1976).

4 *Carrier, O., jr.; Hester, R.K.; Jurevics, H.A., and Tenner, T.E., jr.:* Influence of magnesium on calcium- and potassium-related responses in vascular smooth muscle. Blood Vessels *13:* 321–337 (1976).

5 *Fleisch, J.H.; Mailing, H.M., and Brodie, B.B.:* Beta-receptor activity in aorta: variation with age and species. Circulation Res. *26:* 151–162 (1970).

6 *Fleisch, J.H. and Hooker, C.S.:* The relationship between age and relaxation of vascular smooth muscle in the rabbit and rat. Circulation Res. *38:* 243–249 (1976).

7 *Fujiwara, M.; Kuchii, M., and Shibata, S.:* Differences of cardiac reactivity between spontaneously hypertensive and normotensive rats. Eur. J. Pharmacol. *19:* 1–11 (1972).

8 *Moore, L.; Hurwitz, L.; Davenport, G.R., and Landon, E.J.:* Energy-dependent calcium uptake activity of microsomes from the aorta of normal and hypertensive rats. Biochim. biophys. Acta *413:* 432–443 (1975).

9 *Murakami, W. and Shibata, S.:* Evidence for a dopamine sensitive receptor in the young rat aorta. Res. Commun. Chem. Pathol. Pharmacol. *13:* 349–352 (1976).

10 *Spector, S.; Fleisch, J.H.; Mailing, H.M., and Brodie, B.B.:* Vascular smooth muscle reactivity in normotensive and hypertensive rats. Science, Wash. *166:* 1300–1301 (1969).

11 *Shibata, S.:* Effects of Mn^{++} on ^{45}Ca content and potassium-induced contraction of the aortic strip. Can. J. Physiol. Pharmacol. *47:* 827–829 (1969).

12 *Shibata, S.; Kurahashi, K., and Kuchii, M.:* A possible etiology of contractility impairment of vascular smooth muscle from spontaneously hypertensive rats. J. Pharmac. exp. Ther. *185:* 406–417 (1973).

13 *Shibata, S.; Kuchii, M., and Taniguchi, T.:* Calcium flux and binding in the aortic smooth muscle from the spontaneously hypertensive rat. Blood Vessels *12:* 279–289 (1975).

14 *Triner, L.; Vulliemoz, Y.; Verosky, M., and Manger, W.M.:* Cyclic adenosine monophosphate and vascular reactivity in spontaneously hypertensive rats. Biochem. Pharmacol. *24:* 743–745 (1975).

15 *Webb, R.C. and Bhalla, R.C.:* Altered calcium sequestration by subcellular fractions of vascular smooth muscle from spontaneously hypertensive rats. J. mol. cell. Cardiol. *8:* 651–660 (1976).

16 *Wei, J.W.; Janis, R.A., and Daniel, E.E.:* Calcium accumulation and enzymatic activities of subcellular fractions for aortas and ventricles of genetically hypertensive rats. Circulation Res. *39:* 133–140 (1976).

Dr. *S. Shibata,* University of Hawaii, School of Medicine, Department of Pharmacology, Biomedical Sciences Building, 1960 East West Road, *Honolulu, HI 96822* (USA)

Mechanisms of Vasodilatation. Satellite Symp., 27th Int. Congr. Physiol. Sci.,
Wilrijk 1977, pp. 187–192 (Karger, Basel 1978)

Myogenic Responses in Exercise Hyperemia[1]

Harvey V. Sparks and Robert D. Phair

Department of Physiology, University of Michigan, Ann Arbor, Mich.

The myogenic response to changes in transmural pressure appears to play a role in the vasodilatation of autoregulation and reactive hyperemia (6). The possibility that myogenic responses (defined here as a change in vascular smooth muscle contractile force in response to a change in transmural pressure) play a causal role in exercise hyperemia has been suggested (4, 7), but not extensively evaluated. Before discussing this possibility in more detail, we will outline the responses of vascular smooth muscle to stretch because these responses are probably responsible for whatever myogenic reactions are observed *in vivo.*

1. Responses of Vascular Smooth Muscle to Stretch

Isolated strips of blood vessel wall exhibit the classic relationship between *initial length* and active tension development. The contractile response to a standard stimulus increases with initial length up to an optimum length. Further increases result in a fall in the contractile response (5, 16). This length-tension relationship probably is not responsible *per se* for steady state myogenic autoregulation, in which reduced transmural pressure results in sufficient smooth muscle relaxation to cause dilatation of resistance vessels. That is, reduced contractile force is associated with increased radius (and 'initial length') of the resistance vessels. The length-tension relationship could play a role in reactive hyperemia because the reduced transmural pressure downstream from an arterial occlusion would allow collapse of the resistance vessels resulting in reduced 'initial length' and contractile force. Once flow is restored, the increased radius would result in increased contractile force, bringing resistance back to control.

[1] Supported by USPHS Grant HL 14516.

Quick stretch of some isolated blood vessels results in contraction (5, 14) and quick release in relaxation (5). Rapid changes in strip length have greater effects than slower ones (5, 14). These responses (collectively referred to as the Bayliss response) may well be the basis for the myogenic responses to sudden increases and decreases in transmural pressure associated with autoregulation and reactive hyperemia.

Smieško (13) has described transient vasodilatation of skeletal muscle resistance vessels in response to rapid transients in transmural pressure caused by raising or lowering intravascular pressure for 1 sec. The observation that transient increases in pressure can cause vasodilatation raises the possibility that maintained stretch results in contraction, but transient stretch causes vasodilatation. The latter may be related to the relaxation observed when vascular smooth muscle is subjected to vibratory stretch (8).

2. *Role in Exercise Hyperemia*

What evidence do we have that any of these myogenic responses may play a role in *exercise hyperemia?* In the cases of autoregulation and reactive hyperemia, the myogenic hypothesis has gained credence because situations have been uncovered in which the alternative explanations, especially those relating flow to tissue metabolism, appear to be ruled out (6). It is a little more difficult to separate metabolic and myogenic responses in the case of exercise hyperemia, but it can be done by comparing the time course of changes in metabolism and vascular resistance.

Brief *tetanus* (1 sec) of dog skeletal muscle is followed by transient vasodilatation having a time course faster than the accompanying change in oxidative metabolism (9). This would appear to rule out a role for vessel wall pO_2 or other vasodilators closely linked to oxidative metabolism. To test the possibility that a myogenic response is involved, we compared the vascular response to brief tetanus with the response to external compression caused by inflating a cuff around the muscles (11). A small balloon inserted among the muscle fibers was used to gauge the compression caused by muscle contraction and cuff inflation. Figure 1 shows that the vascular response to cuff inflation is a large fraction of the vascular response to brief tetanus when both stimuli cause the same increase in balloon pressure. External compression does not appear to cause vasodilatation by increasing muscle metabolism because there is no change in venous O_2 saturation at constant flow. Under the conditions of these experiments, an average of approximately 50% of the vasodilatation following brief tetanus appears to be caused by a myogenic response to altered transmural pressure. The rest of the vasodilatation is probably caused by release of potassium ions (K^+) (12).

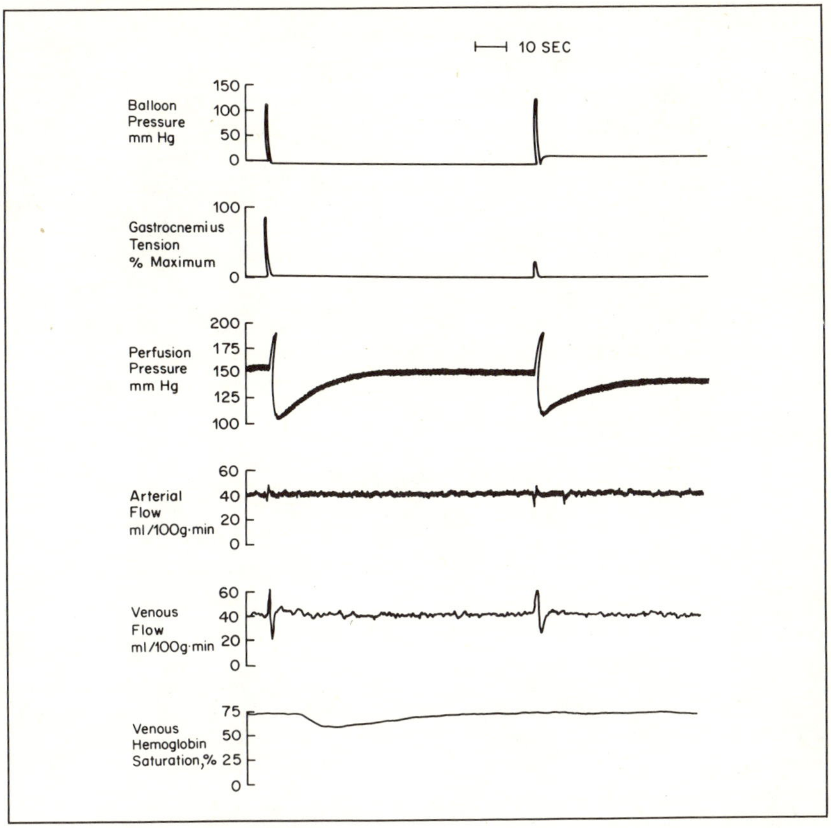

Fig. 1. Curvilinear strip chart records of response to 1-sec tetanus (32 impulses/sec, left) and 1-sec elevated cuff pressure (right). Dog calf muscles; constant flow. Tension response to cuff inflation caused by forcing muscle mass away from force transducer, not by active contraction.

Although natural exercise consists of short bursts of tetanus, most forms of experimental exercise consist of repetitive twitches for many minutes. The following line of reasoning suggests that myogenic hyperemia may be important in *twitch* as well as tetanic exercise. The vascular response to compression is very small at balloon pressures which are associated with the contractile force of twitches. This would appear to rule out a myogenic response during twitch exercise, but this may be an erroneous conclusion. High external compression pressure may mimic the forces within contracting muscle much better than do low external pressures. This is because it appears that resistance vessels are subjected to shearing forces that nip or twist off fairly large arteries (3). This

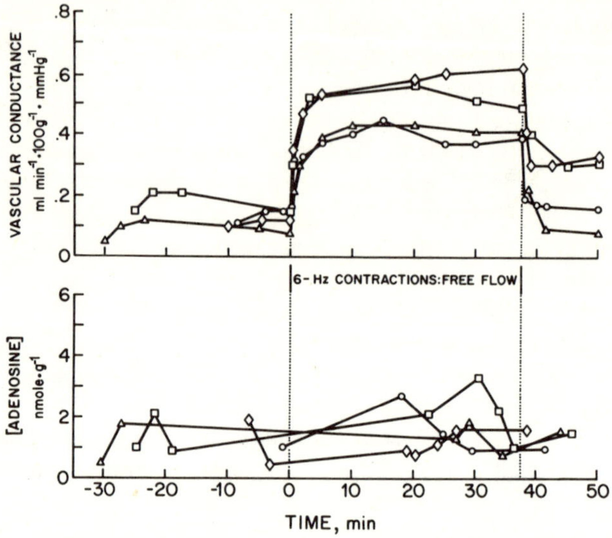

Fig. 2. Vascular conductance and canine skeletal muscle adenosine concentration during rest, twitch exercise at 6 twitches/sec, and rest for 4 dogs, each indicated by a different symbol. Serial tissue samples taken by punch biopsy and quickly frozen in liquid N_2.

means that the arteries distal to the shear points are subjected to rapid and large changes in transmural pressure which are out of proportion to the compressive forces exerted by muscle contraction or external compression. If this is the case, we may well be underestimating the vasodilator effect of a myogenic response associated with twitch exercise.

In the case of brief tetanus, we were able to rule out many metabolic factors by determining that the transient vascular response is more rapid than the transient in oxidative metabolism (9). We attempted to evaluate the dynamic relationship between oxidative metabolism and vascular resistance during continuous twitch work by using sinusoidal variations in twitch frequency (10). When flow was held slightly below the free flow level for the twitch exercise performed, resistance dynamics closely paralleled those of oxidative metabolism. On the other hand, when flow was held at a higher level, resistance dynamics were more rapid than oxidative metabolism. This suggests at least two control mechanisms, one too fast to be closely related to oxidative metabolism. The faster control mechanism is probably not related to K^+ because both release of K^+ and vascular responsiveness to K^+ decline with time, and the sinusoidal stimulations lasted several hours (4). A myogenic mechanism could exert rapid control over vascular resistance and be responsible for the rapid response of vascular resistance at high flow.

Another way to evaluate the potential role for myogenic hyperemia in twitch exercise is to check the list of potential *metabolic controllers* to see if currently available candidates can adequately explain twitch exercise hyperemia. As we have already indicated, K^+ release probably participates in the response to brief tetanus (12), but not the long-term maintenance of exercise hyperemia (4). The importance of increased tissue osmolarity seems to be species-specific. A large exercise hyperemia results from twitch exercise of canine skeletal muscle, but the observed increases in osmolarity and the vascular response to raising blood osmolarity are very modest (4). Vascular smooth muscle may not be sensitive to changes in vessel wall pO_2 occurring *in vivo* (2). In addition,vessel wall pO_2 may well rise in exercise hyperemia during which time flow delivery of O_2 to the vessel wall is increased dramatically (2). Tissue adenosine increases if flow is restricted during twitch exercise (1), but we have found that if flow is allowed to increase, even exercise at 6 twitches per second does not result in a rise in skeletal muscle adenosine (fig. 2). Certainly, other possible mediators remain to be fully evaluated (15), but we would suggest that a myogenic response to alterations in transmural pressure should be on the list of viable candidates.

3. Conclusion

As a working hypothesis, we propose that muscle contractions cause nipping and twisting of arteries which, in turn, results in rapid shifts in transmural pressure. Furthermore, these rapid shifts in transmural pressure result in vasodilatation either via the classic Bayliss response or the rate-sensitive response described by *Smieško.* The myogenic mechanism may account for rapid adjustments in flow as well as longer-term steady-state hyperemia. A substratum of myogenic hyperemia may be modulated by metabolic mechanisms.

References

1 *Bochman, E.L.; Berne, R.M., and Rubio, R.:* Adenosine and active hyperemia in dog skeletal muscle. Am. J. Physiol. *230:* 1531–1537 (1976).

2 *Duling, E.F. and Pittman, R.N.:* Oxygen tension: dependent or independent variable in local control of blood flow? Fed, Proc. Fed. Am. Socs exp. Biol. *34:* 2012–2019 (1975).

3 *Gray, S.D.; Carlsson, E., and Staub, N.C.:* Site of increased vascular resistance during isometric muscle contraction. Am. J. Physiol. *213:* 683–689 (1967).
4 *Haddy, F.J. and Scott, J.B.:* Metabolic factors in peripheral circulatory regulation. Fed. Proc. Fed. Am. Socs exp. Biol. *34:* 2006–2011 (1975).
5 *Johansson, B. and Mellander, S.:* Static and dynamic components in the vascular myogenic response to passive changes in length as revealed by electrical and mechanical recordings from the rat portal vein. Circulation Res. *36:* 76–83 (1975).
6 *Johnson, P.C. and Henrich, H.A.:* Metabolic and myogenic factors in local regulation of the microcirculation. Fed. Proc. Fed. Am. Socs exp. Biol. *34:* 2020–2024 (1975).
7 *Khayutin, V.M.:* Determinants of working hyperemia in skeletal muscle; in *Hudlicka* Circulation in skeletal muscle, pp. 145–157 (Pergamon Press, Oxford 1968).
8 *Ljung, B. and Sivertsson, R.:* Inhibitory effect of vibrations on tension development in vascular smooth muscle. Acta physiol. scand. *85:* 428–430 (1972).
9 *Mohrman, D.E.; Cant, J.R., and Sparks, H.V.:* Time course of resistance and venous O_2 changes following brief tetanus of dog skeletal muscle. Circulation Res. *33:* 323–336 (1973).
10 *Mohrman, D.E. and Sparks, H.V.:* Resistance and VO_2 dynamics during sinusoidal exercise of dog skeletal muscle. Circulation Res. *33:* 337–345 (1973).
11 *Mohrman, D.E. and Sparks, H.V.:* Myogenic hyperemia following brief tetanus of canine skeletal muscle. Am. J. Physiol. *227:* 531–535 (1974).
12 *Mohrman, D.E. and Sparks, H.V.:* Role of K^+ in vascular response to brief tetanus. Circulation Res. *35:* 384–390 (1974).
13 *Smieško, V.:* Unidirectional rate sensitivity component in local control of vascular tone. Pflügers Arch. ges. Physiol. *327:* 324–336 (1971).
14 *Sparks, H.V.:* Effect of quick stretch on isolated vascular smooth muscle. Circulation Res. *15:* I-254–260 (1964).
15 *Sparks, H.V. and Belloni, F.L.:* The peripheral circulation: local regulation; in *Knobil, Sonnenschein and Edelman* Ann. Rev. Physiol., vol. 40 (Annual Reviews, Palo Alto, in press).
16 *Sparks, H.V. and Bohr, D.F.:* The effect of stretch on passive tension and contractility of isolated vascular smooth muscle. Am. J. Physiol. *202:* 835–840 (1962).

H.V. Sparks, Department of Physiology, University of Michigan, *Ann Arbor, MI 48109* (USA)

Mechanisms of Vasodilatation. Satellite Symp., 27th Int. Congr. Physiol. Sci., Wilrijk 1977, pp. 193–199 (Karger, Basel 1978)

Oxygen, Carbon Dioxide, and Hydrogen Ion as Local Factors Causing Vasodilatation[1]

Brian R. Duling[2]

Department of Physiology, School of Medicine, University of Virginia, Charlottesville, Va.

Oxygen, carbon dioxide, and hydrogen ion are all substances which might serve admirably in the regulation of blood flow since they are all influenced by the metabolic activity of the tissue, are freely diffusible and thus can rapidly transmit information between parenchymal cell and vessel wall. All three substances have accordingly been the subject of extensive investigations over the years and it appears that each of them is a strong candidate for a mediator of flow control, at least in certain tissues and during certain types of experimental modifications. It is clear, however, that none of the three alone can prove an adequate and general explanation for local control of blood flow.

Flow control obviously resides with the vascular smooth muscle cells and in particular with the cells of the arterioles. It follows that the problem of quantitating the role of these or any other agent reduces to one of assessing the sensitivity of the smooth muscle cells and determining the relevant ambient concentrations in the environment of the cells. This has proved difficult with all of the agents to be discussed in this symposium, and perhaps especially difficult with these three, because of their high diffusibility in the extracellular fluid and because of the fact that their concentrations are complex functions related to a combination of their production rates and the nature of the oxyhemoglobin dissociation curve. However, the development of microelectrodes and other recent work in this field has increased our understanding of the role of these agents in the regulation of blood flow. The present discussion will be largely

[1] Research reported in this paper carried out by the author was supported by PHS Grant No. 12792.

[2] Established investigator of the American Heart Association.

restricted to recent work in the elucidation of the mechanisms of action of these substances, with special emphasis on determinations of concentrations in the environment of the smooth muscle cells, estimations of the reactivity of the smooth muscle cells, and evaluation of the mechanism of action of the substances in inducing vasodilatation.

1. Carbon Dioxide and Hydrogen Ion

Locally, both hydrogen ion and carbon dioxide are dilators and the smooth muscle relaxing properties of both agents have been shown both *in vitro* and *in vivo* (5, 18, 21, 23, 24, 36, 37). The dilatation induced by CO_2 and H^+ is probably by a direct effect on the vasculature (22, 34). The mechanism by which the two agents alter vascular tone is most probably via an alteration in the membrane calcium flux (36, 38), but this has yet to be conclusively demonstrated, and it should be noted that the contractile proteins themselves are reasonably sensitive to changes in the hydrogen ion activity (27).

In striated muscle, the sensitivity of the vascular bed to hydrogen and carbon dioxide appears to be inadequate to explain most forms of local blood flow control (16, 23), but reactive hyperemia may be an exception (23).

The evidence in support of a role of carbon dioxide and H^+ in blood flow regulation of the cerebral vasculature seems somewhat stronger. The cerebral vasculature is in general quite sensitive to CO_2 (21, 24–26, 33), and the pial vasculature is quite sensitive to both topically applied H^+ and CO_2 (18). The cerebral vasculature is relatively insensitive to changes in plasma pH, however, owing to the presence of the blood-brain barrier (33).

Pittman and Duling (31) have recently shown that the arterioles are sufficiently permeable to CO_2 to permit flux of CO_2 in amounts large enough to alter the oxyhemoglobin dissociation curve in the precapillary vessels. This raises the possibility that significant amounts of CO_2 may shunt between the arterial and venous sides of the circulation and thus minimize the changes in CO_2 experienced by arterioles when the arterial CO_2 is elevated. If this does occur, then the estimates of the reactivity of vascular beds such as that of striated muscle, in which the arterioles and venules are arranged in a parallel fashion, may be too low. This effect would be smaller in tissues such as the brain where the terminal vascular architecture is not arranged with a tight parallelism between arterial and venous sides.

The finding that the cerebral vessels are quite sensitive to changes in H^+ has led to the concept that the regulation of cerebral blood flow is primarily dependent on the pH of the cerebral extracellular space (33). This certainly is compatible with the recent findings by *Kontos et al.* (21) that the effect of CO_2 on the pial vessels is apparently due entirely to the change in pH and that isohydric changes in CO_2 are without effect. In contrast, in the human forearm (23), the hamster cheek pouch (8), and in the heart (29), changes in CO_2 can produce effects independent of changes in extracellular pH.

Knowing that the vessels are sensitive to H^+ or CO_2 does not establish that either substance participates in the regulation of blood flow. To date, it has not been shown unequivocally that the pH or the PCO_2 varies in the environment of the cerebral vessels in the appropriate fashion for any type of physiological regulation. The available literature on the subject remains controversial for both data obtained from measurements of the cerebrospinal fluid pH (12, 19, 20) and for local measurements using microelectrodes (1, 2, 17). Furthermore, in the case of cerebral vasodilatation associated with hypoxia, *Siesjö et al.* (35) have calculated that the rate of cerebral lactate production is inadequate to produce the necessary changes in pH. Thus, it is of great importance to make the appropriate measurements of cerebral pH using currently available measuring techniques.

2. *Oxygen*

The availability of good microelectrodes has made it possible to obtain quantitative measurements of the PO_2 in the immediate environment of the smooth muscles controlling flow. Such measurements have demonstrated that: (1) in most cases, the PO_2 of the vascular smooth muscle is approximately equal to that of the luminal blood in the microvessels (10); (2) that it must lie between approximately 100 and 20 mm Hg in normoxia (8); (3) that it varies with the environmental PO_2 (9); (4) that during occlusion of the vascular supply the PO_2 approaches 0 mm Hg within 1 min (*Lombard and Duling,* unpublished observation), and (5) that the PO_2 on the surface of arterioles in striated muscle remains constant during dilatations induced by striated muscle contraction (14). Thus, to date only during reactive hyperemia and reductions in local oxygen availability, has the PO_2 of the smooth muscle been shown to vary in the appropriate fashion for oxygen to be a regulator of flow by a direct action on the smooth muscle cells. On the other hand, PO_2 of the striated muscle cells does fall during the steady-state phase of muscle contraction in a way which is consistent with the participation of oxygen in the elaboration of some vasodilator metabolite (14).

Reduction of arterial oxygen tension leads to hyperemia and an increase in capillary density in many tissues (3, 9, 14–16, 18, 33). It has been observed that there is frequently no flow increase as arterial PO_2 is reduced, until a level of about 50–60 mm Hg is achieved and this has been the basis for a conclusion that the vasculature is relatively insensitive to blood PO_2 (16, 33). However, *Granger et al.* (15) have shown that the vasculature may respond to changes in PO_2 by an alteration of capillary density rather than a flow change, and thus an evaluation of the O_2 sensitivity of the vasculature based only on observation of flow changes must be made cautiously.

Since the PO_2 of the arteriolar wall usually lies in the range of 100–20 mm Hg, it follows that this is the range in which the smooth muscle must respond to changes in PO_2. We have attempted to demonstrate this sensitivity *in situ* and have been unable to find evidence that it exists (7). However, in that study precise control of the smooth muscle environment was impossible and it is thus necessary to look to *in vitro* studies to fully assess the reactivity of the smooth muscle to O_2.

In vitro, vascular smooth muscle does relax when the environmental PO_2 is lowered (6). However, no quantitative relation between the smooth muscle contractile activity and cell PO_2 can be obtained from simple measurement of the bathing solution PO_2 because of the presence of unstirred layers in the solution and diffusion gradients within the tissues. We attempted to evaluate the importance of these processes by looking at the correlation between O_2 sensitivity and the thickness of the vascular smooth muscle strips studied (32). The minimum PO_2 of the solution which could sustain full contraction of the smooth muscle cells was found to vary with thickness in a way compatible with diffusion theory and a limiting PO_2 of the cells of 2 mm Hg, i.e. far less than the PO_2 range required. *Fay et al.* (11) have recently succeeded in measuring the presence of such gradients in the ductus arteriosus and found that at a solution PO_2 of 148-mm Hg unstirred layers contributed a gradient of 50 mm Hg and an intramural gradient approaching 90 mm Hg was measured. Thus, so far, there are no well-defined quantitative experimental data on the sensitivity of the smooth muscle cell *per se* to oxygen. Furthermore, all *in vitro* studies have been done on the muscle from the larger muscular and conduit arteries, and it is possible that the behavior of the muscle from the arterioles is quite different.

The biochemical basis for the relaxation of smooth muscle which is induced by reduced oxygen is also not known at the present. It is not based on a simple limitation of availability of high energy phosphates following a reduction of oxidative phosphorylation (28). *Coburn* (4) has recently shown that the effect of oxygen on the smooth muscle tone is related to the method of generating tone. He reported that the cells of the K^+ stimulated aorta show a lower sensitivity to reduced oxygen tension than do those of the norepinephrine stimulated aorta. This is in spite of the fact that the oxygen consumption of the vessel strips is affected identically in the two cases. He concluded, on the basis of these findings, that the oxygen-sensitive process controlling contractile strength must be different from oxidative phosphorylation. However, this does not necessarily follow, since there is presumably some chemical intermediate between the alteration in oxygen availability and the vascular relaxation. *Coburn*'s results might simply reflect a different sensitivity of the potassium-induced contraction and the norepinephrine-induced contraction to the vasodilator intermediate. This possibility is made quite reasonable by the fact that both the membrane potential changes and the tension changes in vascular smooth muscle induced by iproveratril, temperature, and external calcium concentration are dependent on whether the cells were stimulated with K^+ or norepinephrine (13, 30).

3. Conclusion

The current status of the problem of the participation of oxygen in the regulation of blood flow is quite different from that for CO_2 and H^+. In the case of CO_2 and H^+ the key fact which is lacking appears to be a demonstration that appropriate variations in extracellular fluid pH occur during the various flow regulation processes. The CO_2 and H^+ effects appear to be induced by a direct effect on the smooth muscle cells. In the case of oxygen, the key question is the sensitivity of the smooth muscle cells to variations in oxygen tension. At present, the available data do not demonstrate that the requisite sensitivity exists and if not, then the participation of O_2 must be via indirect effects on the parenchymal cells, mediated by some metabolic intermediate, perhaps even one of those discussed in this symposium.

References

1 *Astrup, J.; Heuser, D.; Lassen, N.A.; Nilsson, B.; Nornberg, K., and Siesjö, B.K.:* Evidence against H^+ and K^+ as the main factors in the regulation of cerebral blood flow during epileptic discharges, acute hypoxemia, amphetamine intoxication, and hypoglycemia. A microelectrode study; in *Betz* Ioniç actions on vascular smooth muscle, pp. 110–114 (Springer, Berlin 1976).

2 *Betz, E.:* Discussion; in *Ingvar and Lassen* Brain work, pp. 222–227 (Munksgaard, Copenhagen 1975).

3 *Bourdeau-Martini, J.; Odoroff, C.L., and Honig, C.R.:* Dual effect of oxygen on magnitude and uniformity of coronary intercapillary distance. Am. J. Physiol. *226:* 800–810 (1974).

4 *Coburn, R.F.:* Oxygen tension sensors in vascular smooth muscle; in *Reivich, Coburn, Lahiri and Chance* Tissue hypoxia and ischemia, pp. 101–115 (Plenum Publishing, New York 1977).

5 *Daugherty, R.M., jr.; Scott, J.B.; Dabney, J.M., and Haddy, F.J.:* Local effects of O_2 and CO_2 on limb, renal, and coronary vascular resistances. Am. J. Physiol. *213:* 1102–1110 (1967).

6 *Detar, R. and Bohr, D.F.:* Oxygen and vascular smooth muscle contraction. Am. J. Physiol. *214:* 241–244 (1968).

7 *Duling, B.R.:* Oxygen sensitivity of vascular smooth muscle. II. *In vivo* studies. Am. J. Physiol. *227:* 42–49 (1974).

8 *Duling, B.R.:* Changes in microvascular diameter and oxygen tension induced by carbon dioxide. Circulation Res. *32:* 370–376 (1973).

9 *Duling, B.R.:* Microvascular responses to alterations in oxygen tension. Circulation Res. *31:* 481–489 (1972).

10 *Duling, B.R. and Berne, R.M.:* Longitudinal gradients in periarteriolar oxygen tension. A possible mechanism for the participation of oxygen in local regulation of blood flow. Circulation Res. *27:* 669–678 (1970).

11 *Fay, F.S.; Nair, P., and Whalen, W.J.:* Mechanism of oxygen induced contraction of ductus arteriosus; in *Reivich, Coburn, Lahiri and Chance* Tissue hypoxia and ischemia, pp. 123–134 (Plenum Publishing, New York 1977).

12 *Fujishima, M.:* Effect of constricting carotid arteries on cerebral blood flow and on cerebrospinal fluid pH, lactate and pyruvate in dogs. Jap. Heart J. *12:* 467–473 (1971).

13 *Golenhofen, K.; Hermstein, N., and Lammel, E.:* Membrane potential and contraction of vascular smooth muscle (portal vein) during application of noradrenaline and high potassium, and selective inhibitory effects of iproveratril (verapamil). Microvasc. Res. *5:* 73–80 (1973).

14 *Gorczyinski, R.J.:* The microcirculatory basis of functional hyperemia in striated muscle; PhD diss., Charlottesville (1976).

15 *Granger, H.J.; Goodman, A.H., and Granger, D.N.:* Role of resistance and exchange vessels in local microvascular control of skeletal muscle oxygenation in the dog. Circulation Res. *38:* 379–385 (1976).

16 *Haddy, F.J. and Scott, J.B.:* Metabolically linked vasoactive chemicals in local regulation of blood flow. Physiol. Rev. *48:* 688–707 (1968).

17 *Halsey, J.H. and Robinson, C.:* The non role of CSF pH in cerebral autoregulation. Circulation *50:* 89 (1974).

18 *Harris, P.D.; Longnecker, D.E.; Miller, F.N., and Wiegman, D.L.:* Sensitivity of small subcutaneous vessels to altered respiratory gases and local pH. Am. J. Physiol. *231:* 244–251 (1976).

19 *Hernandez-Perez, M.J. and Anderson, D.K.:* Autoregulation of cerebral blood flow and its relation to cerebrospinal fluid pH. Am. J. Physiol. *231:* 929–935 (1976).

20 *Kogure, K.; Scheinberg, P.; Reinmuth, O.M.; Fujishima, M., and Busto, R.:* Mechanisms of cerebral vasodilatation in hypoxia. J. appl. Physiol. *29:* 223–229 (1970).

21 *Kontos, H.A.; Raper, A.J., and Patterson, J.L.:* Analysis of vasoactivity of local pH, PCO_2 and bicarbonate on pial vessels. Stroke *8:* 358–360 (1977).

22 *Kontos, H.A.; Wei, E.P.; Raper, A.J., and Patterson, J.L.:* Local mechanism of CO_2 action on cat pial arterioles. Stroke *8:* 226–229 (1977).

23 *Kontos, H.A.:* Role of hypercapnic acidosis in the local regulation of blood flow in skeletal muscle. Circulation Res. *28/29:* suppl. 1, pp. I–98–I–105 (1971).

24 *Kontos, H.A.; Raper, A.J., and Patterson, J.L., jr.:* Mechanisms of action of CO_2 on pial precapillary vessels. Eur. Neurol. *6:* 114–118 (1971–72).

25 *Lassen, N.A.:* Brief reviews: control of cerebral circulation in health and disease. Circulation Res. *34:* 749–760 (1974).

26 *Lassen, A.N.:* Brain extracellular pH: the main factor controlling cerebral blood flow. Scand. J. clin. Lab. Invest. *22:* 247–251 (1968).

27 *Mrwa, U.; Achtig, I., and Ruegg, J.C.:* Influences of calcium concentration and pH on the tension development and ATPase activity of the arterial actomyosin contractile system. Blood Vessels *11:* 277–286 (1974).

28 *Namm, D.H. and Zucker, J.L.:* Biochemical alterations caused by hypoxia in the isolated rabbit aorta: correlation with changes in arterial contractility. Circulation Res. *32:* 464–470 (1973).

29 *Ng, M.L.; Levy, M.N., and Zieske, H.A.:* Effects of changes of pH and of carbon dioxide tension on left ventricular performance. Am. J. Physiol. *213:* 115–120 (1967).

30 *Peiper, U.; Griebel, L., and Wende, W.:* Different mechanisms of activating the vascular smooth muscle by noradrenaline or potassium-depolarization; in *Betz* Vascular smooth muscle, pp. 7–9 (Springer, Berlin 1972).

31 *Pittman, R.N. and Duling, B.R.:* Effects of altered carbon dioxide tension on intravascular hemoglobin oxygenation. Microvasc. Res. *13:* 211–224 (1977).

32 *Pittman, R.N. and Duling, B.R.:* Oxygen sensitivity of vascular smooth muscle. I. *In vitro* studies. Microvasc. Res. *6:* 202–211 (1973).

33 *Purves, N.J.:* The physiology of the cerebral circulation, p. 420 (University Press, Cambridge 1972).

34 *Severinghaus, J.W. and Lassen, N.:* Step hypocapnia to separate arterial from tissue PCO_2 in the regulation of cerebral blood flow. Circulation Res. *20:* 272–278 (1967).

35 *Siesjö, B.K.; Johannsson, H.; Norberg, K., and Salford, L.:* Brain function, metabolism and blood flow in moderate and severe arterial hypoxia; in *Ingvar and Lassen* Brain work, pp. 101–119 (Munksgaard, Copenhagen 1975).

36 *Breemen, C. van; Farinas, B.R.; Gerba, P., and McNaughton, E.D.:* Excitation-contraction coupling in rabbit aorta studied by the lanthanum method for measuring cellular calcium influx. Circulation Res. *30:* 44–54 (1972).

37 *Wahl, M. and Kuschinski, W.:* Influence of pH on adenosine-induced dilations of pial arteries; in *Betz* Ionic actions on vascular smooth muscle (Springer, Berlin 1976).

38 *Wei, E.P.; Thames, M.D.; Kontos, H.A., and Patterson, J.L., jr.:* Inhibition of the vasodilator effect of hypercapnic acidosis by hypercalcemia in dogs and rats. Circulation Res. *35:* 890–895 (1974).

B.R. Duling, PhD, Department of Physiology, School of Medicine, University of Virginia, *Charlottesville, VA 22901* (USA)

Mechanisms of Vasodilatation. Satellite Symp., 27th Int. Congr. Physiol. Sci., Wilrijk 1977, pp. 200–205 (Karger, Basel 1978)

The Mechanism of Potassium Vasodilatation

F.J. Haddy

Department of Physiology, Uniformed Services University, Bethesda, Md.

Katz and Lindner (18) first observed potassium vasodilatation in 1938; intracoronary administration of small amounts of potassium produced a decrease in coronary vascular resistance. Shortly thereafter, *Dawes* (10) demonstrated and studied more thoroughly the same response in the hindlimb vascular bed. In 1959, we showed that the response also occurs in the forelimb and is not blocked by denervation and adrenergic blockade with phentolamine (12). Since then, potassium vasodilatation has been observed by many investigators in a variety of vascular beds (7, 14, 15).

The mechanism of potassium vasodilatation was for many years an enigma because theory and observations in isolated vascular smooth muscle suggested that the response should be constriction rather than vasodilatation. Both the Nernst equation and Goldman equation indicated that potassium should depolarize the cell and indeed the most frequently reported response of isolated vascular smooth muscle to an increase in bath potassium concentration was contraction – though relaxation is observed when exploring the lower range of concentrations in a preparation having some initial active tension (6, 13).

1. Na^+/K^+ ATPase

In 1971, it occurred to us that potassium vasodilatation might be related to sarcolemmal Na^+, K^+-ATPase and the Na-K pump. We therefore studied the response in the dog gracilis muscle and forelimb before and after the administration of ouabain, a potent Na^+, K^+-ATPase inhibitor. Ouabain blocked potassium vasodilatation and sometime converted it to constriction and this did not occur in the case of acetylcholine and adenosine (8, 9). We therefore proposed (9) that potassium dilates by stimulating an electrogenic Na-K pump (22) in the vascular smooth muscle cell. We hypothesized (9) that there is an ATPase, sensitive to potassium and inhibited by ouabain in the sarcolemma of the vascular smooth muscle cell, and that stimulation of this ATPase by potassium increases Na^+-K^+ pumping across the sarcolemma, that the pump is not neutral, i.e. more sodium ions are pumped out than potassium ions pumped in, that this leads to loss of positive charge in the cell and consequently hyperpolarization of the membrane, and thus relaxation.

Studies since 1971 have been largely supportive of the hypothesis. In that same year, *Wolowyk et al.* (25) in Canada isolated Na^+, K^+-ATPase from a membrane preparation of rabbit carotid smooth muscle. Its presence was subsequently confirmed by *Limas and Cohn* (19). We then decided to see whether ouabain inhibition of the enzyme also blocks the potassium response in a vascular bed other than that in the gracilis muscle. We selected the coronary vascular bed of the dog and found that ouabain does indeed block hypokalemic coronary constriction (5). Just to make sure this has nothing to do with the red cells or changes in vascular transmural pressure, we worked with *Bünger et al.* (7) and examined the diastolic flow responses in the isolated guinea pig heart perfused with a cell free fluid. Here, we found that ouabain not only blocks hyperkalemic diastolic hyperemia but actually converts it to diastolic ischemia (as predicted by the hypothesis). Lowering perfusate temperature from 38 to 26 °C also greatly antagonized potassium vasodilatation.

In 1974, *Gebert and Piekowiak* (13) in Germany also confirmed these results in isolated strips of bovine facial artery. Before ouabain, an increase in bath $[K^+]$ to 10 mEq/l decreased tension and a decrease in bath $[K^+]$ to 1.3 mEq/l increased tension. These responses were reversed in the presence of ouabain. Similar effects of ouabain were observed in isolated cerebral arteries by *Toda* (23, 24). He additionally found that potassium relaxation is also abolished by substitution of LiCl for all of the NaCl and is suppressed by lowering the temperature of the bath or reducing $[Na^+]_o$.

2. Electrogenic Pump

At the same time, *Anderson and Brace* (1, 3–5) in our laboratory examined the hypothesis mathematically and found that the changes in cell potential that occur when extracellular potassium concentration is varied (see below) were predicted mathematically when they used a model of cell transport which includes an electrogenic pump. They were not predicted when they used a model which includes an electroneutral pump. Furthermore, they showed a clear correlation between measured vascular resistance and the cell potential predicted by the model which includes an electrogenic pump.

It is possible to examine the idea mathematically since the processes can be described quantitatively, provided reasonable values for ion permeabilities, intracellular ion concentrations, intracellular volume, and active ion fluxes are available. *Anderson* and *Brace* picked these values for arterial vascular smooth muscle from the literature, inserted them into several different mathematical models, and let the computer calculate membrane potential as a function of extracellular potassium concentration. The mathematical models included one with an electroneutral pump (essentially the Goldman equation) and another with an electrogenic pump. They compared membrane potentials calculated with the two models when potassium concentration is suddenly increased or decreased from normal. They also compared them with measured values from isolated arteries (see below). They found that the model, which includes an electroneutral pump, fails to predict the hyperpolarization which actually occurs as potassium concentration is increased over the low range (∿0 to 10 mEq/l). However, the model which assumes an electrogenic pump, activated by extracellular potassium and intracellular sodium, does correctly predict hyperpolarization with slightly increased extracellular potassium and depolarization with decreased extracellular potassium. It also predicts depolarization with large increases in extracellular potassium (>20 mEq/l), which has long been known to actually occur.

They also compared observed initial change in gracilis muscle vascular resistance and predicted membrane potential (based on the model which includes the electrogenic pump) as a function of changes in extracellular potassium concentration over the range 1–10 mEq/l. The correlation between observed vascular resistance and predicted membrane potential was good. They made a similar comparison of predicted membrane potential with resistance responses observed by *Brecht et al.* (6) in isolated perfused segments of bovine facial artery in response to change in potassium concentration over a wider range (0.5–25 mEq/l). Here too, the correlation was good.

Further evidence that the response of vascular tissue to potassium is consistent with an electrogenic pump is seen in the time course of the responses of forelimb vascular resistance and computed cell potentials to both hyperkalemia and hypokalemia. Calculated potentials correlated qualitatively with observed resistance both during the on response and during the off response.

3. Membrane Potential

Hendrickx and Casteels (16) recently demonstrated that the predicted potential changes actually occur in arterial smooth muscle, thereby providing strong support for the argument which related the contractile responses to an electrogenic pump. In an elegant series of experiments in isolated perfused rabbit ear arteries, they were successful in simultaneously measuring cell potential and vascular resistance. They found that the cell quickly depolarizes during perfusion of the artery with potassium-free perfusate and this is associated with an increase in vascular resistance. The cell then continues to depolarize more slowly. When potassium is reintroduced, the cell hyperpolarizes beyond the resting value (and this hyperpolarization is reversible by ouabain). *Anderson* and *Brace* (3) were able to simulate these potential changes with the model which includes an electrogenic pump. Thus, the initial depolarization appears to result from inhibition of the electrogenic pump and the slower depolarization from changing intracellular ion concentrations. The overshoot on reintroduction of potassium undoubtedly occurs because the internal sodium concentration is now elevated.

More recently, *Hermsmeyer* (17) measured cell potential in isolated vascular smooth muscle and explored the effect of varying potassium concentration over the entire physiological range. He found progressive hyperpolarization on raising potassium concentration and progressive depolarization on lowering the potassium concentration.

Thus, there seems to be little question that potassium stimulates the ouabain-sensitive electrogenic Na-K pump in vascular smooth muscle. It should next be asked whether it is the hyperpolarization which causes the relaxation. Does potassium vasodilatation result from decreased calcium influx subsequent to a hyperpolarization-induced decrease in calcium permeability, or does it result from increased calcium efflux via Na-Ca exchange (21). There are at this time no definitive data bearing on the question. We tend to favor the former mechanism simply because the changes in membrane potential correlate so well with the relaxation.

4. Role of Catecholamines

Do the catecholamines play a role in potassium vasodilatation? There is evidence that uptake of norepinephrine by isolated sympathetic nerve endings is sensitive to the external potassium concentration, low potassium causing decreased uptake and slightly high potassium causing increased uptake (2). Furthermore, in isolated canine saphenous vein strips, increasing the potassium concentration to 10 or 15 mEq/l inhibits the neuronal uptake of catecholamines, but in concentrations over 10 mEq/l potassium causes a concentration-dependent inhibition of the release of norepinephrine evoked by sympathetic nerve stimulation (20). However, variations in the potassium concentration within physiological ranges have little effect on the venomotor responses to sympathetic nerve stimulation or to exogenous norepinephrine, perhaps because the above effects tend to balance one another (*Vanhoutte,* personal commun.). This is compatible with the observation that potassium vasodilatation still occurs in the dog forelimb following adrenergic blockade (12).

5. Other Vasodilator Substances

Do other vasodilator agents act via the ouabain-sensitive electrogenic sodium-potassium pump? We have recently examined the time course of the effect of ouabain on the activity of a number of vasodilators in the dog forelimb (11). Ouabain very quickly abolished the response to intraarterially injected potassium chloride. In contrast, ouabain had little effect on the vasodilator responses to injected magnesium sulfate, acetylcholine, adenosine, and prostaglandin E_1 until late in the experiment when the responses were suppressed but not abolished. Thus, of the vasodilators tested, potassium chloride is the most sensitive to ouabain. Of interest was the observation that ouabain abolished the constrictor response to calcium chloride almost as fast as it blocked the dilator response to potassium chloride. Shortly thereafter, calcium chloride injection actually produced vasodilatation. At this time, norepinephrine still produced constriction. We have no explanation for the effect of ouabain on the action of calcium chloride.

6. Conclusion

Intra-arterial administration of small amounts of potassium produces a decrease in the resistance to blood flow through most vascular beds. This effect is blocked or reversed by ouabain. Studies in isolated vascular smooth muscle indicate that slight increase in $[K^+]_o$ produces hyperpolarization which is also blocked by ouabain. These and other findings suggest that potassium vasodilatation results from stimulation of cell membrane Na^+, K^+-ATPase which activates an electrogenic Na-K pump causing hyperpolarization and relaxation of the vascular smooth muscle cell.

References

1 *Anderson, D.K.:* Cell potential and sodium-potassium pump in vascular smooth muscle. Fed. Proc. Fed. Am. Socs exp. Biol. *35:* 1294–1297 (1976).

2 *Bogdanski, D.F.; Blaszkowski, T.P., and Tissari, A.H.:* Mechanism of biogenic amine transport and storage. IV. Relationship between K^+ and Na^+ requirement for transport and storage of 5-hydroxytryptamine and norepinephrine in synaptosomes. Biochim. biophys. Acta *211:* 521–532 (1970).

3 *Brace, R.A.:* Time course and mechanisms of the acute effects of hypokalemia and hyperkalemia on vascular resistance. Proc. Soc. exp. Biol. Med. *145:* 1389–1394 (1974).

4 *Brace, R.A. and Anderson, D.K.:* Predicting transient and steady state changes in resting membrane potential. J. appl. Physiol. *35:* 90–94 (1973).

5 *Brace, R.A.; Anderson, D.K.; Chen, W.T.; Scott, J.B., and Haddy, F.J.:* Local effects of hypokalemia on coronary resistance and myocardial contractile force. Am. J. Physiol. *227:* 590–597 (1974).

6 *Brecht. K.; Konold, P., and Gebert, G.:* The effect of potassium, catecholamines and other vasoactive agents on arterial segments of the muscular type. Physiol. Bohemoslov. *18:* 15–22 (1969).

7 *Bünger, R.; Haddy, F.J.; Querengässer, A., and Gerlach, E.:* Studies on potassium induced coronary dilation in the isolated guinea pig heart. Pflügers Arch. *363:* 27–31 (1976).

8 *Chen, W.T.; Anderson, D.K.; Scott, J.B., and Haddy, F.J.:* The mechanism of the vasodilator effect of potassium. J. Lab. clin. Med. *78:* 797 (1971).

9 *Chen, W.T.; Brace, R.A.; Scott, J.B.; Anderson, D.K., and Haddy, F.J.:* The mechanism of the vasodilator action of potassium. Proc. Soc. exp. Biol. Med. *140:* 820–824 (1972).

10 *Dawes, G.S.:* The vasodilator action of potassium. J. Physiol., Lond. *99:* 224–238 (1941).

11 *Dobbins, D.; Dabney, J.; Swindall, B., and Haddy, F.:* Blockade of calcium vasoconstriction by ouabain in the canine forelimb. Fed. Proc. Fed. Am. Socs exp. Biol. *36:* 417 (1977).

12 *Emanuel, D.A.; Scott, J.B., and Haddy, F.J.:* Effect of potassium on small and large blood vessels of the dog foreleg. Am. J. Physiol. *197:* 637–642 (1959).

13 *Gebert, G. and Piechowiak, H.:* Effect of potassium and norepinephrine on the tone of the isolated artery: changes by ouabain pretreatment. Experientia *30:* 46–47 (1974).

14 *Haddy, F.J.:* Potassium and blood vessels. Life Sci. *16:* 1489–1498 (1975).

15 *Haddy, F.J. and Scott, J.B.:* Metabolically linked vasoactive chemicals in local regulation of blood flow. Physiol. Rev. *48:* 688–707 (1968).

16 *Hendrickx, H. and Casteels, R.;* Electrogenic sodium pump in arterial smooth muscle cells. Pflügers Arch. *346:* 299–306 (1974).

17 *Hermsmeyer, K.:* Electrogenesis of increased norepinephrine sensitivity of arterial vascular muscle in hypertension. Circulation Res. *38:* 362–367 (1976).

18 *Katz, L.N. and Lindner, E.:* The action of excess Na, Ca and K on the coronary vessels. Am. J. Physiol. *124:* 155–160 (1938).

19 *Limas, C.J. and Cohn, J.N.:* Stimulation of vascular smooth muscle sodium, potassium-adenosinetriphosphatase by vasodilators. Circulation Res. *35:* 601–607 (1974).

20 *Lorenz, R.R. and Vanhoutte, P.M.:* Inhibition of adrenergic neurotransmission in isolated veins of the dog by potassium ions. J. Physiol., Lond. *246:* 479–500 (1975).

21 *Reuter, H.; Blaustein, M.P., and Haeusler, G.:* Na-Ca exchange and tension development in arterial smooth muscle. Phil. Trans. R. Soc. B *265:* 87–94 (1973).

22 *Thomas, R.C.:* Electrogenic sodium pump in nerve and muscle cells. Physiol. Rev. *52:* 563–594 (1972).

23 *Toda, N.:* Responsiveness to potassium and calcium ions of isolated cerebral arteries. Am. J. Physiol. *227:* 1206–1211 (1974).

24 *Toda, N.:* Potassium-induced relaxation in isolated cerebral arteries contracted with prostaglandins $F_{2\alpha}$. Pflügers Arch. *364:* 235–242 (1976).

25 *Wolowyk, M.W.; Kidwai, A.M., and Daniel E.E.:* Sodium-potassium stimulated adenosinetriphosphatase of vascular smooth muscle. Can. J. Biochem. *49:* 376–384 (1971).

F.J. Haddy, MD, PhD, Department of Physiology, Uniformed Services University, 4301 Jones Bridge Road, *Bethesda, MD 20014* (USA)

Mechanisms of Vasodilatation. Satellite Symp., 27th Int. Congr. Physiol. Sci.,
Wilrijk 1977, pp. 206–213 (Karger, Basel 1978)

Vasodilatation Mediated by Hyperosmolarity[1]

S. Mellander and J. Lundvall

Department of Physiology and Biophysics, University of Lund, Lund

It has been known for a long time that infusion of hypertonic solutions into the blood stream can lead to an increase of peripheral blood flow (1, 15, 20). A direct physiologic role of osmolarity in circulatory control was, however, evidenced only 10 years ago when *Mellander et al.* (17) proposed tissue hyperosmolarity to be a mediator of exercise hyperemia in skeletal muscle. Later research has confirmed this hypothesis and, in addition, demonstrated several other important aspects of the 'osmolar circulatory control system' with influences, among others, on the intra-/extravascular fluid distribution, the plasma volume, the cardiac performance, and the central and peripheral autonomic nervous functions.

1. Muscular Exercise and Osmolarity Changes

Let us consider first some general aspects of the intricate problem of the exercise hyperemia response. There is clear experimental evidence today that the reaction is causally linked to muscle metabolism and, further, that *several* metabolic factors contribute to the response by mutual synergistic actions. Since the metabolic pathways in skeletal muscle, and hence the concentrations of individual metabolites, differ with the type of exercise performed, there is every reason to believe that the relative importance of each contributing dilator factor can vary, for instance during different phases of the exercise hyperemia response and under different work loads of the skeletal muscle. It follows that the conditions under which a proposed mediator of exercise hyperemia plays its main role must be investigated and defined, and that experiments aimed at

[1] Supported by Grant 2210-14X from the Swedish Medical Research Council.

revealing its true role in the reaction must be designed so as to reproduce as closely as possible these special conditions. Otherwise, the quantitative importance of the metabolic factor for the vascular reaction might be seriously misinterpreted. Much disagreement in this field of research might be settled if these fundamental matters were considered.

The very precise and graded adjustments of vascular tone and blood flow that occur in skeletal muscle in relation to its varying nutritional demands would seem to require highly complicated feedback control systems. However, once such a biological control system is experimentally defined, we often become aware of the ingenious simplicity by which nature has solved a difficult regulatory problem. Such simplicity seems to be a characteristic of the osmolar control of the exercise hyperemia response. The increased metabolism of working skeletal muscle causes a proportionate production and release of metabolic substances and a consequent graded interstitial hypertonicity which, largely *irrespective* of the particles involved, will cause graded osmotic shrinkage of the adjacent vascular smooth muscle. Such reduction of smooth muscle cell volume leads to graded inhibition of myogenic pacemaker discharge, active vascular relaxation, and an exercise hyperemia response in relation to the increased muscle metabolism. The experimental evidence supporting this concept is briefly outlined below.

The characteristics of the changes in muscle osmolality and in vascular resistance were examined by *Mellander et al.* (17) and *Lundvall* (11) during graded short-term (5 min) exercise of the lower leg muscles in the cat produced by stimulation of the somatomotor fibers in the frequency range up to 4 imp/sec. By such stimulations, graded exercise hyperemia responses ranging from barely discernible to maximal ones can be elicited. Exercise was found always to be associated with a local tissue hyperosmolality, the extent of which was estimated as it was reflected in the venous effluent. During heavy work, venous osmolality could exceed that at rest by as much as 45 mosm/kg H_2O. The hyperosmolality developed almost immediately at the onset of work and reached a maximal level in the venous effluent within less than 3 min. The dilator response of the resistance vessels showed a similar time course, starting almost immediately upon exercise and reaching a maximal level within 1–2 min. The extent of tissue (venous) hyperosmolality showed a highly significant correlation to the magnitude of the concomitant dilator response of the resistance vessels, both these phenomena, in turn, being correlated to the intensity of the work (fig. 1A). The data from more than 80 exercise experiments are here classed with regard to the rate of somatomotor fiber stimulation, where excitation rates less than 1 imp/sec can be considered as light exercise, 1–2 imp/sec as moderate, and 3–4 imp/sec as heavy exercise.

Analysis of the constituents in the venous effluent from muscle during work showed that the hyperosmolality could be ascribed mainly to sodium and lactate

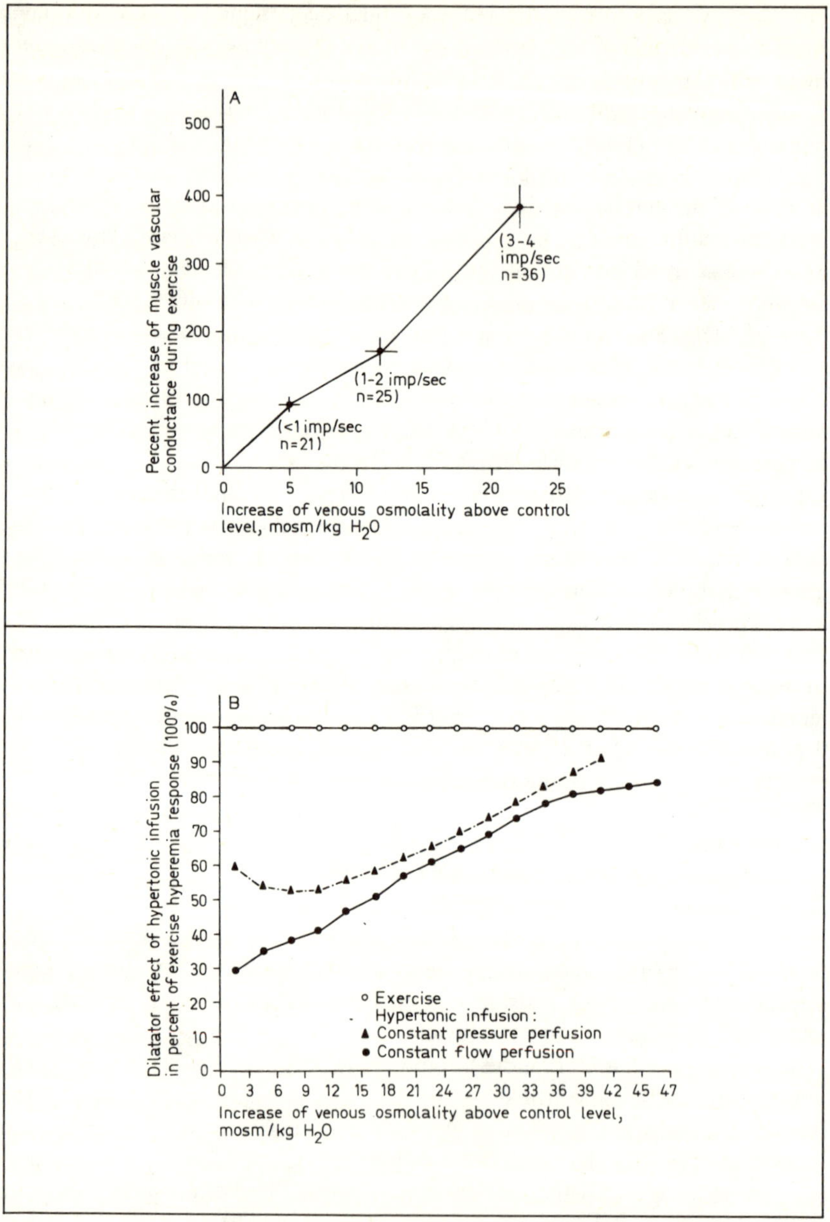

Fig. 1. A Correlation between vasodilator response and tissue (venous) hyperosmolality in cat skeletal muscle during graded exercise. *B* Quantitative evaluation of the role of tissue hyperosmolality for the early exercise hyperemia response in the cat. Reproduced in modification from *Lundvall* (11) by permission.

but also to some other substances e.g. potassium (11). Release of substances from the contracting striated muscle fibers and osmotic water movement into these cells can explain the changed composition of the extracellular fluid.

During more prolonged exercise (15 min duration), venous hyperosmolality remained at its maximal level when exercise was light or moderate (<3 imp/sec), but tended to decline during heavy work and then in parallel with a marked decline of the developed muscle tension (11). The latter type of experimental work therefore seems to be supramaximal; a corresponding heavy work cannot be expected to be performed voluntarily for more than a short period of time. Such an effect of fatigue might at least partly explain the moderate exercise hyperosmolality in the dog reported by some other authors (see below).

The vascular resistance effects of tissue hyperosmolality *per se* were analyzed during experimental hyperosmolality in resting cat skeletal muscle produced by arterial infusion of various hypertonic solutions, such as glucose, sucrose, or sodium lactate (11, 17). These experiments were designed so as to simulate as closely as possible the osmolar changes in exercising muscle with regard to magnitude and time course. Experimental hyperosmolality always evoked a rapidly developing dilator response with an initial marked blood flow increase followed by a somewhat less pronounced but maintained hyperemia level. There was a distinct correlation between the magnitudes of these dilator responses and the concomitantly produced increase of venous osmolality. The vascular response also seemed dependent on the rate at which osmolality was increased (11, 16, 18). The dilator response to hypertonicity could be attributed to a direct active inhibition of vascular smooth muscle tone. Passive or rheological effects, such as infusion artifacts, transcapillary fluid absorption, shrinkage of blood cells, osmotic dehydration of the vascular walls, etc. were shown to be of negligible importance (11).

It was concluded from such exercise and hypertonic infusion experiments in the cat, and from a number of other observations (11), that tissue hyperosmolality is causally linked to the exercise hyperemia response. The importance of this factor seemed to be greatest in early stages of work when the hyperosmolality is pronounced and the rate of the osmolarity increase is high. A quantitative evaluation of the role of tissue hyperosmolality for the early exercise hyperemia response is given in figure 1B. The resistance vessel dilator responses during experimental hyperosmolality in resting muscle are here expressed in percent of the exercise hyperemia response at comparable levels of evoked venous hyperosmolality. The exercise hyperemia at each workload is represented by the 100% line and the lower curves show the extent to which the vascular response can be attributed to tissue hyperosmolality (triangles refer to constant pressure, and squares to constant flow experiments). The diagram indicates that tissue hyperosmolality contributes to some 50% of the early exercise hyperemia response at light and moderate work and that its role is even greater during heavy work characterized by more pronounced hyperosmolality. In later phases of work, tissue hyperosmolality can still be a significant but perhaps not the dominating mediator of the hyperemia response (see below).

The tissue osmolarity hypothesis has also been tested in experiments on man by *Lundvall et al.* (13) by studying the vascular resistance function in the forearm during exercise and arterial hypertonic infusion at comparable venous hyperosmolality values. These results strongly suggested that tissue hyperosmolarity is a causal factor for the exercise hyperemia that develops in our own skeletal muscles during voluntary exercise (cf. 21). As in the cat, this factor seems to be of greater importance in early stages of human voluntary work (13) than during prolonged exercise (2).

Mention should be made in this context that some investigators (3, 22, 23) have reported relatively moderate exercise hyperosmolarity and relatively weak dilator effects of hypertonic infusions in the dog. It is possible, therefore, that the hyperosmolar dilator mechanism is less efficient in the dog than in cat and man. On the other hand, it is doubtful whether those studies can permit any valid conclusion about the quantitative role of tissue hyperosmolarity in exercise hyperemia, since the interstitial osmolar changes produced in resting muscle were slow compared to the rapid osmolar events that occur in exercise.

2. Cellular Mechanism

The cellular mechanisms responsible for the inhibition of vascular tone by hyperosmolarity were investigated by recording electrical and mechanical activity in a spontaneously active single-unit vascular smooth muscle preparation *in vitro* (8, 10, 17). In brief, these studies showed that increased osmolarity caused a pronounced and sustained relaxation of the vascular smooth muscle mainly by inhibiting myogenic pacemaker activity. This effect, in turn, could be ascribed to changes in transmembrane ionic concentration gradients and in membrane permeabilities to ions produced by osmotic reduction of smooth muscle cell volume. Exercise *in vivo* leads to pronounced hyperosmolarity in the interstitium, the immediate environment of the vascular smooth muscle, by production and release of osmotically active particles from the contracting skeletal muscle fibers and by water movement into these cells. It is most likely that this interstitial osmolar change will lead to similar shrinkage of the vascular smooth muscle cells and to the consequent series of electrical and mechanical events that were shown to produce vascular relaxation in the *in vitro* experiments.

In fact, it might be possible that osmotic shrinkage of the smooth muscle cells *in vivo* could lead to inhibition of basal myogenic tone by a similar mechanism to that of a decrease in vascular transmural pressure and, hence, that there is a coupling between the metabolic hyperosmolar stimulus and the myogenic inhibitory pressure stimulus. The myogenic 'receptor' is known to be especially sensitive to the rate of change of transmural pressure, dP_T/dt, and a strong inhibitory rate-dependent stimulus (pressure decrease, i.e. negative dP_T/dt) can cause complete abolition of vascular tone (4, 5, 9). The finding that also the osmotic vasodilator response in skeletal muscle seemed to be rate-dependent strengthens the hypothesis that osmotic and myogenic stimuli might operate via the same myogenic receptor mechanism. Since the different motor units in muscle contract asynchronously in submaximal work, it is likely that local tissue osmolarity in the vicinity of each unit is *oscillating* during work which could cause repetitive strong rate-dependent osmotic stimuli and effective local microvascular dilation each time the motor unit becomes excited. The overall exercise hyperemia response in the whole skeletal muscle (and the venous hyperosmolarity) would still be expected to be steady due to the asynchrony of the different motor-unit contractions.

3. Salivary Glands

Tissue hyperosmolarity apparently contributes to the functional vasodilatation in salivary glands as well, as indicated by *Lundvall and Holmberg* (7, 14). Those studies showed that considerable hyperosmolarity develops during chorda stimulation and that experimental hyperosmolarity of similar magnitude almost can mimic the evoked functional vasodilatation. This suggests that hyperosmolarity acts synergistically with the previously discussed mediators of functional vasodilatation in salivary glands, i.e. via the cholinergic innervation and the bradykinin system.

4. General Hemodynamic Effects

When large muscle groups are active, arterial osmolality in man can increase by some 25 mosm/kg H_2O due to overflow of osmols from the active muscles. This change has been shown to have the following beneficial effects on overall cardiovascular dynamics in exercise. It causes considerable osmotic absorption of intracellular and interstitial fluid from inactive tissues into the circulatory system (12) tending to compensate for the large, mainly osmotic, fluid losses that inevitably occur into the active muscles during exercise (11, 17). Such hyperosmolality can further lead to a decrease of pulmonary vascular resistance (6) and to a positive inotropic action on the heart (24). There is also indirect evidence from ventriculo-cisternal hypertonic perfusion experiments in the dog that cerebral hypertonicity of similar magnitude to what would result from the arterial exercise hyperosmolality can contribute to increased respiration and to the typical alterations of the autonomic nervous activity that characterize whole body exercise (19). It thus appears that the 'osmolar circulatory control system' can contribute in several ways to the homeostasis in exercise.

5. Conclusion

From all the experimental evidence now available, it may be concluded that tissue hyperosmolarity is an important, perhaps dominating, mediator of exercise hyperemia, especially in early stages of work. Other metabolic dilator factors might well contribute to the exercise hyperemia response and could be of special importance for its very initial development and for its maintenance during prolonged work.

References

1 *Binet, L. et Stoicesco, S.:* Les injections intraveineuses de solutions hypertoniques de chlorure de sodium. Paris méd. *73:* 498–503 (1929).

2 *Eklund, B. and Kaijser, L.:* Forearm blood flow after isometric contraction at different loads in relation to potentially vasodilating substances. Scand. J. clin. Lab. Invest. *34:* 23–29 (1974).

3 *Gazitùa, S.; Scott, J.B.; Swindall, B., and Haddy, F.J.:* Resistance responses to local changes in plasma osmolality in three vascular beds. Am. J. Physiol. *220:* 384–391 (1971).

4 *Grände, P.-O.; Lundvall, J., and Mellander, S.:* Evidence for a rate-sensitive regulatory mechanism in myogenic microvascular control. Acta physiol. scand. *99:* 432–447 (1977).

5 *Grände, P.-O. and Mellander, S.:* Characteristics of static and dynamic regulatory mechanisms in myogenic microvascular control. Acta physiol. scand. (in press).

6 *Hauge, A. and Bø, G.:* Blood hyperosmolality and pulmonary vascular resistance in the cat. Circulation Res. *28:* suppl. I, pp. 371–376 (1971).

7 *Holmberg, J. and Lundvall, J.:* Tissue hyperosmolality as a causal factor in vasodilatation following sympathetic stimulation of the submandibular gland. Acta physiol. scand. *98:* 400–406 (1976).

8 *Johansson, B. and Jonsson, O.:* Cell volume as a factor influencing electrical and mechanical activity of vascular smooth muscle. Acta physiol. scand. *72:* 456–468 (1968).

9 *Johansson, B. and Mellander, S.:* Static and dynamic components in the vascular myogenic response to passive changes in length as revealed by electrical and mechanical recordings from the rat portal vein. Circulation Res. *36:* 76–83 (1975).

10 *Jonsson, O.:* Extracellular osmolality and vascular smooth muscle activity. Acta physiol. scand., suppl. 359, pp. 1–48 (1970).

11 *Lundvall, J.:* Tissue hyperosmolality as a mediator of vasodilatation and transcapillary fluid flux in exercising skeletal muscle. Acta physiol. scand., suppl. 379, pp. 1–142 (1972).

12 *Lundvall, J.; Mellander, S.; Westling, H., and White, T.:* Fluid transfer between blood and tissues during exercise. Acta physiol. scand. *85:* 258–269 (1972).

13 *Lundvall, J.; Mellander, S., and White, T.:* Hyperosmolality and vasodilatation in human skeletal muscle. Acta physiol. scand. *77:* 224–233 (1969).

14 *Lundvall, J. and Holmberg, J.:* Role of tissue hyperosmolality in functional vasodilatation in the submandibular gland. Acta physiol. scand. *92:* 165–174 (1974).

15 *Marshall, R.J. and Shepherd, J.T.:* Effect of injections of hypertonic solutions on blood flow through the femoral artery of the dog. Am. J. Physiol. *197:* 951–954 (1959).

16 *McKinley, M.J.; McKenzie, J.S., and Blair-West, J.R.:* Effects of maintained osmolarity changes on rat portal vein spontaneous contractions. Am. J. Physiol. *226:* 718–723 (1974).

17 *Mellander, S.; Johansson, B.; Gray, S.; Jonsson, O.; Lundvall, J., and Ljung, B.:* The effects of hyperosmolarity on intact and isolated vascular smooth muscle. Possible role in exercise hyperemia. Angiologica *4:* 310–322 (1967).

18 *Mellander, S. and Lundvall, J.:* Role of tissue hyperosmolality in exercise hyperemia. Circulation Res. *28/29:* suppl. I, pp. 39–45 (1971).

19 *Mellander, S. and Hillman, J.:* Circulatory and respiratory effects evoked by hypertonic ventriculo-cisternal perfusion. Acta physiol. scand. *94:* 229–235 (1975).

20 *Overbeck, H.W.; Molnar, J.I., and Haddy, F.J.:* Resistance to blood flow through the vascular bed of the dog forelimb. Am. J. Cardiol. *8:* 533–541 (1961).

21 *Overbeck, H.W. and Grega, G.J.:* Response of the limb vascular bed in man to intrabrachial arterial infusions of hypertonic dextrose or hypertonic sodium cloride solutions. Circulation Res. *26:* 717–731 (1970).

22 *Scott, J.B. and Radawski, D.:* Role of hyperosmolarity in the genesis of active and reactive hyperemia. Circulation Res. *28:* suppl. I, pp. 26–32 (1971).

23 *Skinner, N.S., jr. and Costin, J.C.:* Interactions between oxygen, potassium, and osmolality in regulation of skeletal muscle blood flow. Circulation Res. *28:* suppl. I, pp. 73–85 (1971).

24 *Wildenthal, K.; Mierzwiak, D.S., and Mitchell, J.H.:* Acute effects of increased serum osmolality on left ventricular performance. Am. J. Physiol. *216:* 898–904 (1969).

Prof. *S. Mellander,* Department of Physiology and Biophysics, University of Lund, Sölvegatan 19, *S–223 62 Lund* (Sweden)

Mechanisms of Vasodilatation. Satellite Symp., 27th Int. Congr. Physiol. Sci., Wilrijk 1977, pp. 214–221 (Karger, Basel 1978)

Role of Adenosine, Adenosine Triphosphate and Inorganic Phosphate in Resistance Vessel Vasodilatation

Robert M. Berne and Rafael Rubio

Department of Physiology, University of Virginia, School of Medicine, Charlottesville, Va.

Over the past several years, there have been several agents added to the list of possible humoral mediators of vasodilatation associated with either increased metabolic activity or reduced oxygen supply to the tissue. Among these are adenosine, adenosine triphosphate (ATP) and inorganic phosphate (PO_4), substances intimately related to the storage and release of cellular energy.

1. Adenosine

The nucleoside, adenosine, has been known to be a potent vasodilator since 1929 (9), but it was not until 1963 that it was proposed as a mediator of the *coronary vasodilatation* associated with myocardial hypoxia (1). This suggestion was based on the observation that in the effluents of the isolated perfused cat heart and in the coronary sinus blood of the dog heart, inosine and hypoxanthine, degradative products of adenosine, were found when the hearts were subjected to severe hypoxia (1). It was postulated that with myocardial hypoxia, ATP was degraded to adenosine which crossed the sarcolemma and induced relaxation of the vascular smooth muscle of the resistance vessels. However, inosine and hypoxanthine are not vasoactive and their presence could arise from adenosine monophosphate (AMP) by deamination by adenylate deaminase to form inosine monophosphate (IMP), dephosphorylation of IMP by 5′-nucleotidase to form inosine, and splitting of inosine to hypoxanthine and ribose-1-phosphate by nucleoside phosphorylase. If this were the sole degradative pathway in cardiac muscle, adenosine would not be formed and the adenosine hypothesis would have to be discarded. However, if AMP were first dephosphorylated prior to deamination, the adenosine hypothesis would be tenable.

Experiments on isolated perfused guinea pig hearts, in which an adenosine deaminase inhibitor was added to the perfusion medium, showed the presence of adenosine (as well as inosine and hypoxanthine) in the perfusate during periods of cardiac hypoxia (19). Similar findings were obtained in isolated perfused rabbit hearts with anoxia and an uncoupler of oxidative phosphorylation (26). Adenosine was also found in the coronary sinus blood of dog hearts during the hyperemia that followed 30–60 sec of left coronary artery occlusion in concentrations capable of eliciting maximal dilatation of the coronary resistance vessels (31). The findings are in agreement with studies on anoxic unperfused pieces of rat (13) and rabbit (17) myocardium in which adenosine was formed during incubation at 37 °C.

To be a physiological regulator of coronary vascular resistance, adenosine should be present in normal as well as in hypoxic cardiac tissue because overperfusion of the coronary vessels results in coronary vasoconstriction (autoregulation), and would require that the concentration of a vasodilator substance like adenosine be reduced by washout and decreased production, thereby permitting a greater expression of intrinsic basal tone. Experiments in which pericardial perfusion was employed and others in which normal pericardial fluid was examined showed that adenosine is released from the epicardial surface of the well-oxygenated heart (28). Calculation of the myocardial content of adenosine from the concentration in pericardial fluid (assuming equilibration with cardiac interstitial fluid) closely approximated the level measured directly in cardiac muscle by a technique with greatly increased sensitivity over previous methods (22). Although adenosine is present in normal heart muscle and increases when the oxygen supply to the myocardium becomes inadequate, it has not yet been demonstrated that the myocardial adenosine levels are reduced by overperfusion of the coronary vasculature.

Adenosine fulfills several criteria for a physiological mediator of vasodilatation such as a ready source in the form of ATP, a potent vasodilator, swift movement across cell membranes and inactivation by enzymes in tissue and blood. However, the nucleoside must also be formed quickly to be of physiological significance. The rapidity of adenosine formation was determined in rat hearts that were crushed and frozen *in situ* at different periods of time after cross clamping at the atrioventricular groove. Within 5 sec after the onset of ischemia, the adenosine levels increased by more than twofold (3). Recently, adenosine formation was studied in isolated perfused guinea pig hearts during a single cardiac cycle that was prolonged by the use of strontium and a perfusion temperature of 32 °C. Preliminary results show a peak level of adenosine coincident with maximal tension and the nadir of adenosine levels during diastole (unpublished observations).

Since adenosine is quickly deaminated to inactive inosine by adenosine deaminase, and since adenosine deaminase is present in myocardial cells, it was difficult to see how the adenosine formed by dephosphorylation of AMP by 5′-nucleotidase could reach the interstitial fluid intact and elicit dilation of the arterioles. The solution to this problem lay in the location of 5′-nucleotidase in the sarcolemma, the intercalated discs and the transverse tubules of the myocardium (27, 30). Hence, adenosine is formed at the cell margins and can enter the contiguous interstitial fluid without deamination.

Next, it was important to determine the fate of the adenosine, since only a small fraction of it appeared in the cardiac venous effluent as adenosine, inosine or hypoxanthine. Experiments on human red cell ghosts (34), dispersed embryonic chick heart cells (21), isolated perfused rabbit hearts (20), and intact dog hearts (24) revealed that adenosine is rapidly transported into the cells (primarily by facilitated diffusion) and incorporated into intracellular adenine nucleotides. Any adenosine that enters the cell and is not phosphorylated is deaminated to inosine, since even at pharmacological levels of extracellular adenosine, free adenosine was not found within the red cell ghosts or the embryonic chick heart cells (21, 34). Proof that at physiological extracellular concentrations, adenosine is phosphorylated by adenosine kinase and not first deaminated was obtained in isolated guinea pig hearts perfused with Krebs-Henseleit solution containing trace quantities of either uniformly labeled adenosine or inosine (41, 42). The specific activity ratio of nucleoside to purine base for infused adenosine was the same as that of the degraded cardiac adenine nucleotides, indicating direct phosphorylation of adenosine. Similar conclusions were reached in studies on isolated perfused rabbit hearts (20). Preferential phosphorylation of adenosine is attributable to the fact that adenosine kinase has a lower Km for the substrate than does adenosine deaminase and may be located in the cell membrane. The extracellular adenosine in the heart that does not reenter the myocardial cells passes through the capillary endothelium and into the blood were some of it is deaminated to inosine and some of the inosine degraded to hypoxanthine and ribose by nucleoside phosphorylase, which is absent in myocardial cells but present in endothelial cells and pericytes (32). Additional degradation of adenosine occurs in the red cells which contain adenosine deaminase and nucleoside phosphorylase. Hence, most of the adenosine that enters the blood vessels is recovered in coronary sinus blood as inosine and hypoxanthine.

Additional evidence that adenosine plays a role in the regulation of coronary resistance was obtained in two different types of experiments. In one study, isolated guinea pig hearts were perfused with Krebs-Henseleit solution equilibrated with 5% CO_2 and different concentrations of O_2 (33). At 35% O_2 and below, there was a striking increase in coronary flow associated with an increase in the rate of adenosine release and of the adenosine content of the myocardium. The close parallelism between coronary flow and adenosine formation suggests but does not establish a cause and effect relationship. In the second study, after brief occlusions of the left coronary artery, coronary sinus blood of the dog was pumped directly into either a foreleg artery or a renal artery (36). The infused cardiac venous blood elicited vasodilatation of the foreleg vessels and vasoconstriction in the kidney. The only known endogenous substances that can produce such responses are adenosine and AMP; the more likely candidate is adenosine since it has not been clearly established that AMP can cross the intact myocardium cell membrane. Earlier studies (18) of this type in which the cardiac venous blood was oxygenated prior to infusion in a test vessel showed no vasoactivity. The reason for this lack of agreement is most likely accounted for by inactivation of adenosine by the erythrocytes during the 10–15 min required for reoxygenation of the blood.

Most of the studies demonstrating increased adenosine formation are characterized by a reduction in oxygen supply to the myocardium. Recently, however, experiments in which cardiac oxygen requirements have been increased have also been associated with enhanced adenosine formation. Increased cardiac work in the rat, produced by thoracic aortic constriction, resulted in a significant increase in the myocardial adenosine levels (10). In another experiment, a special cannula was introduced into the pericardial sac of the dog, and Krebs-Henseleit solution was injected before and withdrawn after a control period and a period of stellate ganglion stimulation (39). A significant increase in the adenosine content of the pericardial injectate was observed with stellate ganglion stimulation. Using the same type of preparation, the release of adenosine into the pericardial injectate was also observed in the anesthetized dog with stimulation of the motor nerves of the four extremities and in the unanesthetized dog during treadmill exercise (40). These experiments indicate that not only reduced myocardial oxygen supply, but increased need in the absence of cardiac ischemia results in accelerated adenosine production in association with a greater coronary blood flow.

Adenosine may also contribute to the regulation of *cerebral blood flow.* The nucleoside is rapidly produced in ischemic brain (2) and is a potent dilator of cerebral arterioles when applied topically to the pial vessels (2). However, it does not cross the blood-brain barrier at a rate sufficient to produce cerebral arteriolar dilation when administered intraarterially. Furthermore, electrical stimulation of the brain, reduction in arterial pressure, hypoxemia and hypocapnia all elicit a significant increase in brain adenosine levels (29).

Skeletal muscle blood flow may be in part under the influence of adenosine since ischemia plus stimulation of dog muscle (8), dog muscle stimulation alone at constant blood flow (6), and stimulation of rat muscle perfused with Krebs-Henseleit solution (6) result in an increase in skeletal muscle adenosine formation. In addition, it has been observed that during muscle contraction of the isolated dog leg during constant flow perfusion, the specific activity of trace quantities of uniformly labeled adenosine infused intraarterially is reduced and is associated with a parallel decrease in vascular resistance (unpublished observations). The reduction in specific activity of the adenosine in the blood perfusing the contracting muscle must be due to dilution by unlabeled endogenous adenosine.

Mechanism. How adenosine elicits vasodilatation is still unknown. Experiments on isolated guinea pig atria suggest that adenosine blocks calcium uptake (35). Whether this is true for vascular smooth muscle remains to be determined. Another possibility is that adenosine produces an increase in vascular smooth muscle levels of cyclic AMP and that this leads to vasodilatation. This concept is based on the report that relaxation of vascular smooth muscle is associated with increased concentrations of cyclic AMP (37). Studies on the media of hog carotid artery strips incubated with physiological concentrations of adenosine (μM) failed to increase cyclic AMP levels significantly but did so when very high concentrations (m*M*) were used (14). In contrast to results on carotid artery strips, preliminary experiments with cultured vascular smooth muscle cells suggest that physiological concentrations of adenosine greatly increase cyclic AMP levels (unpublished observations). Finally, there is recent evidence that there are specific adenosine receptors in the vasculature of the dog heart (23).

Although adenosine may play a key role in the regulation of blood flow in several organs, it is undoubtedly not the only factor. Furthermore, it is quite likely that the action of adenosine is modulated by other agents, such as oxygen tension, K^+, and pH (12, 38).

2. Adenosine Triphosphate

Whether or not ATP is released from contracting skeletal or heart muscle is quite controversial. It has been reported (7) and denied (8) that isolated frog sartorius muscle releases ATP into the bathing medium during repetitive contraction. It has also been reported in human experiments (11) and denied in dog and rat experiments (5, 8) that contracting perfused skeletal muscle releases ATP into the venous effluent. Since ATP is rapidly destroyed in blood, EDTA was used to prevent its destruction in the human studies. However, EDTA added to ATP-free blood causes release of ATP from the formed elements of the blood and hence no valid conclusions can be drawn from these experiments (5). In the case of contracting rat leg muscle perfused with Krebs-Henseleit solution (5), one would expect to find ATP in the venous effluent if it were released from the muscle. However, none was found, although high concentrations of adenosine and its degradative products were detected in the perfusates.

With respect to the heart, it has recently been reported that isolated perfused guinea pig hearts release ATP when made hypoxic (25). However, the amounts released are below the concentrations required to elicit coronary vasodilatation and, therefore, it is unlikely that ATP plays a role in the regulation of coronary blood flow.

3. Inorganic Phosphate

As in the case of ATP, the role of PO_4 is controversial. PO_4 has been found to increase in the venous blood of stimulated fast (white or low oxidative) muscles of the cat, and when infused intraarterially as Na_2HPO_2 or NaH_2PO_4 to elicit vasodilatation (15). The acid salt is more active and the concentration required to produce maximal dilatation is 20 m*M*, a concentration greatly in excess of that found in venous blood of contracting muscle. Low permeability of the capillary endothelium to PO_4 is thought to be the reason for the high blood levels necessary to obtain vasodilatation (16). Slow (red or high oxidative) muscles are reported to fail to show active hyperemia or PO_4 release with activity (15). However, the absence of active hyperemia may be related to the very high (50 ml/min/100 g/muscle) resting blood flow (15). Furthermore, dog hind leg muscle contains only high oxidative fibers (red and intermediate) yet shows active hyperemia (6) as does cat soleus (red and high oxidative) when control resting blood flow is about 3–5 ml/min/100 g (*Bockman,* personal commun.).

In humans, only a 20% increase in deep forearm venous blood PO_4 levels was observed with forearm exercise, and infusion of PO_4 sufficient to raise blood levels fourfold failed to increase blood flow (4). In one study (8) on blood-perfused dog hind limbs, no increase in venous blood PO_4 was observed after 5 min of muscle contraction during complete ischemia, and in another (6) there was no significant arteriovenous difference in PO_4 during 20 min of contraction of the dog hindlimb perfused with blood at constant flow. At present, no satisfactory explanation can be offered to reconcile these divergent results, but it is obvious that further studies are necessary.

References

1 *Berne, R.M.:* Cardiac nucleotides in hypoxia: possible role in regulation of coronary blood flow. Am. J. Physiol. *204:* 317–322 (1963).

2 *Berne, R.M.; Rubio, R., and Curnish, R.R.:* Release of adenosine from ischemic brain: effect on cerebral vascular resistance and incorporation into cerebral adenine nucleotides. Circulation Res. *35:* 262–271 (1974).

3 *Berne, R.M.; Rubio, R., and Duling, B.R.:* Vasoactive substances affecting the coronary circulation; in *Ross and Hoffman* Myocardial ischemia. Int. Congr. Ser. No. 225, pp. 28–43 (Excerpta Medica, Amsterdam 1971).

4 *Barcroft, H.; Foley, T.H., and McSwiney, R.R.:* Experiments on the liberation of phosphate from the muscles of the human forearm during vigorous exercise and on the action of sodium phosphate on forearm muscle blood vessels. J. Physiol., Lond. *213:* 411–420 (1971).

5 *Bockman, E.L.; Berne, R.M., and Rubio, R.:* Release of adenosine and lack of release of ATP from contracting skeletal muscle. Pflügers Arch. *355:* 229–241 (1975).

6 *Bockman, E.L.; Berne, R.M., and Rubio, R.:* Adenosine and active hyperemia in dog skeletal muscle. Am. J. Physiol. *230:* 1531–1537 (1976).

7 *Boyd, I.A. and Forrester, T.:* The release of adenosine triphosphate from frog skeletal muscle *in vitro.* J. Physiol., Lond. *199:* 115–135 (1968).

8 *Dobson, J.G., jr.; Rubio, R., and Berne, R.M.:* Role of adenine nucleotides, adenosine, and inorganic phosphate in the regulation of skeletal muscle blood flow. Circulation Res. *29:* 375–384 (1971).

9 *Drury, A.N. and Szent-Györgyi, A.:* The physiological activity of adenine compounds with especial reference to their action upon the mammalian heart. J. Physiol., Lond. *68:* 213–237 (1929).

10 *Foley, D.H.; Herlihy, J.T.; Thompson, C.I.; Rubio, R., and Berne, R.M.:* Relationship between cardiac work and levels of myocardial adenosine. Fed. Proc. Fed. Am. Socs exp. Biol. *35:* 348 (1976).

11 *Forrester, T.:* An estimate of adenosine triphosphate release into the venous effluent from exercising human forearm muscle. J. Physiol., Lond. *224:* 611–628 (1972).

12 *Gellai, M.; Norton, J.M., and Detar, R.:* Evidence for direct control of coronary vascular tone by oxygen. Circulation Res. *32:* 279–289 (1973).

13 *Gerlach, E.; Deuticke, B. und Dreisbach, R.H.:* Der Nucleotid-Abbau im Herzmuskel bei Sauerstoffmangel und seine mögliche Bedeutung für die Coronardurchblutung. Naturwissenschaften *50:* 228–230 (1963).

14 *Herlihy, J.T.; Bockman, E.L.; Berne, R.M., and Rubio, R.:* Adenosine relaxation of isolated vascular smooth muscle. Am. J. Physiol. *230:* 1239–1243 (1976).

15 *Hilton, S.M. and Chir, B.:* A new candidate for mediator of functional vasodilatation in skeletal muscle. Circulation Res. *28/29:* suppl. I, pp. I-70–I-72 (1971).

16 *Hilton, S.M.; Hudlická, O., and Jackson, J.R.:* The movement of inorganic phosphate ions across capillaries in skeletal muscle during exercise and at rest. J. Physiol., Lond. *239:* 98–99P (1974).

17 *Imai, S.; Riley, A.L., and Berne, R.M.:* Effect of ischemia on adenine nucleotides in cardiac and skeletal muscle. Circulation Res. *15:* 443–450 (1964).

18 *Jelliffe, R.W.; Wolf, C.R.; Berne, R.M., and Eckstein, R.W.:* Absence of vasoactive and cardiotropic substances in coronary sinus blood of dogs. Circulation Res. *5:* 382–387 (1957).

19 *Katori, M. and Berne, R.M.:* Release of adenosine from anoxic hearts: relationship to coronary flow. Circulation Res. *19:* 420–425 (1966).

20 *Liu, M.S. and Feinberg, H.:* Incorporation of adenosine-8-^{14}C and inosine-8-^{14}C into rabbit heart adenine nucleotides. Am. J. Physiol. *220:* 1242–1248 (1971).

21 *Mustafa, S.J.; Rubio, R., and Berne, R.M.:* Uptake of adenosine by dispersed chick embryonic cardiac cells. Am. J. Physiol. *228:* 62–67 (1975).

22 *Olsson, R.A.:* Changes in content of purine nucleosides in canine myocardium during coronary occlusion. Circulation Res. *26:* 301–306 (1970).

23 *Olsson, R.A.; Davis, C.J.; Khouri, E.M., and Patterson, R.E.:* Evidence for an adenosine receptor on the surface of dog coronary myocytes. Circulation Res. *39:* 93–98 (1976).

24 *Olsson, R.A.; Snow, J.A.; Gentry, M.K., and Frick, G.P.:* Adenosine uptake by canine heart. Circulation Res. *31:* 767–778 (1972).

25 *Paddle, B.M. and Burnstock, G.:* Release of ATP from perfused heart during coronary vasodilatation. Blood Vessels *11:* 110–119 (1974).

26 *Richman, H.G. and Wyborny, L.:* Adenine nucleotide degradation in the rabbit heart. Am. J. Physiol. *207:* 1139–1145 (1964).

27 *Rostgaard, J. and Behnke, O.:* Fine structural localization of adenine nucleoside phosphatase activity in the sarcoplasmic reticulum and the T system of rat myocardium. J. Ultrastruc. Res. *12:* 579–591 (1965).

28 *Rubio, R. and Berne, R.M.:* The release of adenosine by the normal myocardium and its relationship to the regulation of coronary resistance. Circulation Res. *25:* 407–415 (1969).

29 *Rubio, R.; Berne, R.M.; Bockman, E.L., and Curnish, R.R.:* Relationship between adenosine concentration and oxygen supply in rat brain. Am. J. Physiol. *228:* 1896–1902 (1975).

30 *Rubio, R.; Berne, R.M., and Dobson, J.G., jr.:* Sites of adenosine production in cardiac and skeletal muscle. Am. J. Physiol. *225:* 938–953 (1973).

31 *Rubio, R.; Berne, R.M., and Katori, M.:* Release of adenosine in reactive hyperemia of the dog heart. Am. J. Physiol. *216:* 56–62 (1969).

32 *Rubio, R.; Wiedmeier, V.T., and Berne, R.M.:* Nucleoside phosphorylase: localization and role in the myocardial distribution of purines. Am. J. Physiol. *222:* 550–555 (1972).

33 *Rubio, R.; Wiedmeier, V.T., and Berne, R.M.:* Relationship between coronary flow and adenosine production and release. J. mol. cell. Cardiol. *6:* 561–566 (1974).

34 *Schrader, J.; Berne, R.M., and Rubio, R.:* Uptake and metabolism of adenosine by human erythrocyte ghosts. Am. J. Physiol. *223:* 159–166 (1972).

35 *Schrader, J.; Rubio, R., and Berne, R.M.:* Inhibition of slow action potentials of guinea pig atrial muscle by adenosine: a possible effect on Ca^{2+} influx. J. mol. cell. Cardiol. *7:* 427–433 (1975).

36 *Scott, J.B.; Daugherty, R.M., jr.; Dabney, J.M., and Haddy, F.J.:* Role of chemical factors in regulation of flow through kidney, hindlimb, and heart. Am. J. Physiol. *208:* 813–824 (1965).

37 *Triner, L.; Nahas, G.G.; Vulliemox, Y.; Overweg, N.I.A.; Verosky, M.; Habif, D.V., and Ngai, S.H.:* Cyclic AMP and smooth muscle function. Ann. N.Y. Acad. Sci. *185:* 458–476 (1971).

38 *Wahl, M. and Kuschinsky, W.:* Dependency of the dilatatory action of adenosine on the perivascular H^+ and K^+ at pial arteries of cats. Proc. 8th Int. Symp. on Cerebral Blood Flow and Metabolism, Copenhagen 1977, pp. 11.10–11.11.

39 *Watkinson, W.P.; Foley, D.H.; Rubio, R., and Berne, R.M.:* Adenosine production by canine myocardium during stellate ganglion stimulation. Circulation *54:* suppl. II, p. 107 (1976).

40 *Watkinson, W.P.; Foley, D.H.; Rubio, R., and Berne, R.M.:* Adenosine production by myocardium with exercise in the unanesthetized dog. Fed. Proc. Fed. Am. Socs exp. Biol. *36:* 598 (1977).

41 *Wiedmeier, V.T.; Rubio, R., and Berne, R.M.:* Incorporation and turnover of adenosine-U-^{14}C in perfused guinea pig myocardium. Am. J. Physiol. *223:* 51–54 (1972).

42 *Wiedmeier, V.T.; Rubio, R., and Berne, R.M.:* Inosine incorporation into myocardial nucleotides. J. mol. cell. Cardiol. *4:* 445–452 (1972).

R.M. Berne, MD, Department of Physiology, University of Virginia Medical Center, Jordan Hall, Box 449, *Charlottesville, VA 22901* (USA)

Mechanisms of Vasodilatation. Satellite Symp., 27th Int. Congr. Physiol. Sci., Wilrijk 1977, pp. 222–226 (Karger, Basel 1978)

Adenosine and Adenine Nucleotides

Action on Vascular Smooth Muscle

Raymond Verhaeghe

Department of Medical Research, University of Leuven, Leuven

Adenosine and the adenine nucleotides have been known as powerful vasodilators in many vascular beds for almost half a century and their possible role in the physiological regulation of myocardial and skeletal muscle flow has been under intensive investigation (*Berne,* this volume). The mechanism of their local action remains largely unsolved. Although part of the vasodilatation caused *in vivo* by the adenine compounds may be due to an inhibitory effect on sympathetic adrenergic neurotransmission (29), these substances must have a major action on the smooth muscle cells. The present discussion focuses on those direct effects of adenosine and the adenine nucleotides.

Adenosine depresses contractions induced in a variety of isolated blood vessels by norepinephrine, high concentrations of acetylcholine, potassium, barium and prostaglandin $F_{2\alpha}$ (9, 16, 22, 24, 25, 28, 30). The relaxation is rapid in onset and is completely reversible. The concentration required to depress contractions of isolated preparations of larger blood vessels is usually higher than that needed to induce vasodilatation in the intact animal. Adenosine causes a greater relaxation of strips from smaller than from larger coronary vessels (16, 22). In the dog forelimb, adenine nucleotides cause the largest decrease in resistance in the small vessels (7). Thus, the adenine compounds appear to act primarily on smaller resistance vessels.

Adenosine causes only relaxation of isolated blood vessels; in addition, higher concentrations of adenosine diphosphate (ADP) and adenosine triphosphate (ATP) have an excitatory effect which can be demonstrated in resting as well as in contracted vessels (8, 10, 27, 28). The excitation results from an increased influx of extracellular calcium which may be due to an enhanced calcium permeability or to chelation of magnesium (6).

1. Is there an Adenosine Receptor on the Vascular Smooth Muscle Cell Membrane?

Blockade of α-adrenergic, β-adrenergic or muscarinic receptors and inhibition of prostaglandin synthesis do not influence the relaxant effect of adenosine (28, 30). Theophylline, alone or in combination with ethylenediamine (aminophylline), antagonizes the effect of adenosine on the coronary, renal or cerebral circulation in the dog (1, 17, 19); in canine coronary and saphenous vein strips, it significantly attenuates the relaxing effect of the nucleoside (25, 28). The mechanism of this antagonism is not completely understood. Aminophylline does not interfere with the release of adenosine from or its uptake into the myocardial cells; by contrast, inhibition of the latter explains the potentiating effect of dipyridamole on the vasodilatatory action of adenosine (15). Kinetic analysis indicates that the antagonism between theophylline and the nucleoside is competitive (4, 19, 25). When adenosine and theophylline are linked covalently to an oligosaccharide to obtain compounds too large to penetrate cell membranes, a coronary vasodilatation can still be produced by intracoronary infusion of the adenosine-oligosaccharide complex, and this vasodilatation is inhibited by the theophylline-oligosaccharide complex (18). Adenosine is either quickly deaminated or rephosphorylated inside the cell and the nucleoside cannot be detected intracellularly unless the extracellular concentration is extremely high (15, 23). ADP and ATP which do not easily permeate cell membranes also cause relaxation of isolated vessels; since this effect is immediate, it implies that the action of the nucleotides occurs before they are degraded to adenosine (4, 28). These observations all support the concept that the vasodilatation by adenine compounds is triggered by an initial action at the cell membrane level. The existence of a specific receptor for adenosine and adenine nucleotides still has to be proven.

2. Influence on Calcium Metabolism

Even if the initial action of adenine compounds is to bind with the cell membrane, ultimately they must either decrease the intracellular concentration of free activator calcium or inhibit the calcium-activated contractile process. Adenosine reduces the rate of contraction in response to increased concentrations of calcium in depolarized strips from small coronary arteries (22). In hog carotid artery, the nucleoside depresses contractions due to complete depolarization of the cell membrane; it inhibits responses which depend on extracellular calcium (potassium) more than those which depend on both extracellular and intracellular calcium (norepinephrine) (9). Excess calcium abolishes the adenosine-induced relaxation of depolarized coronary artery strips (26). These observations all point to a possible inhibitory effect of adenosine on influx of extracellular calcium. However, the latter interpretation is hard to reconcile with observations in cutaneous and mesenteric veins. In the saphenous vein of the dog, adenosine still depresses contractions induced by norepinephrine in calcium-free solution or after blockade of calcium influx by verapamil; adenosine does not inhibit the spontaneous contractions of longitudinal strips of the same species' portal mesenteric vein although all spontaneous activity ceases when extracellular calcium is removed or verapamil is added (28). Clearly, studies on calcium fluxes are needed to clarify this point.

3. Effect on Cyclic Nucleotide Levels

Adenosine increases the cyclic AMP levels in a variety of tissues and influences the activity of adenylcyclase and phosphodiesterase obtained from different sources (2, 3, 5, 11, 12, 14, 20, 21). Since levels of cyclic 3′,5′-nucleotides could be related to smooth muscle tone, changes in these levels might play a role in the vasodilating properties of adenosine. However, direct measurement of cyclic nucleotide levels in vascular tissue do not confirm this view. No change in cyclic AMP was observed in resting arterial strips exposed to adenosine unless extremely high concentrations were used (9). Acetylcholine contractions in venous strips are accompanied by an increase in cyclic GMP but not in cyclic AMP; these levels remain unaltered during relaxation with adenosine (28). Similarly, the nucleoside does not affect the level of cyclic AMP in intestinal smooth muscle (13).

4. Conclusion

Despite a considerable amount of data, our knowledge of the mechanism of action of the adenine compounds is hampered by the gap in our understanding of relaxation processes in general. Their initial action seems to be located at the cell membrane where they may activate specific receptors. The subsequent intracellular events are a matter of speculation. Whether calcium influx is decreased or its efflux or sequestration enhanced is debatable. One point is clear: cyclic nucleotides are not involved in the relaxation caused by adenine compounds.

References

1 *Afonso, S.:* Inhibition of coronary vasodilating action of dipyridamole and adenosine by aminophylline in the dog. Circulation Res. *26:* 743–752 (1970).

2 *Blume, A.J.; Dalton, C., and Sheppard, H.:* Adenosine-mediated elevation of cyclic

3′,5′-adenosine monophosphate concentrations in cultured mouse neuroblastoma cells. Proc. natn. Acad. Sci. USA *70:* 3099–3102 (1973).

3 *Blume, A.J. and Foster, C.J.:* Mouse neuroblastoma adenylate cyclase. Adenosine and adenine analogues as potent affectors of adenylate cyclase activity. J. biol. Chem. *250:* 5003–5008 (1975).

4 *Bünger, R.; Haddy, F.J., and Gerlach, E.:* Coronary responses to dilating substances and competitive inhibition by theophylline in the isolated perfused guinea pig heart. Pflügers Arch. *358:* 213–224 (1975).

5 *Clark, R.B.; Gross, R.; Su, Y.F., and Perkins, J.P.:* Regulation of adenosine 3′,5′-monophosphate content in human astrocytoma cells by adenosine and the adenine nucleotides. J. biol. Chem. *249:* 5296–5303 (1974).

6 *Daniel, E.E. and Irwin, J.:* On the mechanism whereby certain nucleotides produce contractions of smooth muscle. Can. J. Physiol. Pharmacol. *43:* 89–109 (1965).

7 *Fröhlich, E.D.:* Local effect of adenosine mono-, di- and triphosphate vessel resistance. Am. J. Physiol. *204:* 28–30 (1963).

8 *Furchgott, R.E.:* Metabolic factors that influence contractility of vascular smooth muscle. Bull. N.Y. Acad. Med. *42:* 996–1006 (1965).

9 *Herlihy, J.T.; Bockman, E.L.; Berne, R.M., and Rubio, R.:* Adenosine relaxation of isolated vascular smooth muscle. Am. J. Physiol. *230:* 1239–1243 (1976).

10 *Hrdina, P.; Bonaccorsi, A., and Garattini, S.:* Pharmacological studies on isolated and perfused rat renal arteries. Eur. J. Pharmacol. *1:* 99–108 (1967).

11 *Katz, S. and Tenenhouse, A.:* The relation of adenylcyclase to the activity of other ATP utilizing enzymes and phosphodiesterase in preparations of rat brain; mechanism of stimulation of cyclic AMP accumulation by NaF. Br. J. Pharmacol. *48:* 505–515 (1973).

12 *McKenzie, S.G. and Bär, H.P.:* On the mechanism of adenylcyclase inhibition by adenosine. Can. J. Physiol. Pharmacol. *51:* 190–196 (1973).

13 *McKenzie, S.G.; Frew, R., and Bär, H.P.:* Effects of adenosine and related compounds on adenylate cyclase and cyclic AMP levels in smooth muscle. Eur. J. Pharmacol. *41:* 193–203 (1977).

14 *Mills, D.C.B. and Smith, J.B.:* The influence on platelet agregation of drugs that affect the accumulation of adenosine 3′,5′-cyclic monophosphate. Biochem. J. *121:* 185–196 (1971).

15 *Mustafa, S.J.; Berne, R.M., and Rubio, R.:* Adenosine metabolism in cultured chick-embryo heart cells. Am. J. Physiol. *228:* 1474–1478 (1975).

16 *Norton, J.M.; Gellai, M., and Detar, R.:* Effects of adenosine on isolated coronary vascular smooth muscle. Pflügers Arch. ges. Physiol. *335:* 279–286 (1972).

17 *Oberdörster, G.; Lang, R., and Zimmer, R.:* Influence of adenosine and lowered cerebral blood flow on the cerebrovascular effects of theophylline. Eur. J. Pharmacol. *30:* 197–204 (1975).

18 *Olsson, R.A.; Davis, C.J.; Khouri, E.M., and Patterson, R.E.:* Evidence for an adenosine receptor on the surface of dog coronary arteries. Circulation Res. *39:* 93–98 (1976).

19 *Osswald, H.:* Renal effects of adenosine and their inhibition by theophylline in dogs. Arch. Pharmakol. *288:* 79–86 (1975).

20 *Palmer, G.C.:* Characteristics of the hormonal induced cyclic adenosine 3′,5′-monophosphate response in the rat and guinea-pig lung *in vitro*. Biochim. biophys. Acta *252:* 561–566 (1971).

21 *Sattin, A. and Rall, T.W.:* The effect of adenosine and adenine nucleotides on the cyclic adenosine 3′,5′-phosphate content of guinea pig cerebral cortex slices. Mol. Pharmacol. *6:* 13–23 (1970).

22 *Schnaar, R.L. and Sparks, H.V.:* Response of large and small coronary arteries to nitroglycerin, $NaNO_2$ and adenosine. Am. J. Physiol. *223:* 223–228 (1972).

23 *Schräder, J.; Berne, R.M., and Rubio, R.:* Uptake and metabolism of adenosine by human erythrocyte cells. Am. J. Physiol. *223:* 159–166 (1972).

24 *Toda, N.:* The action of vasodilating drugs on isolated basilar, coronary and mesenteric arteries of the dog. J. Pharmac. exp. Ther. *191:* 139–146 (1974).

25 *Toda, N.; Hojo, M.; Sakae, K., and Usui, H.:* Comparison of the relaxing effect of dopamine with that of adenosine, isoproterenol and acetylcholine in isolated canine coronary arteries. Blood Vessels *12:* 290–301 (1975).

26 *Turnheim, K.; Pittner, H.; Kolassa, N., and Kraupp, O.:* Relaxation of coronary artery strips by adenosine and acidosis. Eur. J. Pharmacol. *41:* 217–220 (1977).

27 *Vanhoutte, P.M. and Shepherd, J.T.:* Effect of temperature on reactivity of isolated cutaneous veins of the dog. Am. J. Physiol. *218:* 187–190 (1970).

28 *Verhaeghe, R.H.:* Action of adenosine and adenine nucleotides on dogs' isolated veins. Am. J. Physiol. *233:* 114–121 (1977).

29 *Verhaeghe, R.H.; Vanhoutte, P.M., and Shepherd, J.T.:* Inhibition of sympathetic neurotransmission in canine blood vessels by adenosine and adenine nucleotides. Circulation Res. *40:* 208–215 (1977).

30 *Walter, P. und Bassenge, E.:* Wirkung von ATP, A-3',5'-MP, Adenosin und Dipyridamol an Streifenpräparaten der A. coronaria, A. renalis und der V. portae. Pflügers Arch. ges. Physiol. *299:* 52–65 (1968).

R. Verhaeghe, MD, Department of Medical Research, University of Leuven, *Leuven* (Belgium)

Mechanisms of Vasodilatation. Satellite Symp., 27th Int. Congr. Physiol. Sci., Wilrijk 1977, pp. 227–240 (Karger, Basel 1978)

Withdrawal of Sympathetic Tone

F.M. Abboud, A.L. Mark, D.D. Heistad and P.G. Schmid

Cardiovascular Division, Department of Medicine and Department of Physiology, Cardiovascular Center, University of Iowa, Iowa City, Iowa

Neurogenic vasodilatation is an important circulatory response which may reduce arterial blood pressure, decrease cardiac afterload, and cause a redistribution of blood flow to specific organs. It can result from activation of sympathetic vasodilator fibers or from the passive withdrawal of sympathetic vasoconstrictor tone. Withdrawal of sympathetic tone occurs in several physiologic and pathophysiologic states. These will be reviewed.

Neurons in the rostral lateral areas of the medulla generate nerve activity along sympathetic adrenergic fibers and modulate vascular tone. The medullary neurons are constantly barraged by afferent impulses that originate in sensory receptor areas in the arterial system, the cardiac and pulmonary regions, the chemoreceptors, in skeletal muscle, and in areas of the central nervous system – particularly the hypothalamus and the cerebellum. Each type of afferent impulse has a selective influence on the medullary centers, and in many clinical situations afferent impulses originating from different receptor areas converge simultaneously on the medullary centers. Afferent impulses originating in arterial and cardiopulmonary baroreceptors exert primarily a restraining influence on the medullary excitatory or pressor centers, causing vasodilatation and bradycardia.

1. Arterial Baroreceptors

Activation of mechanoreceptors or stretch receptors located in the aorta and the carotid sinus region causes a non-uniform response in animals and in man. For example, in the dog electrical stimulation of the carotid sinus nerve causes a much greater vasodilatation in the perfused gracilis muscle than in the perfused paw in the same animal (2a). In the cat, *Little et al.* (26) showed that vasodilator responses in muscle and kidney during elevation of carotid sinus pressure were significant and equivalent, but the reflex bradycardia was rela-

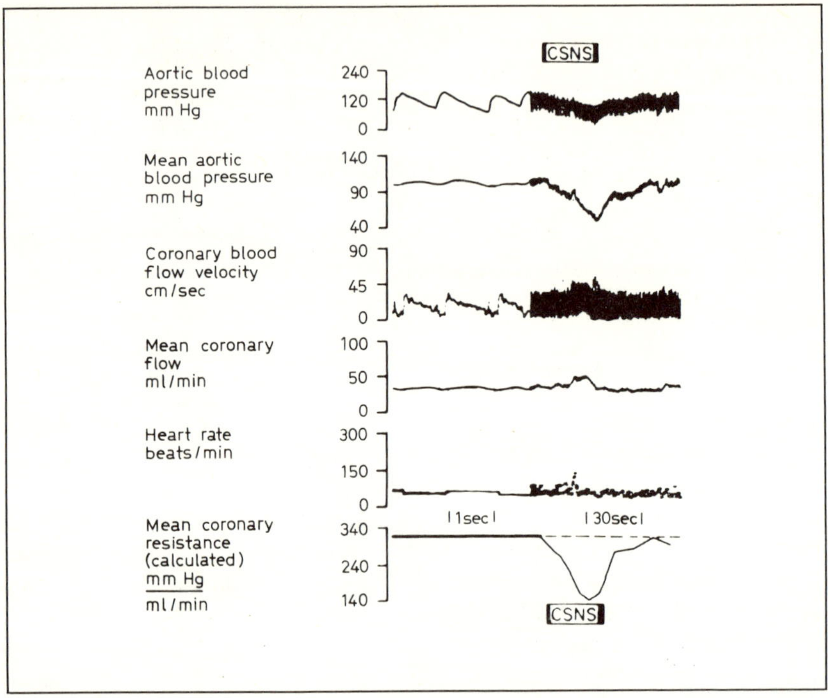

Fig. 1. Effects of carotid sinus nerve stimulation on the coronary circulation of conscious dogs. Despite the hypotension and decreased myocardial work and oxygen demand, there was a coronary vasodilator response caused in part by withdrawal of sympathetic vasomotor tone to the coronary vessels. Reprinted by permission from *Vatner et al.* (38).

tively less intense. We have shown, in the dog coronary, that reflex vasodilatation occurs during carotid nerve stimulation and that it is caused in part by withdrawal of sympathetic tone as well as by activation of vasodilator vagal fibers supplying the coronary vessels (14). Coronary vasodilatation was also demonstrated by *Vatner et al.* (38) during carotid sinus nerve stimulation in the conscious dog (fig. 1). The splanchnic circulation is particularly sensitive to changes in baroreceptor activity; both the splanchnic veins and the splanchnic arteriolar resistance vessels respond vigorously to changes in carotid sinus pressure (7).

In man, carotid sinus nerve stimulation causes reflex bradycardia and hypotension (11). We have stretched the carotid sinus mechanoreceptors by applying negative pressure to the neck of man to simulate hypertension and have noted that reflex bradycardia and hypotension are induced without decreasing vascular resistance of the forearm (3). On the other hand, the withdrawal of sympathetic

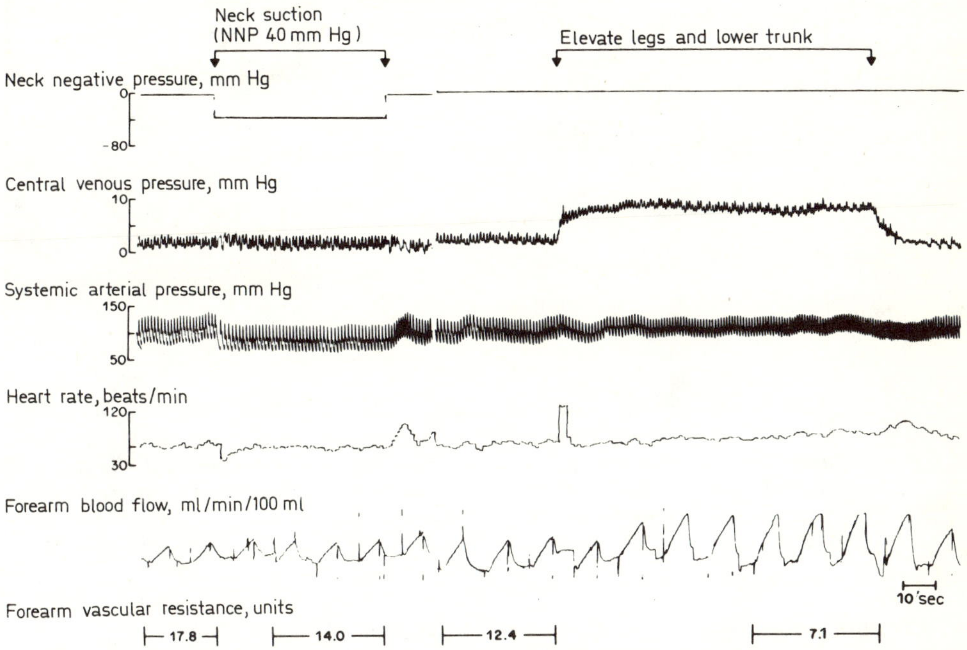

Fig. 2. Comparison of vasodilator responses to stretch of the carotid baroreceptors (neck suction) and the cardiopulmonary baroreceptors (elevation of the legs). Stretch of the carotid baroreceptors causes bradycardia, hypotension, and minimal vasodilation without actual increase in forearm flow, whereas elevation of the legs and stretch of the cardiopulmonary baroreceptors causes a marked increase in forearm blood flow, with a significant reduction in forearm vascular resistance and a slight tachycardia reminiscent of the Bainbridge reflex.

vasoconstrictor tone from forearm vessels can be demonstrated during elevation of the legs, a maneuver that increases central venous pressure and activates cardiopulmonary mechanoreceptors (fig. 2) (3).

Recent electrophysiologic studies on the organization of the central vasopressor pathways by *Gebber et al.* (12) indicate that there are two distinct systems of 'vasopressor pathways': one has a short latency and is rapidly conducting; the other has a longer latency and is slowly conducting. Activation of the baroreceptor reflex blocks the more slowly conducting vasoconstrictor pathways without interfering with the more rapidly conducting system. An exciting possible human corollary of the findings by *Gebber et al.* (12) is our observation that 'carotid neck suction' in man (which activates carotid baroreceptors) prevents the reflex tachycardia and reflex splanchnic vasoconstriction

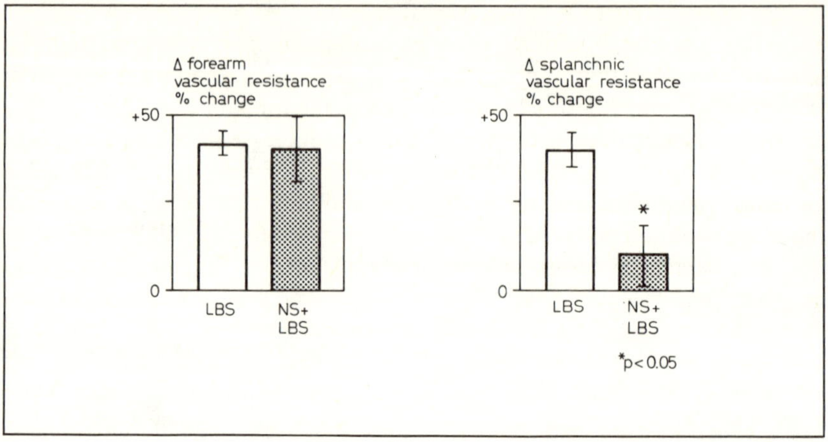

Fig. 3. Effect of carotid sinus reflex on splanchnic vascular resistance. The left panel shows that the simultaneous application of lower body negative pressure (LBS) and neck suction (NS) does not modify the reflex vasoconstrictor response in the forearm, whereas the neck suction inhibits the reflex vasoconstrictor response in the splanchnic bed (right panel).

induced by pooling of blood in the lower extremities without affecting the reflex vasoconstrictor response in the forearm (fig. 3). These findings also suggest that in man, the carotid baroreceptor afferents may be preferentially distributed to splanchnic rather than forearm vasomotor neurons.

2. Cardiopulmonary Receptors

Cardiac non-medullated afferents (C fibers) transmit impulses initiated by the mechanical or chemical activation of cardiac receptors located primarily in the left ventricle and left atrium.

Atrial receptors. The work of *Karim et al.* (22) indicates that stretch of the left atrial receptors may cause an increase in cardiac sympathetic activity, but a selective reduction in sympathetic activity to the kidney; others have found no significant alteration in sympathetic efferent activity to the hind limb. The combination of tachycardia and renal vasodilatation during left atrial stretch might be an effective mechanism by which intravascular volume is reduced and the ventricular volume becomes smaller in response to heart failure (24).

In man, we observed responses to pooling of blood in the lower extremities with low levels of lower body negative pressure. The amount of pooling which was sufficient to decrease right atrial pressure, without altering systemic arterial pressure, triggers reflex vasoconstriction in the forearm without a significant change in heart rate (42). Since arterial pressure does not change, we suggest that reduction of stretch of low pressure cardiopulmonary receptors removes the sympathetic inhibition imposed by the stretch of these receptors and allows the activation of sympathetic vasoconstriction in the forearm.

Renal sympathetic activity may cause an increase in renin release as well as an increase in sodium reabsorption independent of renin release. The inhibition of renal sympathetic activity by 'stretch' and activation of atrial receptors may therefore modulate renin release and sodium excretion (27).

Ventricular receptors. Electrical stimulation of non-medullated vagal afferent C fibers from the heart causes a fall in blood pressure, slowing of the heart rate and vasodilatation (32). A similar response is observed when cardiac receptors are excited with chemicals, such as veratrum alkaloids or nicotine, or with mechanical stimulation when pressure is elevated in isolated cardiac chambers (29). This characteristic cardiocirculatory inhibition is often referred to as the Bezold-Jarisch reflex. The reflex has been ascribed by *Dawes and Comroe* to a 'coronary chemoreflex' rather than to a coronary chemoreceptor reflex because of the lack of convincing evidence that there are chemoreceptors in the coronary arteries. Recent observations by our group (28) indicate that reflex vasodilatation can be triggered by intracoronary injection of nicotine, but not by perfusion of the coronaries with hypoxemic, hypercapnic, and acidotic blood. These observations indicate that inhibitory chemoreflexes can be demonstrated, but the physiologic role of receptors that mediate this chemical reflex does not appear to be one of adjustment to changes in oxygen and carbon dioxide tension of blood. It is possible that they may be activated by distortion of the ventricles.

The studies of *Eckberg et al.* (10) in our laboratories suggest that such a coronary chemoreflex might be triggered in humans during intracoronary injections of contrast medium for coronary arteriography and may represent a clinical counterpart to the Bezold-Jarisch reflex.

Pulmonary receptors. Afferent impulses from the lung also mediate sympathetic restraint. A reflex from the lung affecting systemic blood vessels was demonstrated by *Brodie and Russell* (6) in 1900 and has been more thoroughly analyzed recently by *Daly and Robinson* (8). Inflation of the lungs will activate pulmonary stretch receptors and inhibit the vasomotor center, causing reflex vasodilatation (34).

3. Arterial Chemoreceptors

Afferent impulses are triggered from peripheral arterial chemoreceptors which are sensitive to reductions in arterial PO_2, increases in PCO_2 and decreases in pH. These receptors may also be activated by nicotine or cyanide. Although the predominant responses are sympathetic vasoconstriction in skeletal muscle

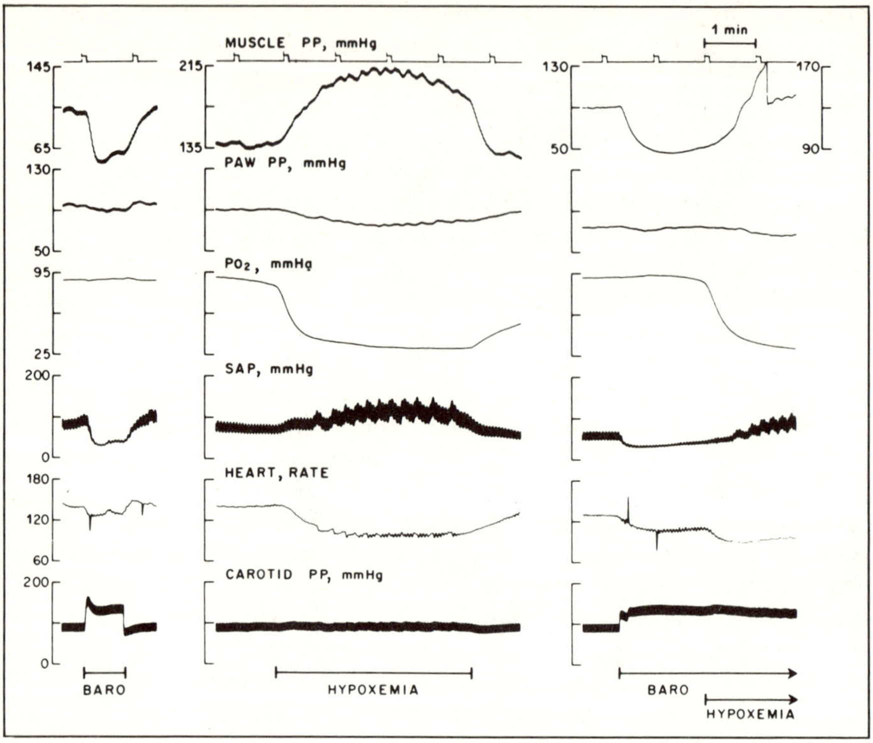

Fig. 4. Effect of stimulation of carotid chemoreceptors by perfusing hypoxemic blood at constant pressure into both carotids (middle panels). The decrease in arterial PO_2 was associated with reflex vasoconstriction in muscle (PP, perfusion pressure at constant flow), vasodilatation in the hindpaw, and bradycardia. The vasoconstrictor response in muscle will override and reverse the dilator effect of carotid hypertension (right panel). From *Abboud et al.* (2b).

and vagal bradycardia, we have observed vasodilatation in the paw (fig. 4) and in coronary vessels (20), and *Pelletier and Shepherd* (35) noted inhibition of sympathetic tone to saphenous veins and venodilatation. These differential responses may contribute to oxygen conservation and perfusion of vital organs.

4. Interference with Reflex Constrictor Responses

Certain stimuli may contribute to a vasodilator response by preventing a compensatory vasoconstriction and by interfering with sympathetic vasoconstrictor tone. In addition to the well-known α-adrenergic blocking drugs, certain

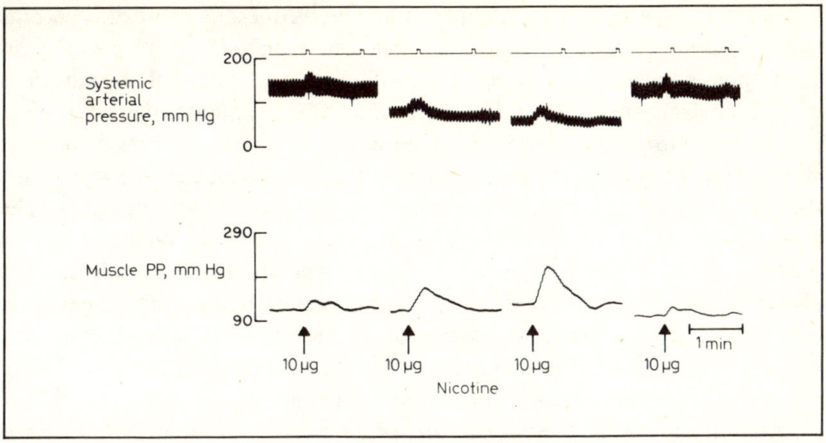

Fig. 5. Interaction between baro- and chemoreceptors. In this experiment, the chemoreceptors were activated by injection of nicotine (10 µg) into one carotid. The reflex vasoconstrictor response in the perfused gracilis muscle was augmented when systemic hypotension was present. Conversely, in other experiments, the elevation of carotid sinus perfusion pressure in one carotid inhibited the reflex chemoreceptor response triggered by injection of nicotine in the opposite carotid, thus supporting the concept of a central nervous system interaction between these two afferent impulses. From *Heistad et al.* (18).

interventions may interfere with adrenergic transmission either at a central or peripheral neural level, at the level of the receptor or at the level of the smooth muscle.

Hypoxemia. Acute hypoxemia in normal subjects inhibits neurogenic reflexes and reduces reflex vasoconstrictor responses to pooling of blood in the lower extremities (15). In animals, severe hypoxemia does not prevent vasoconstrictor responses to norepinephrine and angiotensin in skeletal muscle until the muscle is made to contract (16). Contraction alone or severe hypoxemia alone do not inhibit vasoconstriction in skeletal muscle. In the heart, however, sympathetic nerve stimulation and norepinephrine cause coronary vasodilatation which is reversed to coronary vasoconstriction after blockade of the myocardial responses to sympathetic stimulation by giving practolol, a β_1-blocker. It is possible that substances released by the vigorously contracting myocardium during sympathetic stimulation inhibit the vasoconstrictor effect of the sympathetic adrenergic fibers on the coronary vessels.

Chronic hypoxemia in patients with chronic lung disease inhibits the reflex vasoconstrictor response in the forearm during lower body negative pressure. The acute administration of oxygen to correct the arterial hypoxemia in these patients restores reflex vasoconstriction (17).

Inhibitory arterial baroreceptor influence on the chemoreceptor excitatory reflex. Lesions of the paramedian reticular nucleus abolish or reverse the depressor response to electrical stimulation of myelinated fibers of the carotid sinus nerve (which are predominantly baroreceptor fibers) and attenuate the depressor response to carotid sinus stretch. These lesions augment the pressor response to chemoreceptor stimulation by lobeline. Thus, the baroreceptor projections into the paramedian reticular nucleus appear to buffer the pressor response to chemoreceptor stimulation, making a strong case for a true synaptic interaction between the two afferent inputs. These findings provide electrophysiologic explanation for our observation (18) of an interaction between the chemoreceptor and baroreceptor reflexes: hypertension or elevation of carotid sinus pressure inhibits or buffers the vasoconstrictor response to stimulation of chemoreceptors; conversely, hypotension augments the chemoreceptor reflex (fig. 5).

Inhibitory cardiac-receptor influence on the arterial baroreceptor excitatory response. Acute transient obstruction of the left ventricular outflow causes marked reflex vasodilatation in the perfused gracilis muscle which is abolished by bilateral vagotomy (29). This reflex dilatation occurs at a time when there is a significant fall in systemic arterial pressure which would ordinarily cause reflex vasoconstriction through activation of the arterial baroreceptor reflex. This study indicates that the inhibitory cardiac receptors, when stretched, may override the sympathetic excitatory response to systemic hypotension.

Conversely, recent studies in our laboratories indicate that bilateral vagotomy tends to increase the carotid baroreceptor reflex response to hypotension, suggesting that under resting conditions, vagal afferents exert an inhibitory influence on sympathetic activity.

Effect of heat on reflex vasoconstriction. Total body heating in resting man causes vasoconstriction in skeletal muscle, renal and splanchnic circulations and vasodilatation in arterioles and venules in the skin. When blood flow to the skin is increased by the heat load, cutaneous vessels become minimally responsive to reflex stimuli (36). In addition, the central nervous system interaction between thermal receptors and cardiopulmonary or carotid baroreceptors may predominantly suppress the sympathetic efferent activity to cutaneous beds (19). For example, heating the skin of one arm blocks the reflex vasoconstrictor response in the contralateral hand, but not the forearm, during lower body negative pressure. This inhibition of reflex vasoconstriction favors heat dissipation but attenuates adjustments to other circulatory stresses such as hypovolemia.

5. Interference with Withdrawal of Sympathetic Tone

A rise in systemic arterial pressure or an increase in cardiac size activate arterial and cardiac baroreceptor reflexes, causing withdrawal of sympathetic tone. These reflexes may be inhibited, however, by impulses originating in the hypothalamus, in somatic receptors during exercise, or during activation of chemoreceptors by hypoxemia.

During exercise and emotional stress, arterial blood pressure increases, but despite the rise in pressure there is a sustained tachycardia and vasoconstriction in several vascular beds. This suggests that the arterial baroreceptor reflex is inhibited by afferent impulses originating either in the hypothalamus or in the somatic receptors in skeletal muscle. Thus, the failure of withdrawal of sympathetic tone in these circumstances allows the maintenance of a high output and a high arterial pressure despite the vasodilatation which occurs in skeletal muscle in these two states.

Stimulation of the fastigial nucleus of the cerebellum causes a rise in pressure and responses analogous to the baroreceptor reflex with vasoconstriction and tachycardia (25). *Nathan* (31) has shown that there is an inhibitory interaction between the stimuli originating in the fastigial nucleus and those originating in the arterial baroreceptors with respect to the final efferent output as recorded over the splanchnic nerve.

In a study on the central integration of excitatory chemoreceptor influences and inhibitory arterial and cardiac baroreceptors, *Wennergren et al.* (41) demonstrated that the reflex vasodilator effects elicited via vagal cardiac afferents are effectively suppressed by a concomitant chemoreceptor stimulation. This chemoreceptor suppression of reflex vasodilatation originating from cardiac receptors may be of importance in sustaining vasoconstriction in hypoxic states.

6. Withdrawal of Sympathetic Tone in Pathophysiologic States

Orthostatic hypotension. A dramatic fall in arterial pressure occurs during upright tilt in patients who have orthostatic hypotension on the basis of autonomic denervation, such as patients with peripheral neuropathy, diabetes, and Shy-Drager syndrome. This is caused by the absence of reflex vasoconstriction (1, 4, 23). In patients with idiopathic orthostatic hypotension or diabetic neuropathy, the failure of vasoconstriction is probably related to the loss of peripheral adrenergic innervation of blood vessels (23). On the other hand, in patients with Shy-Drager syndrome, failure of the reflex response is probably related to a central nervous system defect (9). In patients with peripheral autonomic neuropathy, the Valsalva maneuver or apneic facial immersion in water results in vasodilatation of forearm vessels (1, 4, 23). Thus, instead of the vasoconstrictor response seen during these interventions in normal subjects, these patients had a vasodilator response in the extremities which is detrimental in that it tends to accentuate the hypotension and redistributes cardiac output towards skeletal muscle in the extremities.

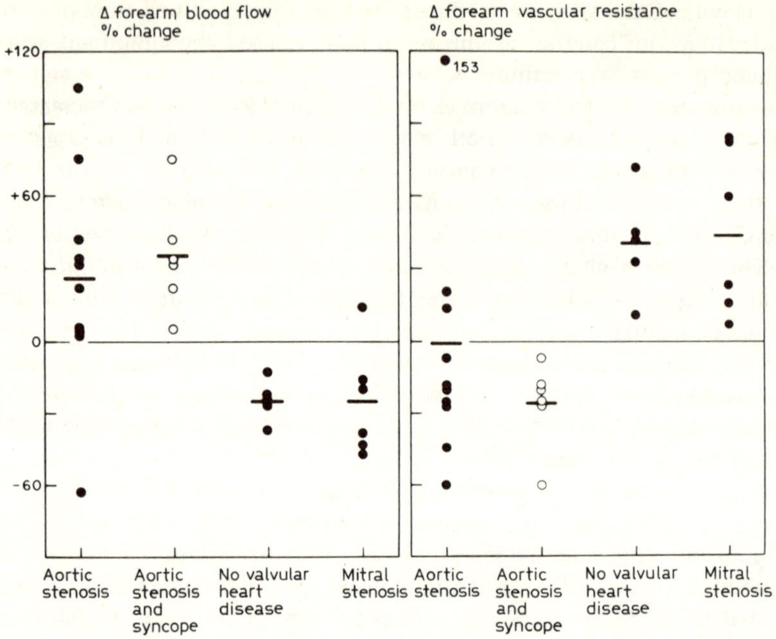

Fig. 6. Effect of leg exercise in aortic stenosis. The reflex vasoconstrictor response seen in the normal subjects and in patients with mitral stenosis is replaced by a vasodilator response in aortic stenosis. From *Mark et al.* (30).

Congestive heart failure. Baroreceptor reflexes are impaired in heart failure; there is less bradycardia during a rise in arterial pressure (21). The impaired reflex may contribute to increased peripheral resistance in heart failure. It has also been suggested that the exaggerated renal sympathetic tone that contributes to sodium retention in heart failure may in part be caused by a failure of inhibitory impulses originating from the stretch of the pulmonary-left atrial region. *Greenberg et al.* (13) suggest that there is an impaired atrial receptor response in dogs with heart failure. One of the actions of digitalis might be to restore responsiveness of atrial receptors to stretch and decrease the sympathetic efferent activities to the kidney favoring natriuresis.

Aortic stenosis. Stretch of left ventricular receptors in patients with aortic stenosis who are subjected to moderate or heavy exercise might contribute to the syncope by preventing the reflex sympathetic vasoconstriction that occurs in the non-active organs and muscles during exercise. This hypothesis is supported by the demonstration that during leg exercise patients with aortic stenosis have an increase in forearm blood flow or vasodilatation, whereas normal subjects and patients with mitral stenosis have a decrease in forearm blood flow or vasoconstriction (fig. 6) (30). The vasodilatation may lead to hypotension and syncope. Three of the patients with aortic stenosis, a history of syncope and forearm vasodilator response to leg exercise were studied after aortic valve replacement. In all three, normal vasoconstriction occurred in the forearm during leg exercise after correction of the aortic stenosis.

Acute myocardial infarction. After acute myocardial infarction, hypotension might be related in part to activation of ventricular mechanoreceptors in the dyskinetic ischemic myocardium which would inhibit sympathetic activity. *Toubes and Brody* (37) demonstrated that following acute myocardial infarction in dogs, peripheral resistance increases only slightly, despite hypotension; but when bilateral vagotomy is performed, there is a marked rise in peripheral resistance and arterial pressure. The findings strongly suggest the presence of an inhibitory cardiac afferent activity mediated through the vagi and originating in the ischemic myocardium.

Bradyarrhythmias are more commonly associated with inferior than anterior myocardial infarction. This phenomenon may be related to a preferential distribution of inhibitory cardiac sensory receptors in the region of the inferior wall of the left ventricle. We demonstrated in dogs that inhibitory cardiac reflexes are activated by injection of nicotine into the circumflex (Cx) and left anterior descending (LAD) coronary arteries. The magnitude of the inhibitory response as reflected in reflex bradycardia and vasodilatation is severalfold greater when nicotine is injected into the Cx than when it is injected into the LAD (39). Cessation of perfusion of the Cx causes bradycardia and more arterial hypotension than cessation of perfusion of the LAD. These results suggest a preferential distribution of inhibitory cardiac receptors in the region of the inferior wall of the canine left ventricle perfused by the circumflex. The results might explain the greater incidence of bradyarrhythmia and hypotension following inferior infarction as contrasted with antero-septal infarction in man (40).

Increased activity of left ventricular receptors during hemorrhage or during occlusion of caval veins. In a recent study, *Öberg and Thoren* (33) showed that a rapid and severe hemorrhage in cats sometimes produces reflex bradycardia similar to that seen in the so-called vasovagal syncope reaction in man. The reflex slowing of the heart is inhibited by section of afferent pathways in the cardiac nerves. Impulse activity is increased in vagal fibers from the left ventricular inhibitory receptors. The vigorous contraction of the ventricle and its distortion during rapid hemorrhage may be a factor in triggering this reflex and may play a role in the clinical syndrome of vasovagal syncope. The protective effect of bradycardia would permit adequate left ventricular filling time.

Hypertension and hyperkinetic states. In hypertension, the baroreceptor reflex is impaired. The threshold of the arterial baroreceptors is increased so that high levels of arterial pressure are necessary to trigger reflex withdrawal of sympathetic tone. This may explain why during exercise the pressor response is much greater in the hypertensive individual. The rise in arterial pressure triggered by activation of somatic efferent receptors during exercise is not buffered properly by the carotid baroreflex and the failure of withdrawal of sympathetic tone becomes much more evident in these patients.

It is known that in some patients with labile hypertension, peripheral resistance may not be increased and the elevated blood pressure may be caused by high cardiac output. In this condition, however, the regional circulations may show abnormalities (5). There may be an increase in skeletal muscle flow and a decrease in splanchnic, cutaneous, and possibly renal blood flow. The high cardiac output and the altered distribution of blood flow are reminiscent of the circulatory adjustment seen in the hypothalamic response during the defense reaction. A similar circulatory adjustment may take place in early labile hypertension and a hypothalamic stimulus may be a factor in maintaining the high output and low resistance. The hypothalamic stimulus may inhibit the baroreceptor reflex and would prevent withdrawal of sympathetic tone and the fall in blood pressure to normal levels through the baroreflex mechanism. Although speculative, these are potential contributing factors to the pathogenesis of hypertension.

References

1 *Abboud, F.M. and Eckstein, J.W.:* Active reflex vasodilatation in man. Fed. Proc. Fed. Am. Socs exp. Biol. *25:* 1611 (1966).

2a *Abboud, F.M.:* Control of the various components of the peripheral vasculature. Fed. Proc. Fed. Am. Socs exp. Biol. *31:* 1226–1239 (1972).

2b *Abboud, F.M.; Heistad, D.D.; Mark, A.L., and Schmid, P.G.:* Differential responses of the coronary circulation and other vascular beds to chemoreceptor stimulation; in *Purves* The peripheral arterial chemoreceptors, pp. 427–447 (Cambridge University Press, London 1975).

3 *Abboud, F.M.; Mark, A.L.; Heistad, D.D., and Schmid, P.G.:* Selectivity of autonomic control of the peripheral circulation in man. Trans. Am. Clin. Climatol. Ass. *86:* 184 (1974).

4 *Bennett, T.; Evans, D.; Hampton, J.R., and Hesking, J.D.:* Abnormal cardiovascular reflexes in subjects with autonomic neuropathy. J. Physiol., Lond. *246:* 41P (1975).

5 *Brod, J.; Fencl, V.; Hejl, Z.; Iirka, J., and Ulrych, M.:* General and regional hemodynamic pattern underlying essential hypertension. Clin. Sci. *23:* 339 (1962).

6 *Brodie, T.G. and Russell, A.E.:* On reflex cardiac inhibition. J. Physiol., Lond. *26:* 92 (1900).

7 *Brooksby, G.A. and Donald, D.E.:* Dynamic changes in splanchnic blood flow and blood volume in dogs during activation of sympathetic nerves. Circulation Res. *29:* 227 (1971).

8 *Daly, M.DeB. and Robinson, B.H.:* An analysis of the reflex systemic vasodilator response elicited by lung inflation in the dog. J. Physiol., Lond. *195:* 387 (1968).

9 *Diamond, M.A.; Murray, R.H., and Schmid, P.G.:* Idiopathic postural hypotension: physiologic observations and report of a new mode of therapy. J. clin. Invest. *49:* 1341 (1970).

10 *Eckberg, D.L.; White, C.W.; Kioschos, J.M., and Abboud, F.M.:* Mechanisms mediating bradycardia during coronary arteriography. J. clin. Invest. *54:* 1455 (1974).

11 *Epstein, S.E.; Beiser, G.D.; Goldstein, R.E.; Stampfer, M.; Wechsler, A.S.; Glick, G., and Braunwald, E.:* Circulatory effects of electrical stimulation of the carotid sinus nerves in man. Circulation *40:* 269 (1969).

12 *Gebber, G.L.; Taylor, D.G., and Weaver, L.C.:* Electrophysiological studies on organization of central vasopressor pathways. Am. J. Physiol. *224:* 470 (1973).

13 *Greenberg, T.T.; Richmond, W.H.; Stocking, R.A.; Gupta, P.D.; Mechon, J.P., and Henry, J.P.:* Impaired atrial receptor responses in dogs with heart failure due to tricuspid insufficiency and pulmonary artery stenosis. Circulation Res. *32:* 424 (1973).

14 *Hackett, J.G.; Abboud, F.M.; Mark, A.L.; Schmid, P.G., and Heistad, D.D.:* Coronary vascular responses to stimulation of chemoreceptors and baroreceptors; evidence for reflex activation of vagal cholinergic innervation. Circulation Res. *31:* 8 (1972).

15 *Heistad, D.D. and Wheeler, R.C.:* Effect of acute hypoxia on vascular responsiveness in man. J. clin. Invest. *49:* 1252 (1970).

16 *Heistad, D.D.; Abboud, F.M.; Mark, A.L., and Schmid, P.G.:* Effect of hypoxemia on responses to norepinephrine and angiotensin in coronary and muscular vessels. J. Pharmac. exp. Ther. *193:* 941 (1975).

17 *Heistad, D.D.; Abboud, F.M.; Mark, A.L., and Schmid, P.G.:* Impaired cardiovascular reflexes in chronically hypoxic patients. J. clin. Invest. *51:* 331 (1972).

18 *Heistad, D.D.; Abboud, F.M.; Mark, A.L., and Schmid, P.G.:* Interaction of baroreceptor and chemoreceptor reflexes. J. clin. Invest. *53:* 1226 (1974).

19 *Heistad, D.D.; Abboud, F.M.; Mark, A.L., and Schmid, P.G.:* Interaction of thermal and baroreceptor reflexes in man. J. appl. Physiol. *35:* 581 (1973).

20 *Heistad, D.D.; Abboud, F.M.; Mark, A.L., and Schmid, P.G.:* Response of muscular and cutaneous vessels to physiologic stimulation of chemoreceptors. Proc. Soc. exp. Biol. Med. *148:* 198 (1975).

21 *Higgins, C.B.; Vatner, S.F.; Eckberg, D.L., and Braunwald, E.:* Alterations in the baroreceptor reflex in conscious dogs with heart failure. J. clin. Invest. *51:* 715 (1972).

22 *Karim, F.; Kidd, C.; Malpus, C.M., and Penna, P.E.:* The effects of stimulation of the left atrial receptors on sympathetic efferent nerve activity. J. Physiol., Lond. *227:* 243 (1972).

23 *Kontos, H.A.; Richardson, D.W., and Norvell, J.E.:* Norepinephrine depletion in idiopathic orthostatic hypotension. Ann. intern. Med. *82:* 336 (1975).

24 *Linden, R.J.:* Function of cardiac receptors. Circulation *48:* 463 (1973).

25 *Lisander, B. and Martner, J.:* Interaction between the fastigial pressor response and the baroreceptor reflex. Acta physiol. scand. *83:* 505 (1971).

26 *Little, R.; Wennergren, G., and Öberg, B.:* Aspects of the central integration of arterial baroreceptor and cardiac ventricular receptor reflexes in the cat. Acta physiol. scand. *93:* 85–96 (1975).

27 *Mancia, G.; Donald, D.E., and Shepherd, J.T.:* Inhibition of adrenergic outflow to peripheral blood vessels by vagal afferents from the cardiopulmonary region in the dog. Circulation Res. *22:* 713 (1973).

28 *Mark, A.L.; Abboud, F.M.; Heistad, D.D.; Schmid, P.G., and Johannsen, U.J.:* Evidence against the presence of ventricular chemoreceptors activated by hypoxia and hypercapnia. Am. J. Physiol. *227:* 178 (1974).

29 *Mark, A.L.; Abboud, F.M.; Schmid, P.G., and Heistad, D.D.:* Reflex vascular responses

to left ventricular outflow obstruction and activation of ventricular chemoreceptors in dogs. J. clin. Invest. *52:* 1147 (1973).

30 *Mark, A.L.; Kioschos, J.M.; Abboud, F.M.; Heistad, D.D., and Schmid, P.G.:* Abnormal vascular responses to exercise in patients with aortic stenosis. J. clin. Invest. *52:* 1138–1146 (1973).

31 *Nathan, M.A.:* Vasomotor projection of the nucleus fastigii to the medulla. Brain Res. *41:* 194 (1972).

32 *Öberg, B. and White, S.:* Circulatory effects of interruption and stimulation of cardiac vagal afferents. Acta physiol. scand. *80:* 383 (1970).

33 *Öberg, B. and Thoren, P.:* Increased acrivity in left ventricular receptors during hemorrhage or occlusion of caval veins in the cat. A possible cause of the vaso-vagal reaction. Acta physiol. scand. *35:* 164–173 (1972).

34 *Ott, N.T. and Shepherd, J.T.:* Vasodepressor reflex from lung inflation in the rabbit. Am. J. Physiol. *221:* 889 (1971).

35 *Pelletier, C.L. and Shepherd, J.T.:* Venous responses to stimulation of carotid chemoreceptors by hypoxia and hypercapnia. Am. J. Physiol. *223:* 97 (1972).

36 *Rowell, L.B.; Brengelmann, G.L.; Blackmon, J.R., and Murray, J.A.:* Redistribution of blood flow during sustained high skin temperature of resting man. J. appl. Physiol. *28:* 415 (1970).

37 *Toubes, D.B. and Brody, M.J.:* Inhibition of reflex vasoconstriction after experimental coronary embolization in the dog. Circulation Res. *26:* 211 (1970).

38 *Vatner, S.F.; Franklin, D.; Citters, R.L. van, and Braunwald, E.:* Effects of carotid sinus nerve stimulation on the coronary circulation of the conscious dog. Circulation Res. *27:* 11–21 (1970).

39 *Walker, J.L.; Abboud, D.M., and Mark, A.L.:* Preferential distribution of inhibitory cardiac receptors in the inferior wall of the left ventricle. Clin. Res. *25:* 261A (1977).

40 *Webb, S.W.; Adgey, A.A.J., and Pantridge, J.F.:* Autonomic disturbance at onset of acute myocardial infarction. Br. med. J. *iii:* 89–92 (1972).

41 *Wennergren, G.; Little, R., and Öberg, B.:* Studies on the central integration of excitatory chemoreceptor influences and inhibitory baroreceptor and cardiac receptor influences. Acta physiol. scand. *96:* 1–18 (1976).

42 *Zoller, R.P.; Mark, A.L.; Abboud, F.M.; Schmid, P.G., and Heistad, D.D.:* The role of low pressure baroreceptors in reflex vasoconstrictor responses in man. J. clin. Invest. *51:* 2967 (1972).

F.M. Abboud, MD, Cardiovascular Division, Department of Medicine and Department of Physiology, Cardiovascular Center, University of Iowa, *Iowa City, IA 52242* (USA)

Mechanisms of Vasodilatation. Satellite Symp., 27th Int. Congr. Physiol. Sci., Wilrijk 1977, pp. 241–247 (Karger, Basel 1978)

Ganglionic Inhibition by Dopamine and Neurogenic Vasodilatation

J.L. Willems and M.G. Bogaert

Heymans Institute of Pharmacology, University of Ghent Medical School, Ghent

Ganglionic blockade has only occasionally been studied as the cause of neurogenic vasodilatation (16). The present discussion focusses on the possibility that dopamine may act at the ganglionic level to inhibit sympathetic outflow to the peripheral vessels, thus causing vasodilatation.

In the dog, intravenous administration of dopamine produces dose-dependent changes in systemic blood pressure. Small doses (1–16 μg/kg) give a decrease of diastolic pressure that has been attributed to a direct vasodilatory effect in renal and mesenteric vascular beds, where dopamine is assumed to activate specific receptors (12, 19, 28). In muscle blood vessels, dopamine can elicit vasodilatation by interfering with neural vascular control: this can be demonstrated using the perfused hindleg technique which allows the separation in time of a direct α-receptor mediated vasoconstrictor effect and the indirect neurogenic vasodilator effect of the drug (2, 4, 23, 24, 26).

1. Dependency upon Sympathetic Tone

Most experimental evidence suggests that the dopamine-induced neurogenic vasodilatation in the hindleg is caused by inhibition of sympathetic vasoconstrictor tone. On administration of doses of dopamine that increase systemic blood pressure and cardiac contractility, baroreceptor or mechanoreceptor reflexes are elicited, which results in a diminution of sympathetic vasoconstrictor activity (18).

The neurogenic vasodilatation induced by dopamine is still present after section of sinoaortic and vagus nerves (2, 23, 24) and blockade of cardiac β-adrenergic receptors (24). In fact, neurogenic dilatation occurs even on administration of small doses of dopamine that give a fall of systemic blood pressure (2, 24, 26). Therefore, inhibition of sympathetic vasoconstrictor tone by dopamine cannot be explained by specific interference with the pressoreceptor reflex. Since the neurogenic vasodilatation in the hindleg induced by several drugs exceeds the sustained vasodilatation produced by sympathectomy, alternative explanations have been proposed which include histamine-mediated dilatation (1, 4) and autoregulation (11). In the case of dopamine, adrenergic antagonists and adrenergic neuron-blocking drugs prevent the neurogenic vasodilatation induced by the drug (23, 24, 26, 34); the sympathetic cholinergic vasodilator system (30) does not participate in the dopamine-induced neurogenic vasodilatation (2, 23, 26, 34).

2. Inhibition of Ganglionic Transmission

The observation that the neurogenic vasodilatation in the isolated perfused hindleg is maximal when dopamine is injected into the abdominal aorta suggests that ganglionic blockade plays a role in the inhibition of the sympathetic vasoconstrictor tone (2). This is also corroborated by the observation that the neurogenic vasodilatation, which is absent after section of the sympathetic chain between L4 and L5, reappears during electrical stimulation of the cranial internodal segment of L5 (34).

Suppression of the sympathetic tone by ganglionic inhibition has been postulated to explain catecholamine-induced dilatation of blood vessels in the dog (6, 17, 20) and relaxation of the cat nictitating membrane (5, 22, 29). It has been shown that noradrenaline and adrenaline inhibit ganglionic transmission and that this effect is mediated by α-receptors (14). Dopamine also inhibits synaptic transmission, as demonstrated in the superior cervical ganglion of cat and rabbit (9, 21, 25) and in the parasympathetic intramural ganglia of the bladder of the cat (8).

The effect of intra-aortic administration of dopamine on synaptic transmission was studied in the paravertebral lumbar ganglia of the dog (31–33, 35). Synaptic transmission in these ganglia was measured by recording the compound action potential from the lower internodal segment or from the ramus communicans griseus of L4 or L5 during supramaximal stimulation (0.3 msec, 1 Hz) of the cranial internodal segment. Dopamine and noradrenaline produced a dose-dependent short-lasting inhibition of synaptic transmission (fig. 1) (31, 32). Dopamine was in that respect equipotent with (–)-noradrenaline and 20 times more potent than (+)-noradrenaline. On the basis of the structure-activity relationship for interaction with α-receptors, one would expect dopamine to be equipotent with (+)-noradrenaline and less potent than (–)-noradrenaline (27). An interaction of dopamine with an α-receptor is therefore unlikely and it was postulated that dopamine interacts in the ganglion on a specific receptor.

The presence of a specific dopamine receptor in the ganglion was further demonstrated by studying the effect of different agonists and antagonists. The

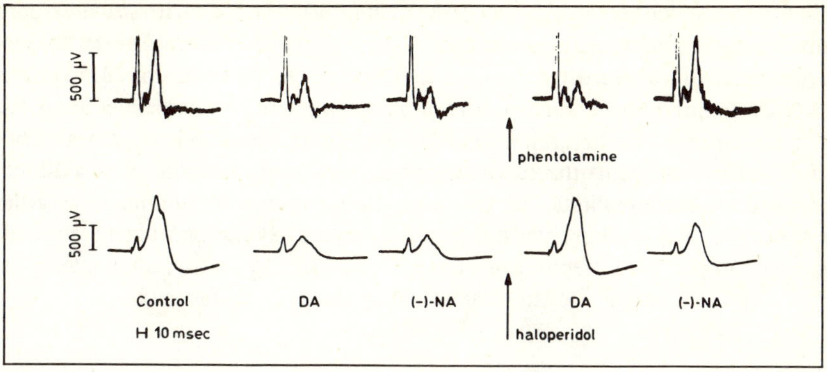

Fig. 1. Lumbar paravertebral ganglion of the dog. The effect of phentolamine (1 mg; upper tracing) and haloperidol (100 µg; lower tracing, different experiment) on the ganglionic inhibition produced by dopamine (DA, 16 µg) and (–)-noradrenaline ([–]-NA, 16 µg). The compound action potential was recorded from the lower internodal segment of L5 upon stimulation of the upper internodal segment. All drugs were injected into the abdominal aorta, 2 min elapsed between DA and (–)-NA, 5 min between the antagonists and the agonists. From *Willems* (31), by permission of the publishers.

dopamine-like agents apomorphine, epinine and piribedil mimick the dopamine-induced ganglionic inhibition. The dopamine antagonists haloperidol, pimozide and (+)-butaclamol selectively antagonize the dopamine effect, while the (–)-noradrenaline effect is selectively blocked by phentolamine (fig. 1) (31, 33).

The effect of dopamine on synaptic transmission is not due to its metabolites since the noradrenaline effect is selectively blocked by phentolamine, while 3-methoxy-4-hydroxyphenylethylamine, 3,4-dihydroxyphenyl acetic acid and tetrahydropapaveroline do not influence ganglionic transmission (32).

Drugs can modulate ganglionic transmission by interfering with excitatory nicotinic or muscarinic receptors, by stimulating adrenergic receptors or by interfering with acetylcholine release from presynaptic nerve endings. For dopamine, as for other catecholamines, a postsynaptic effect (with production of a slow inhibitory potential) and a presynaptic effect (with antagonism of the release of acetylcholine) have been proposed (14).

3. Characterization of the Receptor Mediating Dopamine-Induced Vasodilatation

The characterization of the receptor causing the neurogenic vasodilatation in the isolated perfused hindleg was also attempted by using different agonists and antagonists. The results of these experiments suggest that ganglionic inhibition plays indeed a role in the neurogenic dilatatory effect of dopamine (34).

However, a difficulty remained. It has been published that imipramine antagonizes the dopamine-induced neurogenic vasodilatation (10), whereas the ganglionic inhibition induced by (–)-noradrenaline (21) and dopamine (35) is potentiated by imipramine. Therefore, the dopamine-induced neurogenic vasodilatation was studied in the heteroperfused hindleg of the dog. When imipramine was given intravenously to the receiver, so that it did not reach the perfused hindleg, the neurogenic vasodilatation was potentiated. When imipramine was given to the donor, thus reaching the hindleg, it antagonized the neurogenic effect (35). In the autoperfused preparation, the effect of intravenously administered imipramine in the hindleg abolishes the effect of the drug at the ganglia.

4. Neurogenic Vasodilatation versus Direct Vasoconstrictor Activity

The question was then raised what happens in the dog with intact circulation in which the direct vasoconstrictor effect of the drug (3, 10, 12, 13, 24) opposes the neurogenic vasodilatation. This was studied in the anesthetized dog with intact circulation to the hindleg, under continuous measurement of femoral blood flow and systemic blood pressure (3). In some dogs, dopamine, injected intra-aortically or intravenously, elicited an initial decrease of femoral resistance in the innervated hindleg (fig. 2). This decrease was not due to a baroreceptor mediated reflex. It was blocked by small doses of haloperidol but not by propranolol or atropine. These observations demonstrate that a neurogenic vasodilatation, dominating and preceding the direct vasoconstrictor effect, can occur in the intact animal (3).

It has been suggested that dopamine interferes with sympathetic vasoconstrictor tone by inhibiting noradrenaline release from the sympathetic nerve endings in the vascular wall (13). Experiments designed to study *in vivo* whether dopamine inhibits sympathetic tone by interacting with receptors at vascular adrenergic nerve endings yielded negative results (7, 15).

5. Extrapolation to Man

Dopamine is used in clinical practice in cardiovascular shock. Because of its inotropic action and vasoconstrictor properties in some vascular beds, e.g. femoral arteries, and vasodilatory properties in other beds, e.g. renal and mesenteric arteries, blood pressure can be maintained without important increases in total vascular resistance (12, 13). It is not known whether the dopamine-induced neurogenic vasodilatation occurs in man. In fact, gangrene has been observed in the extremities of patients in shock treated with dopamine (13). Therefore, we studied this problem in a small series of human volunteers. Blood flow in the leg

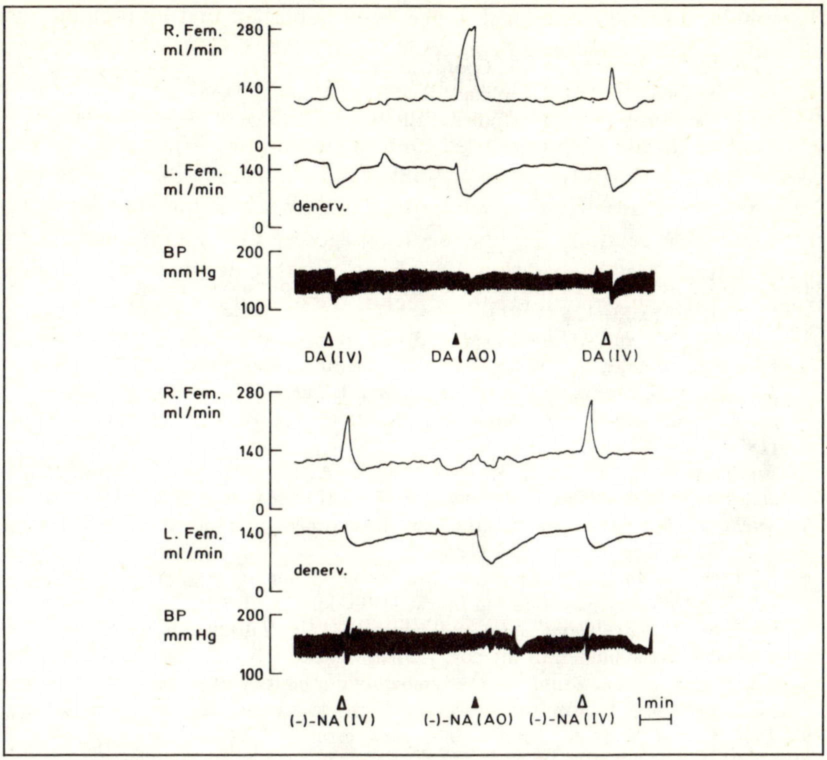

Fig. 2. Femoral blood flow in the dog right (R. Fem.) and left (L. Fem.) femoral artery. The left hindleg has been denervated (denerv.). BP = Systemic blood pressure. Dopamine (DA, 4 μg/kg) and (–)-noradrenaline ([–]-NA, 0.1 μg/kg) were injected intravenously (i.v.) or intra-aortically (ao). From *Bogaert et al.* (3), by permission of the publishers.

was measured plethysmografically in 5 volunteers during infusion of increasing doses of dopamine. An increase in peripheral flow that was not accompanied by an increase in systemic blood pressure was never observed, suggesting that in man an indirect neurogenic effect is not preponderant.

6. Conclusion

The experimental observations summarized show that the dopamine-induced neurogenic vasodilatation in the perfused hindleg of the dog is largely due to a dopamine-induced ganglionic inhibition. This neurogenic vasodilatation contributes to the final cardiovascular effects of dopamine in the anesthetized dog.

Acknowledgments

Supported by grants from the Coordination Committee for Scientific Research (CCAS) and from the Fund for Interdisciplinary Research, Belgium. The collaboration of Dr. *Denis Clement* in the human volunteer studies is kindly acknowledged.

References

1 *Beck, L.:* Active reflex dilatation in the innervated perfused hindleg of the dog. Am. J. Physiol. *201:* 123–128 (1961).

2 *Bogaert, M.G. and Schaepdryver, A.F. de:* Dopamine-induced neurogenic vasodilatation in the hindleg of the dog. Archs int. Pharmacodyn. Thér. *166:* 203–207 (1967).

3 *Bogaert, M.G.; Schaepdryver, A.F. de, and Willems, J.L.:* Dopamine-induced neurogenic vasodilatation in the intact hindleg of the dog. Br. J. Pharmacol. *59:* 283–292 (1977).

4 *Brody, M.J.; Ducharme, D.W., and Beck, L.:* Active reflex vasodilatation induced by veratrine and dopamine. J. Pharmac. exp. Ther. *155:* 84–90 (1967).

5 *Buelbring, E.:* The action of adrenaline on transmission in the superior cervical ganglia. J. Physiol., Lond. *103:* 55–67 (1944).

6 *Buelbring, E. and Burn, J.H.:* An action of adrenaline on transmission in sympathetic ganglia which may play a part in shock. J. Physiol., Lond. *101:* 289–303 (1942).

7 *Buylaert, W.A.; Willems, J.L., and Bogaert, M.G.:* Vasodilatation produced by apomorphine in the hindleg of the dog. J. Pharmac. exp. Ther. *201:* 738–746 (1977).

8 *Groat, W.C. de and Saum, W.R.:* Sympathetic inhibition of the urinary bladder and of pelvic ganglionic transmission in the cat. J. Physiol., Lond. *220:* 297–314 (1972).

9 *Dun, N. and Nishi, S.:* Effects of dopamine on the superior cervical ganglion of the rabbit. J. Physiol., Lond. *239:* 155–164 (1974).

10 *Eble, J.N.:* A proposed mechanism for the depressor effect of dopamine in the anesthetized dog. J. Pharmac. exp. Ther. *145:* 64–70 (1964).

11 *Glick, G.; Wechsler, A.S., and Epstein, S.E.:* Mechanisms of reflex vasodilatation: assessment of the role of neural reuptake of norepinephrine and release of histamine. J. clin. Invest. *47:* 511–520 (1968).

12 *Goldberg, L.I.:* Cardiovascular and renal actions of dopamine. Potential clinical applications. Pharmacol. Rev. *24:* 1–29 (1972).

13 *Goldberg, L.I.; Hsieh, Y.Y., and Resnekov, L.:* Newer catecholamines for treatment of heart failure and shock: an update on dopamine and a first look at dobutamine. Prog. cardiovasc. Dis., vol. 19, pp. 327–340 (Grune & Stratton, New York 1977).

14 *Haefely, W.:* Electrophysiology of the adrenergic neuron; in *Blaschko and Muscholl* Catecholamines. Handbuch der experimentellen Pharmakologie, vol. 33, pp. 661–725 (Springer, Berlin 1972).

15 *Haeusler, G.:* Studies on the possible contribution of a peripheral presynaptic action of clonidine and dopamine to their vascular effects under *in vivo* conditions. Arch. Pharmakol. *295:* 191–202 (1976).

16 *Hancock, J.C. and Hilton, G.:* Control of peripheral vascular resistance in perfused hindquarters of dogs: lack of effect of nicotinic and muscarinic ganglion block. Eur. J. Pharmacol. *31:* 226–231 (1975).

17 *Hartman, F.A.; Kilborn, L.G., and Fraser, L.:* Location of the adrenaline vasodilator mechanisms. Am. J. Physiol. *46:* 168–185 (1918).

18 *Heymans, C. and Neil, E.:* Reflexogenic areas of the cardiovascular system (Churchill, London 1958).

19 *Higgins, C.B.; Millard, R.W.; Braunwald, E., and Vatner, S.F.:* Effects and mechanisms of action of dopamine on regional hemodynamics in the conscious dog. Am. J. Physiol. *225:* 432–437 (1973).

20 *Jones, J.J.:* Inhibition of the vasomotor system of the anaesthetized dog by epinephrine and norepinephrine. Circulation Res. *10:* 156–160 (1962).

21 *Kadzielawa, K.; Gawecka, I., and Kadzielawa, R.:* The potentiating influence of imipramine on ganglionic effects of catecholamines. Int. J. Neuropharmacol. *7:* 517–521 (1968).

22 *Kewitz, H. und Reinert, H.:* Prüfung pharmakologischer Wirkungen am oberen sympathischen Halsganglion bei verschiedenen Erregungszuständen. Naunyn-Schmiedebergs Arch. exp. Path. Pharmak. *215:* 547–555 (1952).

23 *Li, D.M.F. and Bentley, G.A.:* Reflex vascular responses to vasoactive substances. Eur. J. Pharmacol. *12:* 203–214 (1970).

24 *McDonald, R.H. and Goldberg, L.I.:* Analysis of the cardiovascular effects of dopamine in the dog. J. Pharmac. exp. Ther. *140:* 60–66 (1963).

25 *McIsaac, R.J.:* Ganglionic blocking properties of epinephrine and related amines. Int. J. Neuropharmacol. *5:* 15–26 (1966).

26 *Osswald, W.:* Über die durch Sympathicomimetica ausgelöste neurogene Vasodilatation. Arch. Pharmakol. *265:* 67–80 (1969).

27 *Patil, P.N.; Miller, D.D., and Trendelenburg, U.:* Molecular geometry and adrenergic drug activity. Pharmacol. Rev. *26:* 323–392 (1975).

28 *Setler, P.E.; Pendleton, R.G., and Finlay, E.:* The cardiovascular actions of dopamine and the effects of central and peripheral catecholaminergic receptor blocking drugs. J. Pharmac. exp. Ther. *192:* 702–712 (1975).

29 *Trendelenburg, U.:* Modification of transmission through the superior cervical ganglion of the cat. J. Physiol., Lond. *132:* 529–541 (1956).

30 *Uvnas, B.:* Cholinergic vasodilator nerves. Fed. Proc. Fed. Am. Socs exp. Biol. *25:* 1618–1622 (1966).

31 *Willems, J.L.:* Dopamine-induced inhibition of synaptic transmission in lumbar paravertebral ganglia of the dog. Arch. Pharmakol. *279:* 115–126 (1973).

32 *Willems, J.L.:* Experimentele bijdrage tot de farmakologie van dopamine; PhD thesis, Ghent (1974).

33 *Willems, J.L.; Bekaert, I.E., and Buylaert, W.A.:* Effects of piribedil, S584, (+)and (–)-butaclamol on ganglionic transmission: further evidence for a ganglionic dopamine receptor. J. Pharmacol. (in press, 1978).

34 *Willems, J.L. and Bogaert, M.G.:* Dopamine-induced neurogenic vasodilatation in isolated perfused muscle preparation of the dog. Arch. Pharmakol. *286:* 413–428 (1975).

35 *Willems, J.L. and Bogaert, M.G.:* The influence of imipramine on dopamine-induced ganglionic inhibition and neurogenic vasodilatation in the dog. Arch. Pharmakol. *286:* 429–436 (1975).

J.L. Willems, MD, Heymans Institute, De Pintelaan 135, *B–9000 Ghent* (Belgium)

Mechanisms of Vasodilatation. Satellite Symp., 27th Int. Congr. Physiol. Sci., Wilrijk 1977, pp. 248–257 (Karger, Basel 1978)

Vasodilatation Caused by Peripheral Inhibition of Adrenergic Neurotransmission

M.A. McGrath and P.M. Vanhoutte

Department of Physiology, Mayo Clinic, Rochester Minnesota, and Department of Internal Medicine, Universitaire Instelling Antwerpen, University of Antwerp, Wilrijk

The main pathway for central nervous control of vascular tone is the sympathetic system. The activity of the post-ganglionic adrenergic neurones can be altered in the blood vessel wall itself, because a number of vasoactive substances modulate the amount of transmitter liberated by the adrenergic nerve terminals (10, 24, 28, 30, 35, 36). The present discussion will focus on these substances which inhibit adrenergic neurotransmission at the local level, an effect which helps explain why they cause vasodilatation.

1. Neuro-Humoral Modulation

Acetylcholine. The inhibitory effect of acetylcholine on the response to nerve stimulation has been demonstrated in many different arteries and veins of the dog, as well as in cutaneous arteries and veins of the rabbit, the mesenteric vessels of the rat, and the penile artery of the bull (4, 6, 8, 14, 25, 29, 34, 38, 45). In these vessels, the response to exogenous norpinephrine is much less depressed by acetylcholine; therefore, the latter must inhibit the release of adrenergic transmitter. Direct evidence to support this interpretation comes from experiments performed on preparations incubated with ^{3}H-norepinephrine where the overflow of labeled transmitter is measured (1, 29, 34, 37, 39–41, 45). Acetylcholine, in concentrations which have no direct effect on the vascular smooth muscle cells, or only moderately relaxes them, causes a marked inhibition of the release of norepinephrine evoked by nerve stimulation (fig. 1). This inhibitory effect has also been confirmed in the intact animal (38).

The inhibitory effect of acetylcholine on adrenergic neurotransmission is blocked by atropine, and can be mimicked by different muscarinic agonists;

thus, a muscarinic receptor is involved. Acetylcholine also depresses the release of ^{3}H-norepinephrine evoked by depolarizing solutions (high K^+) but not that caused by tyramine or Ca^{++} ionophores (34, 37, 38–40, 41, 43, 45); it is likely that the muscarinic inhibition of adrenergic neurotransmission is due to hyperpolarization of the sympathetic nerve terminals (35, 40).

Histamine. The inhibitory effect of histamine on adrenergic neurotransmission is demonstrated best in the dog saphenous vein, because the agent has no direct dilatory effect on the smooth muscle of this preparation. In concentrations which do not affect basal tone of the veins, histamine causes a marked relaxation during sympathetic nerve stimulation; the relaxation is paralleled by a decrease in the evoked release of ^{3}H-norepinephrine (fig. 1) (21). Histamine also depresses the release of transmitter evoked by depolarizing solution, but not its displacement by tyramine. The inhibitory effect on adrenergic neurotransmission can also be demonstrated in tibial arteries, and is due to activation of prejunctional H_2 receptors, since it is blocked by metiamide but not by the histamine H_1 antagonist pyrilamine (21).

5-Hydroxytryptamine. As for acetylcholine and histamine, the inhibition by 5-hydroxytryptamine of adrenergic neurotransmission has been convincingly demonstrated in the dog cutaneous vein, where the autacoid augments the responses to exogenous norepinephrine, but depresses those to sympathetic nerve stimulation; the latter relaxation is paralleled by a decreased release of transmitter (fig. 1) (19). The agent does not affect the displacement of ^{3}H-norepinephrine caused by tyramine, but reduces K^+-evoked release. The exact pharmacological nature of the mechanism mediating the prejunctional action of 5-hydroxytryptamine remains undefined since it is not affected by methysergide, phentolamine or morphine (19).

Prostaglandins. In certain, but not all blood vessels, prostaglandins E_1 inhibit the vasoconstriction caused by sympathetic nerve stimulation to a greater extent than those caused by exogenous norepinephrine; to judge also from studies in other adrenergically innervated organs, the inhibition of the responses to nerve stimulation must be due to inhibition of the release of norepinephrine (2, 16–18, 28, 30–32).

Enkephalin. In the rabbit ear artery, enkephalin has been shown to specifically depress the responses to sympathetic nerve stimulation, which suggests that it inhibits adrenergic neurotransmission (9).

Conclusion. Do these inhibitory receptors (fig. 2) on adrenergic nerve endings play a physiological role? Whereas the exact role of the inhibitory responses noted on exposure to enkephalin and prostaglandins during nerve stimulation is hard to define, that of those to acetylcholine, histamine and 5-hydroxytryptamine is easier to envisage. With these substances, which all seem to affect the exocytotic release process, the prejunctional inhibitory action is evident at lower concentrations than those needed to directly affect the vascular

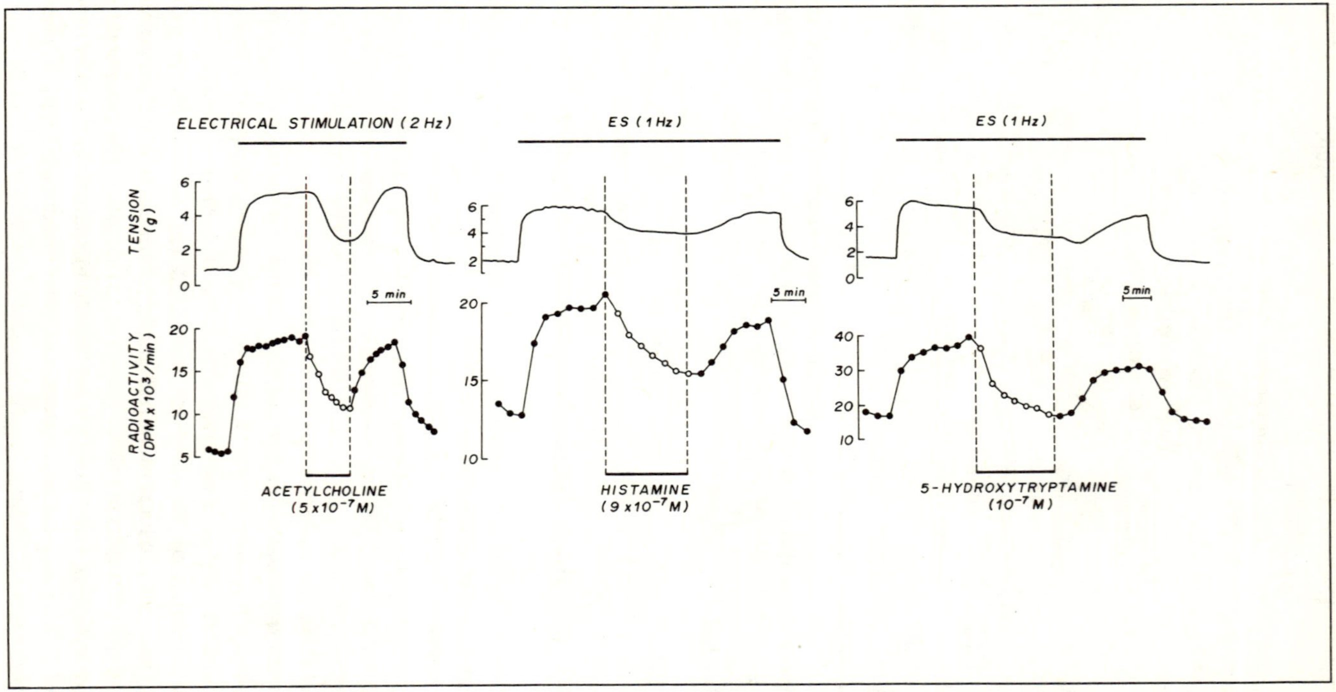

Fig. 1. Concentrations of acetylcholine (left), histamine (middle), and 5-hydroxytryptamine (right), which do not cause changes in tension in unstimulated preparations or during contractions to exogenous norepinephrine, cause a pronounced relaxation during sympathetic nerve stimulation (upper); the relaxation is due to a decrease in the evoked release of adrenergic neurotransmitter (lower). Experiments performed on three different saphenous vein strips previously incubated with ^{3}H-norepinephrine; the overflow of the latter is a qualitative measurement of the amount of transmitter liberated. From *McGrath* (19), *McGrath and Shepherd* (21), and *Vanhoutte et al.* (37).

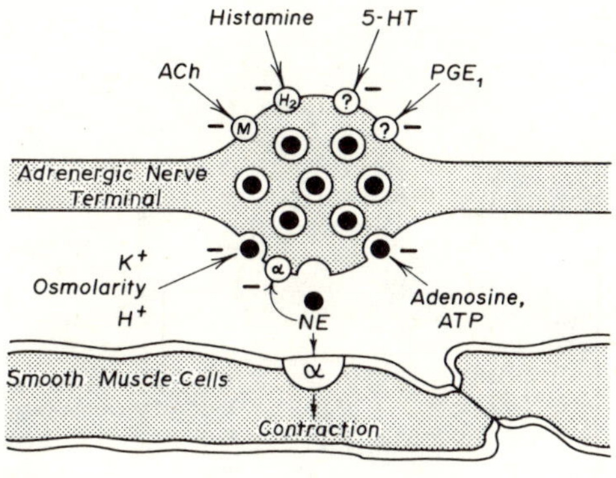

Fig. 2. Schematic representation of different inhibitory influences on adrenergic neurotransmission in the blood vessel wall. The effects of norepinephrine itself are not discussed in the text. NE = Norepinephrine; M = muscarinic; Ach = acetylcholine; 5HT = 5-hydroxytryptamine; H = histamine; PGE$_1$ = prostaglandin E$_1$; α = α-adrenergic receptor; ? = unknown mechanism.

smooth muscle cells. Hence the inhibition they exert on adrenergic neurotransmission helps to explain the potent vasodilator effects that these drugs have in certain vascular beds and should be considered every time they are administered to blood vessels under sympathetic control. In the case of acetylcholine, it is known in the heart that vagal stimulation reduces the release of norepinephrine evoked by sympathetic stimulation (11, 12); in the canine gastric artery, atropine augments and physostigmine depresses the response to field stimulation which activates both adrenergic and cholinergic nerve endings in the wall of that vessel (33). Thus, it is tempting to assume that muscarinic inhibition of adrenergic neurotransmission is part of the normal regulatory mechanisms not only in the heart but also in blood vessels subserved by cholinergic terminals. This would mean that during cholinergic vasodilatation, the acetylcholine released in the vicinity of the adrenergic nerve endings switches off the existing sympathetic tone. Such a local modulating role, which obviously also applies for histamine and 5-hydroxytryptamine, makes the concept of the peripheral summation of the opposing effects of two autonomic mediators on the effector cells obsolete, and results in a more efficient and more economical way to obtain a vasodilator response (35, 36).

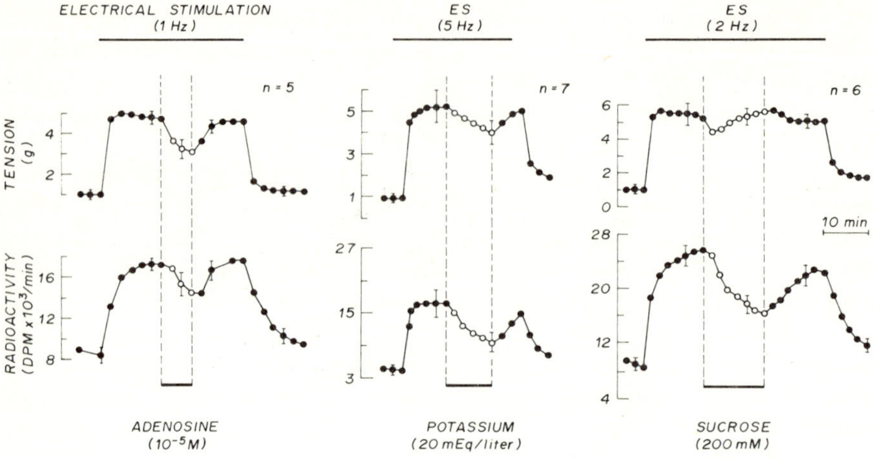

Fig. 3. A concentration of adenosine (left), and increases in K^+ concentration (middle) or in osmolarity (right) which do not inhibit responses to norepinephrine, cause relaxation (upper) and a significant depression of the release of ^{3}H-norepinephrine (lower) evoked by nerve stimulation, in three groups of five saphenous veins each. From *Lorenz and Vanhoutte* (13), *McGrath and Shepherd* (20), and *Verhaeghe et al.* (46).

2. Metabolic Modulation

Adenosine and adenine nucleotides. Increasing concentrations of adenosine and of adenosine triphosphate (ATP) depress the response of the rat mesenteric artery and of canine vascular strips to sympathetic nerve stimulation more than that to norepinephrine (15, 46); a similar qualitative difference can be demonstrated for adenosine in the intact dog, using the perfused lateral saphenous vein as a model of an adrenergically innervated blood vessel (46). Concentrations of adenosine and ATP, lower than those required to inhibit the response to exogenous norepinephrine, significantly decrease the release of ^{3}H-norepinephrine evoked by sympathetic nerve stimulation (fig. 3) (46). In the rabbit kidney and in canine adipose tissue, adenosine also inhibits the release of ^{3}H-norepinephrine evoked by nerve stimulation (5). In addition, adenosine inhibits the release of tritiated transmitter evoked by potassium depolarization, but not its displacement by tyramine. Theophylline antagonizes the prejunctional effect of adenosine, which may imply that the latter involves the cyclic nucleotides system and interferes with an intracellular step of the excitation-secretion coupling (46).

Potassium. With increases in potassium concentration from 5.9 to 20 mEq/l, there is a concentration-dependent decrease in the contractile response to supra-maximal sympathetic nerve stimulation; the decrease in tension is paralleled by a decreased overflow of adrenergic neurotransmitter (fig. 3) (13, 44). The tyramine-induced release of ^{3}H-norepinephrine is unaffected by the ion.

Hyperosmolarity. In the isolated canine saphenous vein, hyperosmolar sucrose solutions depress the contractile response to sympathetic nerve stimulation; this depression is paralleled by a decreased efflux of neurotransmitter (fig. 3) (20). No depression of the responses to nerve stimulation is observed with hyperosmolar urea solutions which suggests that the inhibitory effect on the adrenergic neuroeffector interaction is related to changes in intracellular volume. Hyperosmolar sucrose depresses the response to tyramine to a greater extent than that to exogenous norepinephrine and that to nerve stimulation; in resting preparations, hyperosmolarity reduces the spontaneous overflow of norepinephrine and its metabolites (20). In these two respects, the inhibitory effect of hyperosmolarity on adrenergic nerve endings differs from the other inhibitory influences discussed so far, which suggests that hyperosmolarity does not interfere specifically with the exocytotic process.

Hydrogen ions. In the dog's cutaneous vein, moderate metabolic acidosis (from pH 7.4 to 7.1) depresses the frequency-response curve to nerve stimulation without affecting the dose-response curve to exogenous norepinephrine (41); this degree of acidosis decreases the evoked release of radiolabeled neurotransmitter to the same extent (30%) as it depresses the contractile response (42).

Conclusion. Adenosine, adenine nucleotides, potassium ions, hyperosmolarity, and hydrogen ions have all been proposed as mediators of the vasodilatation occurring in metabolically active tissues. In metabolically active tissues, and in particular in contracting skeletal muscle, the response of the resistance vessels to sympathetic nerve stimulation is reduced (3, 7, 26, 27). The studies summarized above indicate that the mediators of metabolic hyperemia are capable of inhibiting adrenergic neurotransmission (fig. 2); they do so in concentrations lower than those required for their direct effects on the vascular smooth muscle cells themselves. When extrapolated to the intact organism, this means that changes in the composition of the extracellular fluid of metabolically active tissues disconnect the blood vessels subserving them from the homeostatic sympathetic vasomotor control by their inhibitory effect on adrenergic neurotransmission and thus achieve a functional sympatholysis at the local level which is graded to the degree of tissular activity. As for the neurohumoral vasodilators, such a mechanism optimalizes the vasodilator impact of these changes in activity and helps direct compensatory increases in sympathetic outflow towards the resting areas of the body.

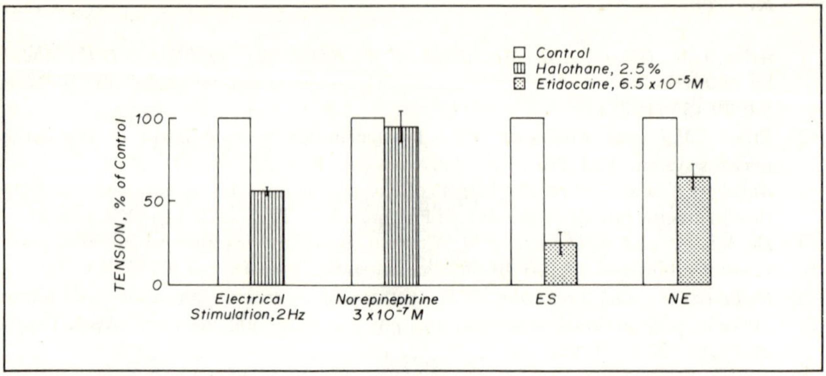

Fig. 4. Both the general anesthetic agent halothane (left) and the local anesthetic drug etidocaine (right) depress the response of dog saphenous veins to sympathetic nerve stimulation significantly more than that to exogenous norepinephrine. This demonstrates the inhibitory effect that both agents have on adrenergic neurotransmission. From *Muldoon et al.* (22, 23).

3. Pharmacological Modulation

Substances such as bretylium tosylate or guanethidine, which have been used because they inhibit adrenergic neurotransmission, need not be discussed here. Other well-known vasodilators also affect the adrenergic nerve terminals. Halothane depresses the responses of cutaneous veins to a much greater extent than those to exogenous norepinephrine; the inhibition of sympathetic neurotransmission that the agent causes explains the pronounced venodilatation occurring during halothane anesthesia (fig. 4) (22). Likewise, local anesthetic agents inhibit the release of norepinephrine evoked by nerve stimulation in isolated blood vessels of the dog, which explains part of their potent vasodilator effect (fig. 4) (23). These examples illustrate the need to assess whether or not a prejunctional effect is involved when investigating vasodilatation caused by therapeutic agents.

Acknowledgments

The authors thank Dr. *John T. Shepherd* for his continuing interest and stimulating comments, and *L. Van den Eynde* for typing the manuscript. Our own work discussed was supported in part by Grant HL 5883 from the National Heart and Lung Institute and by Grant 2.09.010.3–1977 of the NFGWO; it would not have been possible without the collaboration of *M.G. Collis, J. De Mey, R.R. Lorenz, G.M. Tyce, R. Van Hee, T.J. Verbeuren* and *R. Verhaeghe*.

References

1 *Allen, G.S.; Glover, A.B.; McCulloch, M.W.; Rand, M.J., and Story, D.F.:* Modulation by acetylcholine of adrenergic transmission in the rabbit ear artery. Br. J. Pharmacol. *54:* 49–53 (1975).

2 *Brody, M.J. and Kadowitz, P.J.:* Prostaglandins as modulators of the autonomic nervous system. Fed. Proc. Fed. Am. Socs exp. Biol. *33:* 48–60 (1974).

3 *Burcher, E. and Garlick, D.:* Effects of exercise metabolites on adrenergic vasoconstriction in gracilis muscle of the dog. J. Pharmac. exp. Ther. *192:* 149–156 (1975).

4 *De Mey, J. and Vanhoutte, P.M.:* Comparison of the reactivity of isolated saphenous veins of rabbit and dog. Archs int. Pharmacodyn. Thér. (in press, 1977).

5 *Hedqvist, P. and Fredholm, B.B.:* Effects of adenosine on adrenergic neurotransmission: prejunctional inhibition and postjunctional enhancement. Arch. Pharmakol. *293:* 217–223 (1976).

6 *Hume, W.R.; Lande, I.S. de la, and Waterson, J.G.:* Effect of acetylcholine on the response of the isolated rabbit ear artery to stimulation of the perivascular sympathetic nerves. Eur. J. Pharmacol. *17:* 227–233 (1972).

7 *Kjellmer, I.:* On the competition between metabolic vasodilatation and neurogenic vasoconstriction in skeletal muscle. Acta physiol. scand. *63:* 450–459 (1965).

8 *Klinge, E. and Sjöstrand, N.O.:* Contraction and relaxation of the retractor penis muscle and the penile artery of the bull. Acta physioll scand. suppl. 420, pp. 1–88 (1974).

9 *Knoll, J.:* Neuronal peptide (enkephalin) receptors in the ear artery of the rabbit. Eur. J. Pharmacol. *39:* 403–407 (1976).

10 *Langer, S.Z.:* Presynaptic regulation of catecholamine release. Biochem. Pharmacol. *23:* 1793–1800 (1974).

11 *Levy, M.N. and Blattberg, B.:* Effect of vagal stimulation on the overflow of norepinephrine into the coronary sinus during cardiac sympathetic nerve stimulation in the dog. Circulation Res. *38:* 81–85 (1976).

12 *Löffelholz, K. and Muscholl, E.:* Inhibition by parasympathetic nerve stimulation of the release of the adrenergic transmitter. Arch. Pharmakol. *267:* 181–184 (1970).

13 *Lorenz, R.R. and Vanhoutte, P.M.:* Inhibition of adrenergic neurotransmission in isolated veins of the dog by potassium ions. J. Physiol., Lond. *246:* 479–500 (1975).

14 *Malik, K.U. and Ling, G.M.:* Modification by acetylcholine of the response of rat mesenteric arteries to sympathetic stimulation. Circulation Res. *25:* 1–9 (1969).

15 *Malik, K.U. and McGiff, J.C.:* Relationship of glucose metabolism to adrenergic transmission in rat mesenteric arteries. Effects of glucose deprivation, glucose metabolites, and changes in ionic composition on adrenergic mechanisms. Circulation Res. *35:* 553–574 (1974).

16 *Malik, K.U. and McGiff, J.C.:* Modulation by prostaglandins of adrenergic transmission in the isolated perfused rabbit and rat kidney. Circulation Res. *36:* 599–609 (1975).

17 *Malik, K.U. and McGiff, J.C.:* Cardiovascular action of prostaglandins; in *Karim* Prostaglandins (MTP Press, Lancaster 1976).

18 *McGiff, J.C.; Malik, K.U., and Terragno, N.A.:* Prostaglandins as determinants of vascular reactivity. Fed. Proc. Fed. Am. Socs exp. Biol. *35:* 2382–2387 (1976).

19 *McGrath, M.A.:* 5-Hydroxytryptamine and neurotransmitter release in canine blood vessels: inhibition by low and augmentation by high concentrations. Circulation Res. (in press, 1977).

20 *McGrath, M.A. and Shepherd, J.T.:* Hyperosmolarity: effects on nerves and smooth muscle of cutaneous veins. Am. J. Physiol. *231:* 141–147 (1976).

21 *McGrath, M.A. and Shepherd, J.T.:* Inhibition of adrenergic neurotransmission in canine vascular smooth muscle by histamine. Mediation by H_2-receptors. Circulation Res. *39:* 566–573 (1976).

22 *Muldoon, S.M.; Vanhoutte, P.M.; Lorenz, R.R., and Dyke, R.A. van:* Venomotor changes caused by halothane acting on the sympathetic nerves. Anesthesiology *43:* 41–48 (1975).

23 *Muldoon, S.M.; Verbeuren, T.J., and Vanhoutte, P.M.:* Effects of amide-liked local anesthetics on adrenergic neuroeffector junction in cutaneous veins of dog. J. Pharmac. exp. Ther. *196:* 723–736 (1976).

24 *Rand, M.J.; McCulloch, M.W., and Story, D.F.:* Pre-junctional modulation of noradrenergic transmission by noradrenaline, dopamine and acetylcholine; in *Davies and Reid* Central action of drugs in blood pressure regulation, pp. 94–132 (University Park Press, Baltimore 1975).

25 *Rand, M.J. and Varma, B.:* The effects of cholinomimetic drugs on responses to sympathetic nerve stimulation and noradrenaline in the rabbit ear artery. Br. J. Pharmacol. *38:* 758–770 (1970).

26 *Rowlands, D.J. and Donald, D.E.:* Sympathetic vasoconstrictive responses during exercise or drug induced vasodilation. Circulation Res. *23:* 45–60 (1968).

27 *Skinner, N.S. jr. and Costin, J.C.:* Interactions between oxygen, potassium and osmolality in regulation of skeletal muscle blood flow. Circulation Res. *28/29:* suppl. I, pp. 73–85 (1971).

28 *Starke, K.; Endo, T.; Taube, H.D., and Borowski, E.:* Presynaptic receptor systems on noradrenergic nerves; in *Almgren, Carlesson and Engel* Chemical tools in catecholamine research, vol. II, pp. 193–200 (North-Holland, Amsterdam 1975).

29 *Steinsland, O.S.; Furchgott, R.F., and Kirpekar, S.M.:* Inhibition of adrenergic neurotransmission by parasympathomimetics in the rabbit ear artery. J. Pharmac. exp. Ther. *184:* 346–356 (1973).

30 *Stjärne, L.:* Basic mechanisms and local feedback control of secretion of adrenergic and cholinergic neurotransmitters; in *Iverson, Iverson and Snyder* Handbook of psychopharmacology, pp. 179–233 (Plenum Publishing, New York 1975).

31 *Stjärne, L. and Brunding, J.:* Prostaglandin E_2 inhibits nerve stimulation-induced smooth muscle contraction in isolated human omental vein by depressing secretion of noradrenaline. Acta physiol. scand. *97:* 526–528 (1976).

32 *Stjärne, L. and Gripe, K.:* Prostaglandin-dependent and -independent feedback control of noradrenaline secretion in vasoconstrictor nerves of normotensive human subjects. Arch. Pharmakol. *280:* 441–446 (1973).

33 *Hee, R.H. van and Vanhoutte, P.M.:* Cholinergic inhibition of adrenergic neurotransmission in dogs' gastric arteries. Archs int. Pharmacodyn. Thér. *220:* 348 (1976).

34 *Vanhoutte, P.M.:* Inhibition by acetylcholine of adrenergic neurotransmission in vascular smooth muscle. Circulation Res. *34:* 317–326 (1974).

35 *Vanhoutte, P.M.:* Cholinergic inhibition of adrenergic neurotransmission. Fed. Proc. Fed. Am. Socs exp. Biol. *36:* 2444–2449 (1977).

36 *Vanhoutte, P.M.:* Heterogeneity in vascular smooth muscle; in *Kaley and Altura* Microcirculation, vol. II (Academic Press, New York, in press).

37 *Vanhoutte, P.M.; Lorenz, R.R., and Tyce, G.M.:* Inhibition of norepinephrine-^{3}H-release from sympathetic nerve endings in veins by acetylcholine. J. Pharmac. exp. Ther. *185:* 386–394 (1973).

38 *Vanhoutte, P.M. and Shepherd, J.T.:* Venous relaxation caused by acetylcholine acting on the sympathetic nerves. Circulation Res. *32:* 259–267 (1973).

39 *Vanhoutte, P.M. and Verbeuren, T.J.:* Inhibition by acetylcholine of ^{3}H-norepineph-

rine release in cutaneous veins after alpha-adrenergic blockade. Archs int. Pharmacodyn. Thér. *221:* 160–162 (1976).

40 *Vanhoutte, P.M and Verbeuren, T.J.:* Inhibition by acetylcholine of the norepinephrine release evoked by potassium in canine saphenous veins. Circulation Res. *39:* 263–269 (1976).

41 *Vanhoutte, P.M. and Verbeuren, T.J.:* Effect of pH changes on adrenergic neuroeffector interaction in isolated cutaneous veins. Circulation Res. *53/54:* suppl. II, p. 42 (1976).

42 *Vanhoutte, P.M. and Verbeuren, T.J.:* Proc. Int. Union of Physiological Sciences, vol. 13, p. 777 (1977).

43 *Vanhoutte, P.M.; Verbeuren, T.J., and Collis, M.G.:* Muscarinic inhibition of adrenergic neurotransmission is not due to direct interference with Ca^{++} entry in the adrenergic nerve endings. J. Pharmacol. (in press).

44 *Vanhoutte, P.M.; Verbeuren, T.J., and Lorenz, R.R.:* Effects of potassium ions on nerve endings and effector cells in adrenergically innervated vascular smooth muscle. Coll. INSERM *50:* 425–442 (1975).

45 *Verbeuren, T.J. and Vanhoutte, P.M.:* Acetylcholine inhibits potassium-evoked release of ^{3}H-norepinephrine in different blood vessels of the dog. Archs int. Pharmacodyn. Thér. *221:* 347–350 (1976).

46 *Verhaeghe, R.H.; Vanhoutte, P.M., and Shepherd, J.T.:* Inhibition of sympathetic neurotransmission in canine blood vessels by adenosine and adenine nucleotides. Circulation Res. *40:* 208–215 (1977).

Dr. *P.M. Vanhoutte,* Department of Internal Medicine, Universitaire Instelling Antwerpen, Universiteitsplein 1, *B–2610 Wilrijk* (Belgium)

Mechanisms of Vasodilatation. Satellite Symp., 27th Int. Congr. Physiol. Sci., Wilrijk 1977, pp. 258–265 (Karger, Basel 1978)

β-Adrenergic Receptor-Mediated Vasodilatation

John A. Bevan, Barbara L. Pegram, John L. Prehn and Raymond J. Winquist

Brain Research Institute and Department of Pharmacology, School of Medicine, University of California, Los Angeles, Calif.

1. The Vascular β-Receptors

The presence of β-adrenergic receptors in the smooth muscle of arteries and veins has been shown by studies of many types of vessels from a variety of species. Current interest in the vascular β-adrenergic receptors centers upon their location, classification, neurogenic activation and excitation-response coupling.

Moran (25, 26), building on the work of *Cannon and Rosenblueth* (8), *Ahlquist* (1), *Powell and Slater* (29), *Ariens* (3), *Buckner and Patil* (6), and others, has defined the characteristics of the various adrenergic receptors and their subtypes. The agonist potency ratios for the β_1-receptor are isoproterenol $>$ norepinephrine $\geqslant$ epinephrine, and for the β_2-receptors, isoproterenol $>$ epinephrine $\gg$ norepinephrine. *Furchgott* (13) proposed strict criteria for the classification of adrenergic receptors. All factors which might influence the concentration of the agonists at the receptor and/or modify its response should be accounted for. In figure 1, the potency ratios of isoproterenol, epinephrine, and norepinephrine for β_1-receptors in non-vascular tissues – the heart and gastrointestinal tract – and for β_2-receptors in tracheal muscle – determined under conditions that approximate those laid down by *Furchgott* – are shown. The studies of *Bohr* (5) and *de la Lande et al.* (12) suggest that the β-receptors in the coronary arteries and possibly skeletal muscle arteries should be classified as β_1-receptors. Measurements on the rabbit thoracic aorta (14) suggest that this vessel contains β_2-receptors.

The evidence that β-adrenergic receptors in blood vessels are activated by locally released adrenergic transmitter is derived from experiments in which either α- or β-adrenergic receptors were blocked. Neurogenic vasodilatation attributable to norepinephrine release has been demonstrated after α-adrenergic receptor blockade – in the skeletal muscle vasculature of the dog (38) and in the hepatic and systemic circulation of the cat (16, 17). β-Adrenergic blockade has been shown to enhance vasoconstrictor responses to epinephrine (7, 9, 15–18, 28, 33, 35).

Hughes and Vane (20, 21) found that the β-adrenergic receptor component of neurogenic vasodilatation in the rat portal vein could be demonstrated only with stimulation frequencies greater than 6 Hz. In the dog femoral artery, *Gerova and Gero* (15) found that the propranolol-sensitive vasodilatation component of the neurogenic response became apparent only after 10 sec of sympathetic stimulation. Vascular β-receptors in adipose tissue seem unrelated to the adrenergic nerve terminal (32). These data suggest that the β-adrenergic receptors are somewhat distant from sites of transmitter release. This is consistent with the observation that vascular β-receptors are activated by circulating not locally released catecholamines (34). However, *Viveros et al.* (38) and *Greenway and Lawson* (17) have observed β-mediated decreases in vascular resistance at low

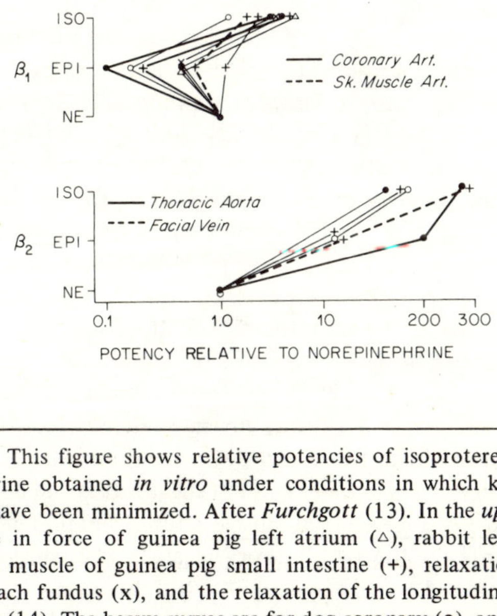

Fig. 1. This figure shows relative potencies of isoproterenol, epinephrine relative to norepinephrine obtained *in vitro* under conditions in which known factors that influence sensitivity have been minimized. After *Furchgott* (13). In the *upper* group, values are shown for increase in force of guinea pig left atrium (△), rabbit left atrium (•), relaxation of longitudinal muscle of guinea pig small intestine (+), relaxation of circular muscle of the rabbit stomach fundus (x), and the relaxation of the longitudinal muscle of the rabbit small intestine (○) (14). The heavy curves are for dog coronary (•), and skeletal muscle arteries (x) (5), and for rabbit coronary artery (+) (12). In the *lower* group, relative potencies are shown for relaxation of guinea pig trachea (○) (14), relaxation of calf trachea (x) (3), and relaxation of calf iris (•) (6). The heavy curves are for relaxation of the rabbit thoracic aorta (•) (14), and the relaxation of the rabbit facial vein (x).

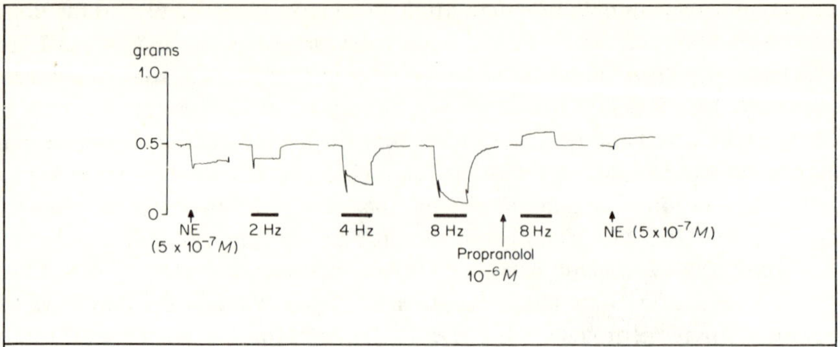

2

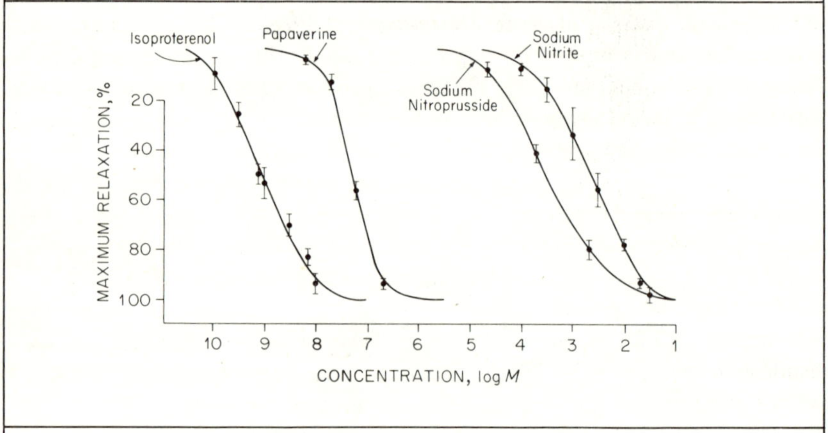

3

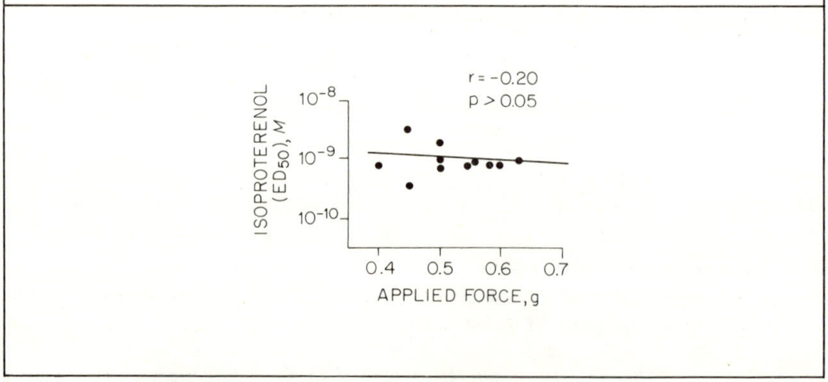

4

Fig. 2. Part of isometric force record of changes following transmural nerve stimulation at different frequencies and the addition of norepinephrine (NE) before and after propranolol. Bars indicate duration of transmural nerve stimulation. From ***Pegram et al.*** (27).

Fig. 3. Dose-relaxation response curves from buccal segment of facial vein of the rabbit to isoproterenol, papaverine, sodium nitroprusside, and sodium nitrite.

Fig. 4. Relationship between isoproterenol ED_{50} and initial tension applied to rabbit facial vein segment.

frequencies of sympathetic stimulation in striated muscle, suggesting that the β-receptor is found close to the neuronal varicosities. It is of interest that a difference has been found between the ratio of the size of the α- and β-mediated responses, and also the nature of the β-receptor in large compared with small coronary arteries (5, 34). The differing conclusions reached in the above studies may be, in part, related to the size of the blood vessel studied.

2. *β-Receptor in Rabbit Facial Vein*

In 1976, *Pegram et al.* (27) described some of the more important properties of the buccal segment of the rabbit facial vein. This comparatively short length of vessel has an adrenergic innervation that ramifies throughout the thickness of the tunica media and develops intrinsic tone spontaneously in the tissue bath. It dilates to both endogenous and exogenous norepinephrine. These three properties are all absent in more proximal and distal segments, which show more conventional vascular properties. The relaxation response of the buccal segment was potentiated by an α- and reversed by a β-adrenergic receptor-blocking agent. The properties of this segment suggest that the sympathetic response of the vessel is dominated by the β-adrenergic receptor (fig. 2).

On the basis of agonist potency ratios, the β-receptor in the rabbit facial vein appears to be β_2 in type (fig. 1). The facial vein exhibits a neurogenic response to sympathetic stimulation at 0.1 Hz, suggesting that the β-receptors are not at a significant distance from an adrenergic varicosity (4).

There may be some functional connection between the occurrence of β-receptors and the ability to develop myogenic tone in response to stretch (see below). When placed under passive tension, the facial vein develops maintained tone (40). There is a highly significant correlation between the equilibrium level of developed tone and the initially applied tension ($r = 0.885$; $p < 0.001$). The magnitude of the maximum inhibition of this tone by isoproterenol is similar to that produced by papaverine, sodium nitrite, and sodium nitroprusside (fig. 3). The mean ED_{50} for the relaxant effect of these drugs are 8.6×10^{-10} *M*, 5.26×10^{-8} *M*, 3.6×10^{-4} *M*, and 2.6×10^{-3} *M*, respectively. The ED_{50} for isoproterenol is independent of initial stretch and level of intrinsic tone ($r = -0.20$ and -0.08, respectively) (fig. 4). This implies that the response to a particular concentration of isoproterenol is proportional to the magnitude of the tone, and due to the activation of the same population of receptors. The relationship also intimates a possible interaction between the mechanism of myogenic tone and the β-receptor. In contrast, there was a positive correlation between the ED_{50} for papaverine and the level of tone [$r = 0.73$ (9); $p < 0.02$]. Thus, the sensitivity and magnitude of the response to papaverine varies with the prevailing level of intrinsic tone.

Our observations show that the development of myogenic tone in the facial vein is an active process, dependent on external calcium but not dependent on circulating vasoactive substances, sympathetic innervation, nor on the intramural synthesis of substances previously shown to influence vascular tone (22, 23). The properties of the tone are similar to those previously described for veins (2) and the resistance vessels in the skeletal muscle (37). The drugs that inhibit this tone are potent vasodilators in the intact animal.

3. Electrophysiology of Vascular β-Receptor Activation

The electrophysiological correlates of the β-adrenergic receptor-mediated vasodilatation were studied using intracellular microelectrode recording (30). The average maximum resting membrane potential of single smooth muscle cells from 10 veins was -46.0 ± 13.5 mV (31). The actual membrane potential may be closer to the maximum recorded potential in some experiments: -60 mV. Upon exposure to isoproterenol (10^{-8} to 10^{-7} *M*), the average maximum membrane potential increased to -68.0 ± 9.8 mV. Again, the maximum recorded in some experiments was in excess of -80 mV. Sodium nitrite (3×10^{-2} *M*) produced a considerably smaller increase in membrane potential.

Field stimulation of intramural nerves (using 0.25–0.3 msec duration pulses) also produced hyperpolarization along with relaxation (fig. 5). This hyperpolarization was blocked by propranolol (10^{-6} *M*) and potentiated by phentolamine (5×10^{-7} *M*). Hyperpolarization was directly dependent on frequency of stimulation, reaching a maximum at 16–32 Hz. A stimulation frequency of 4–8 Hz usually produced a hyperpolarization and relaxation 80% or more of the maximum (27). Neurogenic hyperpolarization also depends on the level of the recorded membrane potential of the impaled cell. Cells having resting membrane potentials of -50 mV or more hyperpolarized less (4–8 mV) than those with lower recorded membrane potentials, which hyperpolarized by up to 24 mV. Both hyperpolarization and relaxation occurred with a latency of about 0.7 sec following onset of even the highest frequency of nerve stimulation used (32 Hz). Thus, hyperpolarization and relaxation began just before the end of a 1-sec stimulus train at 16 Hz.

Elevation of external (K^+) by a factor of 4–5 caused rhythmic oscillations (about 0.5 Hz) in membrane potential and tension. The potential waves occurred in all cells impaled and were always phase-locked with the tension changes, implying strong electrical coupling between the smooth muscle cells in the rabbit facial vein. Under these conditions, transmural nerve stimulation (4–8 Hz) produced hyperpolarization and relaxation and abolished the oscillations in both membrane potential and tension. Following cessation of stimulation, or with low frequency (1–2 Hz) stimulation, the waves increased in amplitude and period.

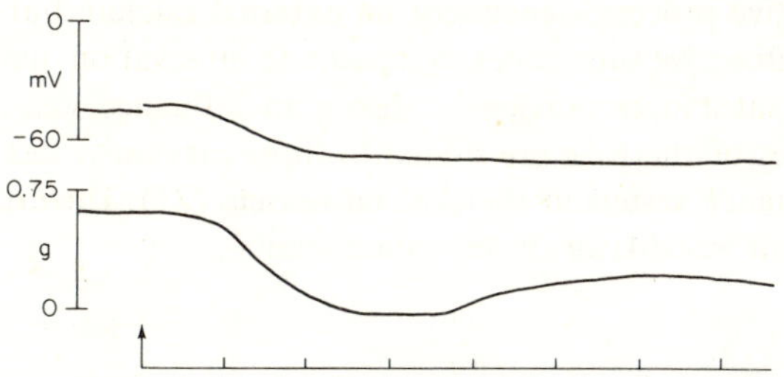

Fig. 5. Simultaneous recording of membrane potential (upper record) of an impaled cell and tension (lower record) in a segment of the rabbit facial vein during transmural nerve stimulation at 32 Hz. Stimulation starts at the beginning of the time marker which shows 1-sec intervals. The impaled cell hyperpolarizes from a resting potential of -44 to -64 mV. The increase in tension from the maximum relaxation is transient.

Further elevation of external (K^+) prevented both hyperpolarization and relaxation.

Hyperpolarization in response to isoproterenol has been recorded in the rabbit portal vein (19), and decrease in action potential frequency in rabbit anterior mesenteric vein (10) using the sucrose gap technique. Hyperpolarization to isoproterenol was recorded in guinea pig portal vein using microelectrodes (39). In rat uterine muscle, decreased Na conductance may contribute to the β-adrenergic receptor-mediated hyperpolarization (11), although isoproterenol still produces relaxation in Na-free solution. Electrogenic Ca pumping has implicated (24) but cannot be entirely responsible for the effect. In the rabbit main pulmonary artery, isoproterenol caused increased membrane resistance; it was concluded that electrogenic Na pumping is mainly responsible for the hyperpolarization (36).

In the rabbit facial vein, preliminary evidence tends to rule out electrogenic Na pumping, since relaxation still occurs in response to isoproterenol (10^{-9} *M*) in the presence of ouabain (10^{-4} *M*), and both relaxation and hyperpolarization occur in K-free solution, which should also block the pump. Other possible mechanisms remain to be tested.

Acknowledgments

New research presented in this chapter was supported by USPHS Grant HL 08359 and AHA-GLAA 408 IG.

It is a pleasure to thank *David Nichols* for carrying out some of the experiments reported in this chapter.

References

1 *Ahlquist, R.P.:* A study of the adrenotropic receptors. Am. J. Physiol. *153:* 586–600 (1948).

2 *Alexander, R.S.:* Role of calcium in the plasticity of venous smooth muscle. Am. J. Physiol. *213:* 287–294 (1967).

3 *Ariens, E.J.:* The structure-activity relationships of beta-adrenergic drugs and beta-adrenergic blocking drugs. Ann. N.Y. Acad. Sci. *139:* 606–631 (1967).

4 *Bevan, J.A. and Bevan, R.D.:* Response of blood vessels to sympathetic nerve stimulation. Proc. Symp. Molecular and Cellular Aspects of Vascular Smooth Muscle in Health and Disease (Karger, Basel, in press, 1978).

5 *Bohr, D.F.:* Adrenergic receptors in coronary arteries. Ann. N.Y. Acad. Sci. *139:* 799–807 (1967).

6 *Buckner, C.K. and Patil, P.N.:* Steric aspects of adrenergic drugs. XVI. Beta-adrenergic receptors of guinea-pig atria and trachea. J. Pharmac. exp. Ther. *176:* 634–649 (1971).

7 *Burks, T.F. and Cooper, T.:* Enhancement of peripheral alpha-receptor stimulation by blockade of 'silent' beta receptors. Circulation Res. *21:* 703–715 (1967).

8 *Cannon, W.B. and Rosenblueth, A.:* Autonomic neuroeffector systems (MacMillan, New York 1937).

9 *Celander, O.:* The range of control exercised by the 'sympathicoadrenal system'. Acta physiol. scand. *32:* suppl. 116, pp. 1–132 (1954).

10 *Cuthbert, A.W. and Sutter, M.C.:* The effects of drugs on the relations between the action potential discharge and tension in a mammalian vein. Br. J. Pharmacol. Chemother. *25:* 592 (1965).

11 *Daniel, E.E.; Paton, D.M.; Taylor, G.S., and Hodgson, B.J.:* Adrenergic receptors for catecholamine effects on tissue electrolytes. Fed. Proc. Fed. Am. Socs exp. Biol. *29:* 1410–1425 (1970).

12 *Lande, I.S. de la; Harvey, J.S., and Holt, S.:* Response of the rabbit coronary arteries to autonomic agents. Blood Vessels *11:* 319–337 (1974).

13 *Furchgott, R.F.:* The classification of adrenoceptors (adrenergic receptors). An evaluation from the standpoint of receptor theory; in *Blaschko and Muscholl* Catecholamines, pp. 283–335 (Springer, Berlin 1972).

14 *Furchgott, R.F.:* The pharmacological differentiation of adrenergic receptors. Ann. N.Y. Acad. Sci. *139:* 553–570 (1967).

15 *Gerova, M. and Gero, J.:* The effect of prolonged sympathetic stimulation on conduit vessel diameter. Experientia *24:* 1134–1135 (1968).

16 *Greenway, C.V. and Lawson, A.E.:* Beta-adrenergic receptors in the hepatic arterial bed of the anesthetized cat. Can. J. Physiol. Pharmacol. *47:* 415–420 (1969).

17 *Greenway, C.V. and Lawson, A.E.:* The effects of adrenaline and noradrenaline on venous return and regional blood flows with special reference to intestinal blood flow. J. Physiol., Lond. *186:* 579–599 (1966).

18 *Greenway, C.V.; Lawson, A.E., and Starke, R.D.:* Vascular responses of the spleen to nerve stimulation during normal and reduced blood flow. J. Physiol., Lond. *194:* 421–433 (1968).

19 *Holman, M.E.; Kasby, C.B.; Suthers, M.B., and Wilson, J.A.F.:* Some properties of the smooth muscle of rabbit portal vein. J. Physiol., Lond. *196:* 111 (1968).

20 *Hughes, J. and Vane, J.R.:* An analysis of the responses of the isolated portal vein of the rabbit to electrical stimulation and to drugs. Br. J. Pharmacol. Chemother. *30:* 46 (1967).

21 *Hughes, J. and Vane, J.R.:* Relaxations of the isolated portal vein of the rabbit induced by nicotine and electrical stimulation. Br. J. Pharmacol. Chemother. *39:* 476 (1970).

22 *Kalsner, S.:* Intrinsic prostaglandin release: a mediator of anoxia-induced relaxation in an isolated coronary artery preparation. Blood Vessels *13:* 155–166 (1976).

23 *Malik, K.U. and Nasjletti, A.:* Facilitation of adrenergic transmission by locally generated angiotensin II in rat mesenteric arteries. Circulation Res. *38:* 26–30 (1976).

24 *Marshall, J.M. and Kroeger, E.A.:* Adrenergic influences on uterine smooth muscle. Phil. Trans. R. Soc. B *265:* 135–148 (1973).

25 *Moran, N.C.:* Pharmacological characterization of adrenergic receptors. Pharmacol. Rev. *18:* 503–512 (1966).

26 *Moran, N.C.:* Adrenergic receptors; in *Blaschko, Sayers and Smith* Handbook of physiology, vol. VI, chap. 29, pp. 447–472 (Am. Physiological Society, Washington 1975).

27 *Pegram, B.L.; Bevan, R.D., and Bevan, J.A.:* Facial vein of the rabbit: neurogenic vasodilatation mediated by β-adrenergic receptors. Circulation Res. *39:* 854–860 (1976).

28 *Pitt, B.; Elliot, E.C., and Gregg, D.E.:* Adrenergic receptor activity in the coronary arteries of the unanesthetized dog. Circulation Res. *21:* 75–86 (1967).

29 *Powell, C.E. and Slater, I.H.:* Blocking of inhibitory adrenergic receptors by a dichloro analog of isoproterenol. J. Pharmac. exp. Ther. *122:* 480–499 (1967).

30 *Prehn, J.L. and Bevan, J.A.:* Facial vein of the rabbit: intracellulary recorded hyperpolarization of smooth muscle cells induced by beta-adrenergic stimulation (in preparation, 1977).

31 *Prehn, J.L. and Bevan, J.A.:* Isoproterenol hyperpolarization of smooth muscle cells in rabbit facial vein. Fed. Proc. Fed. Am. Socs exp. Biol. *36:* 523 (1977).

32 *Rosell, S. and Belfrage, E.:* Adrenergic receptors in adipose tissue and their relation to adrenergic innervation. Nature, Lond. *253:* 738–739 (1975).

33 *Ross, G.:* Effects of epinephrine and norepinephrine on the mesenteric circulation of the cat. Am. J. Physiol. *212:* 1037–1042 (1967).

34 *Ross, G.:* The regional circulation. Annu. Rev. Physiol. *33:* 445–478 (1971).

35 *Ross, G. and Kurrasch, M.:* Adrenergic responses of the hepatic circulation. Am. J. Physiol. *216:* 1380–1385 (1969).

36 *Somlyo, A.P.; Somlyo, A.V., and Smiesko, V.:* Cyclic AMP and vascular smooth muscle; in Adv. cyclic Nucl. Res., vol. 1, pp. 175–194 (Raven Press, Hewlett 1972).

37 *Uchida, E. and Bohr, D.F.:* Myogenic tone in isolated perfused resistance vessels from rats. Am. J. Physiol. *216:* 1343–1350 (1969).

38 *Viveros, O.H.; Garlic, D.G., and Renkin, E.M.:* Sympathetic beta-adrenergic vasodilation in skeletal muscle of the dog. Am. J. Physiol. *215:* 1218–1225 (1968).

39 *Loh, D. von:* The effect of adrenergic drugs on spontaneously active vascular smooth muscle studied by long-term intracellular recording of membrane-potential. Angiologica *8:* 144 (1971).

40 *Winquist, R.J. and Bevan, J.A.:* Development of intrinsic muscle tone in the rabbit facial vein. Proc. West. Pharmacol. Soc. *20:* 149–152 (1977).

J.A. Bevan, MD, Brain Research Institute and Department of Pharmacology, School of Medicine, University of California, *Los Angeles, CA 90024* (USA)

Mechanisms of Vasodilatation. Satellite Symp., 27th Int. Congr. Physiol. Sci., Wilrijk 1977, pp. 266–277 (Karger, Basel 1978)

Histaminergic and Cholinergic Vasodilator Systems

Michael J. Brody

Department of Pharmacology and Cardiovascular Center, University of Iowa, Iowa City, Iowa

Relaxation of vascular smooth muscle induced by changes in neurogenic activity is achieved by two mechanisms acting either independently or simultaneously. The first of these mechanisms involves withdrawal of vasoconstrictor activity of the adrenergic sympathetic nerves. The second mechanism involves the release of a substance which promotes vascular smooth muscle relaxation.

The purpose of this paper is to review the current status of two major neurogenic vasodilator systems. These systems are the sympathetic cholinergic and the histaminergic vasodilator systems. The systems are similar in that their function is to regulate vascular smooth muscle caliber in skeletal muscle primarily. Despite a similar anatomic distribution, the systems subserve uniquely different physiological functions.

1. Sympathetic Cholinergic System

This system has been the subject of several excellent reviews by *Uvnäs* (72, 73), and the reader is referred to these for the details and historical development. Although it had been known for many years that stimulation of autonomic nerves could give rise to vasodilatation, it was not until 1935 that *Bülbring and Burn* (24) and *Rosenblueth and Cannon* (58) demonstrated that the vasodilatation produced in the hindlimbs of experimental animals by stimulation of the sympathetic nerves was mediated by a cholinergic mechanism. These early studies were significant in that they demonstrated that the sympathetic innervation to the blood vessels was organized into distinct vasoconstrictor and vasodilator pathways (25). It is interesting to note that the functional and anatomical separation of sympathetic vasoconstrictor and vasodilator fibers is now being revised. Adrenergic discharge appears to suppress the sympathetic cholinergic nerves (28) and acetylcholine is among the most potent inhibitors of transmitter release from noradrenergic nerve terminals (76).

Segmental vascular and organ distribution. From both functional and anatomical standpoints, it appears that sympathetic cholinergic vasodilator fibers distribute almost exclusively to small arterial and arteriolar precapillary vessels (72). Thus, the major function of the system is to produce large changes in vascular resistance without greatly affecting capacitance vessels. Although veins are not ordinarily innervated by sympathetic cholinergic terminals, the existence of inhibitory muscarinic receptors on noradrenergic terminals innervating venous smooth muscle emphasizes potential interactions between adrenergic and cholinergic innervation when both are present in the same vascular segment (76).

Largely on the basis of studies by *Uvnäs* (72, 73) and co-workers, it had been felt that sympathetic cholinergic innervation was limited to a functional role only in skeletal muscle. A number of studies have suggested that this is not the case and that other organs receive a cholinergic vasodilator innervation which can be unmasked at least by direct electrical stimulation of the anatomic sympathetic nerves. For example, sympathetic cholinergic vasodilator innervation has been described in skin (23) and uterus (5, 12, 61). Important vascular beds, such as the renal and cerebral circulations, are innervated by cholinergic fibers which in the case of brain appear to be non-sympathetic, while in the case of kidney may well belong to the anatomic sympathetic nerves (48, 50). The function of the proposed renal cholinergic sympathetic nerves is not well established although there have been several suggestions that these nerves may participate in visceral renal reflexes (43) and intra-renal reflexes promoting redistribution of renal blood flow (49).

Myers et al. (49) have recently challenged the well-accepted concept that sympathetic cholinergic innervation is purely post-ganglionic. They have identified histologically ganglion cells in skeletal muscle, which persist after total denervation of the muscle. They suggest that the sympathetic cholinergic vasodilator response elicited by hypothalamic stimulation is produced by release of an unknown transmitter from terminals originating from ganglion cells which are innervated by the classical post-ganglionic cholinergic sympathetic nerves. They envision the site of cholinergic blockade of sympathetic vasodilatation to be the synapse between the post-ganglionic sympathetic terminal and the ganglionic cell and not the neuroeffector junction. The pharmacological evidence supporting this hypothesis is intriguing; however, the study does not rule out the direct functional link between post-ganglionic sympathetic nerves and skeletal muscle vessels, nor does it provide a quantitative histologic description of the fate of cholinergic terminals following denervation of the muscle.

Species. Cholinergic sympathetic vasodilatation has been demonstrated in dog, cat, and rabbit (73) as well as in man (1, 65). It is of interest that sympathetic cholinergic vasodilatation has not been found in other widely used experimental animals such as rat (18, 20) and primates (73). By carefully identifying central sites from which cholinergic vasodilatation could be elicited in dogs and cats, *Schramm et al.* (63) found that a system homologous to the cholinergic sympathetic vasodilator system was present in a sub-human primate, the squirrel monkey. The chemical mediator of this homologous vasodilator pathway remains to be determined but it is clearly not acetylcholine.

Central representation. The early studies which identified the central component of the sympathetic cholinergic vasodilator system (32, 33, (72) have been refined and extended by the classic study of *Schramm and Bignall* (64). The cholinergic sympathetic vasodilator pathway (and its homologous subhuman primate system) originates in motor cortex and is a discrete descending pathway which relays in hypothalamus and continues through ventral mesencephalon descending through medullary lateral spinothalamic tract. Of particular importance to vasomotor mechanisms is the confirmation (64) that the sympathetic cholinergic pathway did not pass through pontine or medullary reticular 'vasomotor centers'. (See discussion below on role of cholinergic vasodilatation in baroreceptor reflexes.)

Until recently, electrophysiological evidence for activation of specific sympathetic vasodilator pathways by central stimulation has been rather scant. *Folkow and Gernandt* (34) first observed increased neural activity of the peroneal nerve on excitation of the hypothalamic cholinergic vasodilator area; however, these recordings were not correlated with muscle blood flow changes nor were the fibers clearly distinguished from those subserving a vasoconstrictor function. Two recent studies (39, 45) document the existence of specific vasodilator neurones in the sympathetic innervation of the skeletal muscle. These fibers appear to be ordinarily silent, are resistant to somatic reflex activation, and increase their activity upon activation of hypothalamic areas which induce vasodilatation mediated by a cholinergic mechanism.

Baroreceptor reflex. The question of whether the sympathetic cholinergic vasodilator system plays a role in reflex regulation of the circulation, once agreed by most workers to be answered negatively, has been reopened. A large number of studies have demonstrated that activation of the baroreceptor reflex, by increasing arterial pressure or by direct stimulation of the carotid sinus nerves, produces vasodilatation which is not affected by cholinergic blockade (9, 10, 19, 35, 36, 77). Recently, others have found evidence for reduction of the baroreceptor reflex by cholinergic blockade in the cat (67, 68), rabbit (26), and dog (54). We recently studied (66) the influence of low doses of atropine (0.2 mg/kg) on reflex vasodilatation induced in the perfused hindquarters of the cat by several pressor doses of norepinephrine. No effect was seen on the magnitude or duration of the reflex responses while responses to intra-arterial acetylcholine were abolished. It is our opinion that only small doses of atropine, sufficient to produce total cholinergic blockade, provide the specificity needed to test for the role of cholinergic vasodilatation in the baroreceptor reflex. The results of others (26, 54, 67, 68) were obtained with doses of cholinergic blocker well in excess of those needed to produce minimum blockade. It may be concluded that cholinergic mechanisms probably play no role in the baroreceptor reflex, or at best, make a very minor contribution (26, 54, 67, 68) to the total response.

Physiological role. The best evidence for a normal physiological function of the cholinergic vasodilator system is that it plays a role in changing muscle blood flow in association with behavioral reactions such as rage or anxiety (3, 4). Although activation of specific hypothalamic areas producing defense reactions and sham rage are clearly associated with large increases in muscle blood flow mediated by the cholinergic system, it is less clear whether this effect represents a significant effect metabolically since activation of the cholinergic system appears to shunt flow away from nutritive channels and reduces oxygen uptake (17, 56, 57). A recent report (13) may help resolve the discrepancy. Sympathetic vasodilatation appears to activate muscle phosphorylase and may thus help prepare skeletal muscle for anaerobic metabolism. Another behavioral correlate involving cholinergic vasodilatation is the conditioned response in the trained limbs of cats (79). It is clear that a substantial amount of additional work is required before the precise physiological roles of the cholinergic system are established.

2. Histaminergic System

Since the early original descriptions of its profound pharmacologic effects (27) and endogenous distribution (2), histamine has been among the most widely investigated of the naturally occurring biogenic amines. There now exists an extensive literature on the role of histamine as a chemical mediator of physiologic and pathophysiologic phenomena such as inflammation, asthma, anaphylaxis, salivary secretion, and gastric hydrochloric acid secretion. Details about these proposed roles for histamine may be found in several monographs (30, 41, 46).

Histaminergic nature of reflex vasodilatation. The proposed role of histamine as a transmitter in the peripheral sympathetic nervous system is intimately involved with its function as a physiologic regulator of vascular resistance. A key observation was made in 1961 by *Beck* (7), who observed that the vasodilatation produced in the hindlimbs by activation of the baroreceptor reflex could not be accounted for solely on the basis of withdrawal of adrenergic vasoconstrictor tone, as was traditionally accepted (74). These experiments indicated that an endogenous vasodilator material must have been responsible for that component of reflex dilatation which exceeded withdrawal of sympathetic tone. Using classic pharmacologic techniques, it was determined (10) that histamine, and not a large number of potential endogenous vasodilator mediators, was responsible for that component of reflex vasodilatation produced by the action of a vasodilator. The role of histamine in reflex vasodilatation has been confirmed many times with the use of antihistamines and the specificity of histamine blockade has also been confirmed (15, 16). Although most of the work on histamine-mediated active vasodilatation has been performed on dogs (6, 8, 23, 37, 62, 78), the role of histamine in this vascular response has also been demonstrated in monkeys (42), cats (70), and rats (69).

Formal demonstration of a possible neurotransmitter function of histamine requires much more than the pharmacologic demonstration that the neurogenic response is amenable to specific receptor blockade. In dogs and monkeys (59), histamine was found to be concentrated in post-ganglionic sympathetic nerves relative to other portions of the autonomic nervous system. The uptake of radiolabeled histamine, like that of norepinephrine, was also concentrated at the post-ganglionic sympathetic nerve (59). Perfused muscle 'prelabeled' with isotopic histamine released the label (and metabolites) during reflex vasodilatation but not during vasodilatation produced by nitroglycerin or during sympathetic nerve stimulation (19). Similar release of labeled histamine has been demonstrated in dog (11), cat (71) and during activation of specific spinal ventral outflow pathways (44). These data strongly suggested that histamine was elaborated from specific histaminergic fibers activated during reflex vasodilatation.

Nature of histamine release. The effects of post-ganglionic sympathetic denervation on histamine content of skeletal muscle were unexpected (60). If histamine were a classical neurotransmitter, released from specific post-ganglionic sympathetic nerves, the amine should have disappeared upon nerve degeneration. Histamine levels in muscle were unaffected, however. This led to the postulate (60) that the release of histamine during reflex vasodilatation was from a non-neurogenic store probably under the reciprocal control of adrenergic nerves. Recently, direct functional evidence has been provided for this postulated mechanism of release using a model system for active reflex vasodilatation (38). Vasoconstrictor tone was established in a decentralized perfused muscle using electrical stimulation of the cut lumbar sympathetic chain. Upon cessation of nerve stimulation (analogous to reflex withdrawal of sympathetic tone) active vasodilatation occurred. The dilatation was blocked by antihistamine and was also prevented by α-adrenergic receptor blockade. The latter observation demonstrates that an adrenergic link exists between the post-ganglionic sympathetic nerve and the proposed non-neurogenic store of histamine. On the basis of these data, it was proposed (38) that the normal tonic release of norepinephrine from sympathetic nerve terminals acts upon a non-neurogenic cellular store of histamine to stabilize the membrane and prevent elaboration of the amine, through an α-adrenergic receptor mechanism. Upon cessation of the release of norepinephrine, the stabilizing effect is lost and vasodilatation ensues, in part from the local vascular action of histamine and in part from disappearance of the vasoconstrictor effect of norepinephrine.

Location of vascular histamine. Since considerable insight about the nature of the storage and release of non-neurogenic histamine could be obtained from a demonstration of the localization of the proposed histamine store, studies on the cellular localization of vascular histamine have been carried out. Earlier studies demonstrated that histamine is found in high concentrations in the walls of both arteries and veins (40, 60). On the basis of indirect evidence, the storage site for histamine did not appear to be associated with nerves since the histamine content of the vessel wall was not reduced by careful stripping of adventitia.

In general, histamine could be stored in either a mast cell or nonmast cell pool. Much of the earlier work on tissue histamine content was performed on rodents in whom a substantial fraction of histamine is found in mast cells stored in association with specific granules (75). These cells can be stained metachromatically with toluidine blue and similar stains, and are easily identified and visualized. In addition, storage of histamine in such cells can be disrupted with the use of agents such as Compound 48/80 which disrupt the mast cell membrane and allow the storage granules to be discharged (55). Mast cell storage of histamine did not appear to be important in dog since the content in skeletal muscle and other organs was not affected by enormous doses of Compound 48/80 (60).

Using a variety of histologic and fluorescence histochemical techniques, we found that in all species studied, save cattle, no mast cells were associated with the blood vessel wall (31). This provided direct evidence that the histamine involved in mediating reflex vasodilatation was not associated with a mast cell pool. Despite the fact that histamine was observed in large concentration in the vessel wall (60), it could not be visualized using a sensitive and specific fluorescence histochemical technique (21). It was proposed that histamine, unlike norepinephrine, might be diffusely distributed across the vessel wall and thus not reach sufficient local concentration to be visualized histochemically.

Histidine, the immediate precursor amino acid for histamine, was infused into muscle in large concentrations along with inhibitors of histamine methyltransferase and diamine oxidase, the enzymes primarily responsible for histamine degradation. Under these conditions, it was anticipated that synthesis of histamine would be accelerated by the presence of large amounts of precursor and that the amine would accumulate because metabolic degradation would be blocked (21). No fluorescence was observed in preparations receiving histidine alone, histamine alone, or histamine in combination with the inhibitors of histamine degradation. The only conditions which produced specific fluorescence in the vessel wall were infusion of histidine along with the enzyme inhibitors. Using microspectrophotofluorometry, the fluorophore was found to possess the spectral qualities of authentic histamine. Aside from specific histamine-containing cells associated with gastric mucosa in the rat, these results represent the first direct demonstration of a non-mast cell store of histamine. Specific cells in blood vessel walls appear to have the capacity to synthesize and store histamine. Ordinarily, the storage of the amine is of insufficient concentration to be visualized with available histochemical techniques. However, under the special conditions described, the amine has been readily observed.

Receptor mechanisms. The nature and location of receptors for histamine in the cardiovascular system were investigated using new specific histamine antagonists (14) and agonists (29). On the arterial side, it was found that histamine interacts with two specific receptors, both of which induce vasodilator responses. Burimamide and metiamide, new antagonists of histamine which interact with type H_2 receptors (14), block that component of the histamine response not antagonized by the classical type H_1 receptor antagonists such as mepyramine. The contraction produced in veins by histamine was demonstrated to involve H_1 receptors only. Using specific receptor agonists – pyridyl-ethylamine (PEA) for H_1 receptors and 4-methyl histamine for H_2 receptors – the results obtained with H_1 and H_2 receptor antagonists were confirmed. These agents each produced vasodilatation which was antagonized in relatively specific fashion by the respective antagonists. Details of these studies are summarized elsewhere (51–53).

The observation that the vasodilator effects of histamine could be sequentially antagonized in a specific manner with the use of H_1 and H_2 receptor antagonists led to an evaluation of the role of histamine as mediator of several physiological vasodilator responses (53). Several forms of neurogenically induced active vasodilatation were demonstrated to involve both H_1 and H_2 receptors. These included active reflex vasodilatation, vasodilatation induced by sympathetic nerve stimulation following adrenergic neuronal blockade, and the vasodilatation which follows cessation of sympathetic nerve stimulation. Although there have been frequent suggestions that histamine may be involved in reactive hyperemia and the hyperemia associated with exercise, no evidence for a role of histamine could be identified with the combined use of H_1 and H_2 receptor antagonists.

Histaminergic vasodilator mechanisms. Figure 1 schematically depicts the mechanisms by which endogenous and exogenous stores of histamine as well as adrenergic nerve terminals are involved in producing relaxation of vascular smooth muscle. Experimental details supporting this schema are discussed in this and in another review (22). Histamine is contained in two storage locations; the mast cell depicted at the bottom of figure 1 and a non-mast cell pool depicted at the top of the blood vessel. Histamine administered into the blood stream appears to have the capacity to interact on the luminal side of the vessel with two types of histamine receptors, H_1 and H_2. In distinction, histamine liberated from a mast cell pool appears to interact selectively with an H_2 receptor which

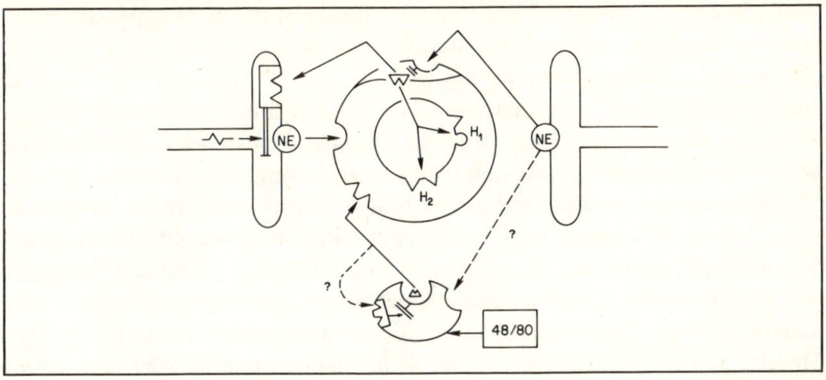

Fig. 1. Histaminergic vasodilator mechanisms. Norepinephrine interacts with α-adrenergic receptors blocked by phentolamine. H_1 = Mepyramine; H_2 = metiamide. Details are found in the text – from (22).

may well be on the adventitial side of the vessel (53). Reflex vasodilatation appears to involve release of histamine from the non-mast cell pool which now can be visualized with fluorescence histochemistry following histidine loading. The mechanism of release appears to involve disappearance of the anti-release or membrane-stabilizing activity of norepinephrine liberated from special non-vasoconstrictor sympathetic fibers. These are depicted on the right side of figure 1. Reflex vasodilatation occurs simultaneously from the disappearance of the vasoconstrictor effects of norepinephrine release from the terminal shown on the left and from a release of histamine from the non-mast cell vascular muscle pool. Although not definitely established as part of reflex vasodilatation, histamine itself through an H_2 receptor mechanism, inhibits the release of norepinephrine from adrenergic nerve terminals (47).

3. Conclusion

Two vasodilator systems, innervating primarily skeletal muscle and subserving distinct physiological functions, have been identified and characterized. The neurotransmitters subserving these sympathetic efferent pathways are histamine and acetylcholine, respectively. The histaminergic system is involved as the mediator of active vasodilatation evoked by the baroreceptor reflex. Histaminergic vasodilatation exceeds in magnitude the maximum vasodilatation achieved by inhibition of adrenergic vasoconstrictor tone. Biochemical and fluorescence histochemical studies of histamine distribution in sympathetic nerves and vascular smooth muscle indicate that histamine is stored in a non-neuronal, non-mast cell pool in media of blood vessels. Release of this pool appears to be under reciprocal regulation of non-vasoconstrictor adrenergic fibers, i.e. adrenergic discharge inhibits histamine release while inhibition of adrenergic firing leads to discharge of histamine from its vascular storage site. Both H_1 and H_2 vascular histamine receptors are involved in the vasodilator responses. The histaminergic system can be activated centrally from brain stem and hypothalamic sites. The cholinergic vasodilator system probably does not play an important role in the baroreceptor reflex and its central representation appears to be anatomically distinct from the histaminergic system. Acetylcholine is released from classical post-ganglionic sympathetic cholinergic nerves, although evidence exists for inhibition of the cholinergic vasodilator mechanism by the adrenergic system. The sympathetic cholinergics appear to be involved physiologically in emotional reactions.

References

1 *Abboud, F.M. and Eckstein, J.W.:* Reflex vasoconstrictor and vasodilator responses in man. Circulation Res. *18:* suppl. I, pp. I-96–I-103 (1966).

2 *Abel, J.J. and Kubota, S.:* Presence of histamine (4-imidazole ethylamine) in the hypophysis cerebri and other tissues of the body and its occurrence among the hydrolytic decomposition products of proteins. J. Pharmac. exp. Ther. *13:* 243–299 (1919).

3 *Abrahams, V.C. and Hilton, S.M.:* Active muscle vasodilatation and its relation to the flight and fight reactions in the conscious animal. J. Physiol., Lond. *140:* 16–17 (1958).

4 *Abrahams, V.C.; Hilton, S.M., and Zbrozina, A.:* Active muscle vasodilatation elicited by mesencephalic stimulation. Its relation to the defence reaction. J. Physiol., Lond. *148:* 32P (1959).

5 *Adham, N. and Schenk, E.:* Autonomic innervation of the rat vagina, cervix and uterus and its cyclic variation. Am. J. Obstet. Gynec. *104:* 508–516 (1969).

6 *Aoki, V.S. and Brody, M.J.:* Medullary control of vascular resistance: an electrophysiological analysis. Circulation Res. *18:* suppl. 1, pp. 75–85 (1966).

7 *Beck, L.:* Active reflex vasodilatation in the innervated perfused hind limb of the dog. Am. J. Physiol. *201:* 123–128 (1961).

8 *Beck, L.:* Histamine as the potential mediator of active reflex vasodilatation. Fed. Proc. Fed. Am. Socs exp. Biol. *24:* 1298–1310 (1965).

9 *Beck, L.:* A new concept of autonomic interactions in the peripheral sympathetic nervous system. Tex. Rep. Biol. Med. *22:* 375–409 (1964).

10 *Beck, L. and Brody, M.J.:* The physiology of vasodilatation. Angiology *12:* 202–221 (1961).

11 *Beck, L.; Schon, D.; Spalding, C.T., and Wise, D.:* Evidence for greater release of histamine during active reflex dilatation than during passive reflex dilatation. Proc. West. Pharmacol. Soc. *14:* 63–67 (1971).

12 *Bell, C.:* Autonomic nervous control of reproduction: circulatory and other factors. Pharmacol. Rev. *24:* 657–736 (1972).

13 *Berdina, N.A. and Rodionov, I.M.:* Activation of anaerobic metabolism in cat skeletal

muscle during cholinergic sympathetic vasodilatation. Pflügers Arch. *267:* 37–42 (1976).
14 *Black, J.W.; Duncan, W.A.M.; Durane, C.V.; Ganellin, C.L., and Parsons, E.M.:* Definition and antagonism of histamine H_2-receptors. Nature, Lond. *236:* 385–390 (1972).
15 *Boerth, R.C.; Ryan, M.J., and Brody, M.J.:* Pharmacologic blockade of reflex vasodilatation: effects on postulated neurohumoral mechanisms. J. Pharmac. exp. Ther. *172:* 52–61 (1970).
16 *Boerth, R.C.:* Pharmacologic blockade of reflex vasodilatation: peripheral actions of antihistamines. Eur. J. Pharmacol. *210:* 312–320 (1972).
17 *Bolme, P. and Novotny, J.:* Oxygen uptake in skeletal muscle of the anesthetized dog during sympathetic vasodilatation. Acta physiol. scand. *77:* 333–343 (1969).
18 *Brody, M.J.:* Cardiovascular responses following immunological sympathectomy. Circulation Res. *15:* 161 (1964).
19 *Brody, M.J.:* Neurohumoral mediation of active reflex vasodilatation (symp.) Fed. Proc. Fed. Am. Socs exp. Biol. *25:* 1583–1592 (1966).
20 *Brody, M.J. and Dixon, R.L.:* Vascular reactivity in experimental diabetes mellitus. Circulation Res. *14:* 494 (1964).
21 *Brody, M.J.; Hakanson, R.; Owman, C., and Sundler, F.:* An improved method for the histochemical demonstration of histamine and other compounds producing fluorophores with OPT. J. Histochem. Cytochem. *20:* 945 (1972).
22 *Brody, M.J.; Knuepfer, M.; Strait, M.R., and Shaffer, R.A.:* Histamine receptors in vascular smooth muscle: mechanisms of vasodilatation; in *Yellin* Histamine receptors (Spectrum Press, New York, in press).
23 *Brody, M.J. and Shaffer, R.:* Distribution of vasodilator nerves in the canine hindlimb. Am. J. Physiol. *218:* 470–474 (1970).
24 *Bülbring, E. and Burn, J.H.:* The sympathetic dilator fibre in the muscles of the cat and dog. J. Physiol., Lond. *83:* 483 (1935).
25 *Burn, J.H.:* Sympathetic vasodilator fibre. Physiol. Rev. *18:* 137 (1938).
26 *Chiariello, M.; Condorelli, M.; Caprio, L. de; Rengo, F.; Sacca, L., and Trimarco, B.:* The cholinergic component in the reflex vasodilatation elicited by stimulation of the depressor nerves in the rabbit. Br. J. Pharmacol. *57:* 375–379 (1976).
27 *Dale, H.H. and Laidlow, P.P.:* The physiological action of β-iminazolylethylamine. J. Physiol., Lond. *41:* 318–344 (1910).
28 *Dorr, L.D. and Brody, M.J.:* Functional separation of adrenergic and cholinergic fibers to skeletal muscle vessels. Am. J. Physiol. *208:* 417 (1965).
29 *Durant, G.J.; Ganellin, C.R., and Parsons, M.E.:* Chemical differentiation of histamine H_1- and H_2-receptor agonists. 168th Am. Chem. Soc. Natn. Meet., MEDI-29 (1974).
30 *Eichler, O. und Farah, A. (eds.):* Handbuch der experimentellen Pharmakologie (Springer, Berlin 1966).
31 *El-Ackad, T.M. and Brody, M.J.:* Evidence for non-mast cell histamine in the vascular wall. Blood Vessels *12:* 181–191 (1975).
32 *Eliasson, S.; Lindgren, P., and Uvnas, B.:* Representation in the hypothalamus and the motor cortex in the dog of the sympathetic vasodilator outflow to the skeletal muscles. Acta physiol. scand. *27:* 18 (1952).
33 *Eliasson, S.; Lindgren, P., and Uvnas, B.:* The hypothalamus, a relay station of the sympathetic vasodilator tract. Acta physiol. scand. *31:* 290 (1954).
34 *Folkow, B. and Gernandt, B.E.:* An electrophysiological study of the sympathetic vasodilator fibers of the limb. Am. J. Physiol. *169:* 622 (1952).
35 *Folkow, B. and Uvnäs, B.:* The distribution and functional significance of sympathetic vasodilators to the hindlimbs of the cat. Acta physiol. scand. *15:* 389 (1948).

36 *Frumin, M.J.; Ngai, S.H., and Wang, S.C.:* Evaluation of vasodilator mechanisms in the canine hind leg; question of dorsal root participation. Am. J. Physiol. *173:* 428 (1953).

37 *Graham, B.H. and Lioy, F.:* Histaminergic vasodilatation in the hindlimb of the dog. Pflügers Arch. *342:* 307–318 (1973).

38 *Heitz, D.C. and Brody, M.J.:* Possible mechanism of histamine release during active vasodilatation. Am. J. Physiol. *228:* 1351 (1975).

39 *Horeyseck, G.; Janig, W.; Firchner, F., and Thamer, V.:* Activation and inhibition of muscle and cutaneous postganglionic neurones to hindlimb during hypothalamically induced vasoconstriction and atropine-sensitive vasodilation. Pflügers Arch. *361:* 231–240 (1976).

40 *Howland, R.D. and Spector, S.:* Disposition of histamine in mammalian blood vessels. J. Pharmac. exp. Ther. *182:* 239–245 (1972).

41 *Kahlson, G. and Rosengren, E. (eds.):* Biogenesis and physiology of histamine (Williams & Wilkins, Baltimore 1971).

42 *Levin, J.A.; Bartlett, J.D., and Beck, L.:* Active reflex vasodilatation induced by intravenous epinephrine or norepinephrine in primates. J. Pharmac. exp. Ther. *161:* 262–270 (1968).

43 *Liang, C.C.:* A possible sympathetic cholinergic mechanism in the renal reflex elicited by stimulation of abdominal viscera in the dog. Clin. exp. Pharmacol. Physiol. *2:* 103–117 (1975).

44 *Lioy, F. and White, K.P.:* ^{14}C-histamine release during vasodilatation induced by lumbar ventral root stimulation. Pflügers Arch. *342:* 319–324 (1973).

45 *Lopes, O.U. and Palmer, J.F.:* Electrical activity in sympathetic fibres to hind limb muscles of the cat produced by hypothalamic stimulation. Q. Jl exp. Physiol. *62:* 41–53 (1977).

46 *Maslinski, C. (ed.):* Histamine (Dowden, Hutchinson & Ross, Stroudsburg 1973).

47 *McGrath, M.A. and Shepherd, J.T.:* Inhibition of adrenergic neurotransmission in canine vascular smooth muscle by histamine. Mediation by H_2-receptors. Circulation Res. *39:* 566–573 (1976).

48 *McKenna, O.A. and Angelakos, E.T.:* Acetylcholinesterase-containing nerve fibres in the canine kidney. Circulation Res. *23:* 645–651 (1968).

49 *Myers, H.A.; Schenk, E.A., and Honig, C.R.:* Ganglion cells in arterioles of skeletal muscle: role in sympathetic vasodilatation. Am. J. Physiol. *229:* 126–138 (1975).

50 *Owman, C.; Edvinsson, L., and Nielsen, K.C.:* Autonomic neural receptor mechanisms in brain vessels. Blood Vessels *11:* 2–13 (1974).

51 *Powell, J.R. and Brody, M.J.:* Identification and specific blockade of two receptors for histamine in the cardiovascular system. J. Pharmac. exp. Ther. *196:* 1–14 (1976).

52 *Powell, J.R. and Brody, M.J.:* Identification and blockade of vascular H_2 receptors (symp.). Fed. Proc. Fed. Am. Socs exp. Biol. *35:* 1935–1941 (1976).

53 *Powell, J.R. and Brody, M.J.:* Participation of H_1 and H_2 histamine receptors in physiological vasodilator responses. Am. J. Physiol. *231:* 1002–1009 (1976).

54 *Rengo, F.; Sacca, L.; Trimarco, B.; Caprio, L. de; Vitale, D.; Caridi, A. de e Condorelli, M.:* Study sul sistema colinergico vasodilatante muscolare. Minerva cardioangiol. *24:* 366–378 (1976).

55 *Riley, J.F. and West, G.B.:* The presence of histamine in tissue mast cells. J. Physiol., Lond. *120:* 528–549 (1953).

56 *Rosell, S. and Uvnäs, B.:* Vasomotor control of oxygen consumption in skeletal muscle. Acta physiol. scand. *50:* suppl. 175, pp. 129–130 (1960).

57 *Rosell, S. and Uvnäs, B.:* Vasomotor nerve activity and oxygen uptake in skeletal muscle of the anesthetized cat. Acta physiol. scand. *54:* 209–222 (1962).

58 *Rosenblueth, A. and Cannon, W.B.:* The chemical mediation of sympathetic vasodilator nerve impulses. Am. J. Physiol. *112:* 33 (1935).

59 *Ryan, M.J. and Brody, M.J.:* Distribution of histamine in the canine autonomic nervous system. J. Pharmac. exp. Ther. *174:* 123–132 (1970).

60 *Ryan, M.J. and Brody, M.J.:* Neurogenic and vascular stores of histamine in the dog. J. Pharmac. exp. Ther. *181:* 83–91 (1972).

61 *Ryan, M.J.; Clark, K.E., and Brody, M.J.:* Neurogenic and mechanical control of canine uterine vascular resistance. Am. J. Physiol. *227:* 547–555 (1974).

62 *Sakuma, A. and Beck, L.:* Pharmacological evidence for active reflex dilatation. Am. J. Physiol. *201:* 129–133 (1961).

63 *Schramm, L.P.; Honig, C.R., and Bignall, K.E.:* Active muscle vasodilation in primates homologous with sympathetic vasodilation in carnivores. Am. J. Physiol. *221:* 768–777 (1971).

64 *Schramm, L.P. and Bignall, K.E.:* Central neural pathways mediating active sympathetic muscle vasodilation in cats. Am. J. Physiol. *221:* 754–767 (1971).

65 *Shepherd, J.T.:* Physiology of the circulation in human limbs in health and disease, p. 70 (Saunders, Philadelphia 1963).

66 *Strait, M.R. and Brody, M.J.:* Unpublished observation.

67 *Takeuchi, T. and Manning, J.W.:* Muscle cholinergic dilators in the sinus baroreceptor response in cats. Circulation Res. *29:* 350–357 (1971).

68 *Takeuchi, T. and Manning, J.W.:* Hypothalamic mediation of sinus baroreceptor evoked muscle cholinergic vasodilator response. Am. J. Physiol. *224:* 1280–1287 (1973).

69 *Tobia, A.J.; Adams, M.D.; Miya, T.S., and Bousquet, W.F.:* Histamine and reflex vasodilatation in the rat. Life Sci. *8:* 745–750 (1969).

70 *Tuttle, R.S.:* Histaminergic component in the baroreceptor reflex of the pyramidal cat. Fed. Proc. Fed. Am. Socs exp. Biol. *25:* 1593–1595 (1966).

71 *Tuttle, R.S.:* Physiological release of histamine-C^{14} in the pyramidal cat. Am. J. Physiol. *213:* 620–624 (1967).

72 *Uvnäs, B.:* Sympathetic vasodilator system and blood flow. Physiol. Rev. *40:* suppl. 4, p. 69 (1960).

73 *Uvnäs, B.:* Cholinergic vasodilator nerves. Fed. Proc. Fed. Am. Socs exp. Biol. *25:* 1618–1622 (1966).

74 *Uvnäs, B.:* Central cardiovascular control; in *Field, Magoun and Hall* Handbook of physiology, vol. II, sect. 1 (Am. Physiological Society, Washington 1960).

75 *Uvnäs, B.:* Histamine storage and release. Fed. Proc. Fed. Am. Socs exp. Biol. *33:* 2172–2176 (1974).

76 *Vanhoutte, P.M. and Shepherd, J.T.:* Venous relaxation caused by acetylcholine acting on the sympathetic nerves. Circulation Res. *32:* 259–267 (1973).

77 *Vatner, S.F.; Franklin, D.; Citters, R.L. van, and Braunwald, E.:* Effects of carotid sinus nerve stimulation on blood-flow distribution in conscious dogs at rest and during exercise. Circulation Res. *27:* 495–503 (1970).

78 *Vatner, S.F. and McRitchie, R.J.:* Reflex limb dilatation following norepinephrine and angiotensin II in conscious dogs. Am. J. Physiol. *230:* 557–563 (1976).

79 *Zanchetti, A. and Ellison, G.D.:* Specific appearance of sympathetic cholinergic vasodilatation in muscles during conditioned movements. Nature, Lond. *232:* 124–125 (1971).

M.J. Brody, PhD, Department of Pharmacology and Cardiovascular Center, University of Iowa, *Iowa City, IA 52242* (USA)

Mechanisms of Vasodilatation. Satellite Symp., 27th Int. Congr. Physiol. Sci., Wilrijk 1977, pp. 278–284 (Karger, Basel 1978)

An Assessment of Purinergic Nerve Involvement in Vasodilatation

G. Burnstock

Department of Anatomy and Embryology, University College, London

Classically, the autonomic nervous system has been considered to consist of two components, namely adrenergic and cholinergic nerves. However, experiments of the past decade have revealed a third component in the autonomic nervous system which is neither adrenergic nor cholinergic (3, 9). These nerves are present in the gastro-intestinal tract of a wide range of vertebrate species and have also been identified in lung, trachea, retractor penis, bladder, oesophagus, eye, seminal vesicle and in some parts of the central nervous system. Evidence has been presented that the principal active substance released by some, at least, of these nerves is a purine nucleotide, probably adenosine triphosphate (ATP), and they have therefore been termed 'purinergic' (4, 5). The evidence includes: synthesis and storage of ATP in nerves characterised by a predominance of 'large opaque vesicles'; release of ATP from the nerves when they are stimulated; mimicry by exogenously applied ATP of the action of nerve-released transmitter; the presence of Mg^{++}-activated ATPase, 5′-nucleotidase and adenosine deaminase, enzymes which inactivate ATP; and finally, the similar blocking and potentiating effects produced by drugs on the responses to exogenously applied ATP and nerve stimulation. A model of the synthesis, storage, release and inactivation of ATP during purinergic nerve transmission has been proposed (fig. 1). The ultrastructure of these nerves and the electrophysiology of transmission to smooth muscle have been described (1, 26). Their physiological roles are beginning to be understood, at least in the gastro-intestinal tract, and the possibility that they also supply parts of the vascular system has been considered (4, 5, 19, 28, 29).

1. Purine Nucleotides and Nucleosides and Vasodilatation

The high sensitivity of blood vessels to adenine nucleotides and nucleosides was first recognised by *Drury and Szent-Györgyi* (12). Since then, there have

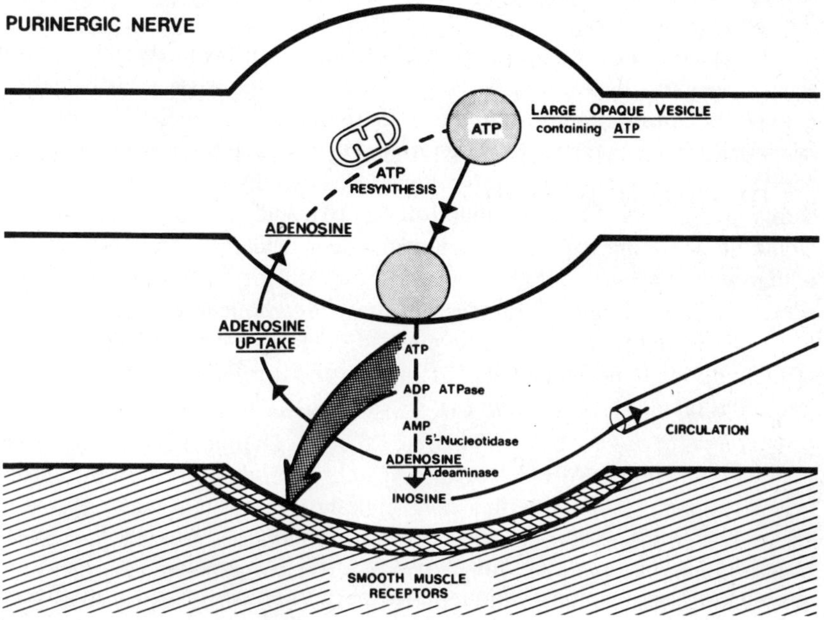

Fig. 1. Schematic representation of synthesis, storage, release and inactivation of ATP at the purinergic neuromuscular junction. From *Burnstock* (4).

been many reports of their actions on various vessels (4, 5, 16, 18). Adenyl compounds cause dilatation of most blood vessels but they have been variably reported to dilate or constrict vessels in the lung and kidney. The doses required to produce vasodilatation *in vivo* are usually lower than the threshold doses which produce slowing of the heart. The high potency and wide-spread action of adenine nucleotides and nucleosides on vascular smooth muscle could indicate a neurohumoral role by analogy with the potent responses of vessels to noradrenaline and acetylcholine released from established autonomic nerves, although high potency alone cannot be taken as evidence for release of purine nucleotides or nucleosides from nerves.

Adenine nucleotides and nucleosides have been collected in the venous effluent of several vascular beds during hypoxia or in other situations and have been considered as physiological regulators of blood flow (2, 24). Evidence for involvement of purine nucleotides and nucleosides in the regulation of blood flow in particular vascular beds has been reviewed in detail recently (6). Release of these compounds during hypoxia or during reflex or direct stimulation of non-adrenergic, non-cholinergic nerves and their potent mimicking vasomotor actions have been demonstrated in various vessels including those in the heart, kidney, skin, brain, salivary glands, lung, mesentery, intestine, uterus and skeletal muscle.

In general, adenosine rather than ATP has been favoured as the vasoactive agent in reactive hyperaemia in various vascular beds, despite the fact that ATP is usually 10 or more times as active in these vessels. One reason for this is that it has usually been assumed that adenosine, but not ATP because of its charge, can pass across cell membranes (15); however, it seems likely that ATP can pass across membranes of hypoxic muscle cells (10), and a specific release mechanism would be operating if ATP was released from purinergic nerves. Another reason is that with the techniques used in early experiments, adenosine was the major adenyl compound collected in the perfusate; however, in experiments using the highly sensitive luciferin-luciferase method, ATP has been detected in the venous effluent during hyperaemia (11, 13, 24, 25); and it is well known that extracellular ATP is rapidly broken down (4).

2. Purinergic Receptors

Evidence has been presented recently to support the view that purinergic receptors can be separated into two types (7). This would be analogous to the established classifications of adrenergic receptors into α- and β-adrenoceptors, cholinergic receptors into muscarinic and nicotinic, and histamine receptors into H_1 and H_2 receptors (30). The main basis for distinguishing these 'receptors' is the selective actions of agonists and antagonists: P_1 purinoceptors are competitively blocked by methylxanthines such as theophylline, caffeine and aminophylline, while P_2 purinoceptors are blocked by quinidine, high concentrations of 2-substituted imidazolines and 2′,2-pyridilisatogen; the relative potencies of agonists for P_1 receptors are AD$\geqslant$AMP$>$ADP$\geqslant$ATP, while the reverse is the case for P_2 receptors. There is also evidence to suggest that P_1, but not P_2, purinoceptors involve activation of adenylate cyclase leading to the production of cyclic AMP. In general, P_1 receptors seem to be predominant in most cardiovascular beds, in the trachea and brain, while P_2 receptors seem to be predominant in the gastro-intestinal tract and urinogenital system. An apparent anomaly in analysing the distribution of purinoceptors is that some of the actions of ATP are blocked by methylxanthines (e.g. brain, and trachea, where P_1 receptors predominate). However, careful examination of the time course of the agonist responses and production of cyclic AMP suggest that ATP is acting after prior hydrolysis to adenosine. The relative effects of P_1 and P_2 antagonists on the responses of various tissues to purinergic nerve stimulation suggest that ATP rather than adenosine is the transmitter released from purinergic nerves in the taenia coli and bladder, while adenosine appears to be the purinergic transmitter in the trachea. The primary responses of the coronary and pulmonary circulations to hypoxia appear to be associated with ATP rather than adenosine as was originally thought, since whereas the vasodilator responses to adenosine are blocked by P_1 antagonists, most workers report that the responses of the pulmonary and coronary circulations to hypoxia and ATP (vasoconstrictor in the lung; vasodilator in the heart) are unaffected (7).

3. Purinergic Nerves

A fluorescence histochemical method has recently been developed which appears to be selective for purinergic nerves (20, 23). We have been applying this method to whole mounts of perivascular plexus and the preliminary results are

Table I. Quinacrine fluorescence in nerves around blood vessels

Quinacrine-positive	Quinacrine-negative
Rabbit	
Ear artery	renal artery
Renal vein	vessels in:
Portal vein	iris
Vessels in:	anococcygeus
Gallbladder	rectococcygeus
Urinary bladder	atrium
Pia mater	dura mater
Choroid plexus	
Caecum	
Guinea pig	
Mesenteric artery	renal artery
Vessels in:	femoral artery
Colon	portal vein
Caecum	renal vein
On surface of	saphenous vein
vas deferens	femoral vein
	vessels in:
	iris
	urinary bladder
	atrium
Rat	
Portal vein	vessels in:
Vessels in:	iris
Anococcygeus	meninges
Cat	
Vessels in:	
Atrium	

Courtesy of *R. Crowe.*

Fig. 2. Rabbit portal vein. Tissue incubated in 5×10^{-7} *M* quinacrine in Krebs' solution for 60 min at 37 °C and gassed with 95% oxygen and 5% carbon dioxide. A plexus of fine varicose fibres can be seen, as well as a bundle of non-terminal fibres. Fluorescent micrograph of whole mount. × 200. Courtesy of *R. Crowe.*

summarised in table I. An example of the appearance of the quinacrine-positive plexus in the rabbit portal vein is illustrated in figure 2. Quinacrine-positive plexuses persist after sympathectomy and have a different appearance from perivascular plexuses stained for acetylcholinesterase, if present. In addition 'large opaque vesicles' comparable to those identified in purinergic nerves in the gastrointestinal tract and lung, have been observed in perivascular plexuses in skin and brain.

Neuromodulation, defined as the regulation of release of transmitter from nerves by the action of neurohumoral agents on prejunctional receptors, is a relatively recent concept which is rapidly becoming recognised to be of physiological significance (27). Apart from inhibition of transmitter output by negative feedback of noradrenaline mediated through prejunctional α-adrenoceptors (21), some inhibition of vasoconstrictor activity of adrenergic nerves appears to be due to the prejunctional actions of ATP and adenosine (17, 31). It seems likely that modulation of cholinergic vasodilator nerve activity by the prejunctional action of adenine nucleotides and nucleosides also occurs (14, 32).

The discovery that ATP is a potent inducer of prostaglandin synthesis (22) led *Burnstock et al.* (8) to suggest that ATP released from purinergic nerves may be linked with prostaglandins in the mechanism of peristalsis, and the possibility that adenine nucleotides are linked with prostaglandins in the physiological regulation of blood flow has also been raised (6).

4. Conclusion

These various pieces of evidence, considered collectively, lead to the conclusion that adenine nucleotides and/or nucleosides are serious contenders for a physiological role in the regulation of blood flow. Whether the source of these compounds is non-nervous or nervous tissues (or both) is still not resolved, but it seems possible that the rapid actions of ATP released from purinergic nerves (acting on P_2 purinoceptors) may be superimposed in some vascular beds on the slower and longer-lasting effects of adenosine released as a result of local metabolic activity (acting on P_1 purinoceptors).

References

1 *Bennett, M.; Burnstock, G., and Holman, M.E.:* Transmission from intramural inhibitory nerves to the smooth muscle of the guinea-pig taenia coli. J. Physiol. Lond. *182:* 541–558 (1966).

2 *Berne, R.M.:* Regulation of coronary blood flow. Physiol. Rev. *44:* 1–29 (1964).

3 *Burnstock, G.:* Evolution of the autonomic innervation of visceral and cardiovascular systems in vertebrates. Pharmacol. Rev. *21:* 247–324 (1969).

4 *Burnstock, G.:* Purinergic nerves. Pharmacol. Rev. *24:* 509–581 (1972).

5 *Burnstock, G.:* Purinergic transmission; in *Iversen, Iversen and Snyder* Handbook of psychopharmacology, vol. 5, pp. 131–194 (Plenum Publishing, New York 1975).

6 *Burnstock, G.:* Cholinergic and purinergic regulation of blood vessels; in *Bohr, Somlyo and Sparks* Handbook of physiology (vascular smooth muscle) (Williams & Wilkins, Baltimore, in press).

7 *Burnstock, G.:* A basis for distinguishing two types of purinergic receptors. Proc. Int. Conf. on Biological Membranes, 1977 (Raven Press, New York, in press).

8 *Burnstock, G.; Cocks, T.; Paddle, B.M., and Staszewska-Barczak, J.:* Evidence that prostaglandin is responsible for the 'rebound contraction' following stimulation of non-adrenergic, non-cholinergic ('purinergic') inhibitory nerves. Eur. J. Pharmacol. *31:* 360–362 (1975).

9 *Campbell, G.:* Autonomic nervous supply to effector tissues; in *Bülbring, Brading, Jones and Tomita* Smooth muscle, pp. 451–495 (Arnold, London 1970).

10 *Chaudry, I.H. and Gould, M.K.:* Effect of externally added ATP on glucose uptake by isolated rat soleus muscle. Biochim. biophys. Acta *196:* 327–337 (1970).

11 *Chen, W.T.; Chou, C.C.; Scott, J.B., and Haddy, F.J.:* Evidence for participation of ATP in active hyperaemia of heart during stellate ganglion stimulation. Physiologist *15:* 104–109 (1972).

12 *Drury, A.N. and Szent-Györgyi, A.:* The physiological activity of adenine compounds with especial reference to their action upon the mammalian heart. J. Physiol., Lond. *68:* 213–327 (1929).

13 *Forrester, T. and Lind, A.R.:* Identification of adenosine triphosphate in human plasma and the concentration in the venous effluent of forearm muscles before, during and after sustained contraction. J. Physiol., Lond. *204:* 347–364 (1969).
14 *Ginsborg, B.L. and Hirst, G.D.:* Theophylline and adenosine at the neuromuscular junction. Br. J. Pharmacol. *43:* 432–433 (1971).
15 *Glynn, I.M.:* The ionic permeability of the red cell membrane; in Prog. Biophysics, vol. 8, pp. 241–307 (Pergamon Press, London 1957).
16 *Haddy, F.J. and Scott, J.B.:* Metabolically linked vasoactive chemicals in local regulation of blood flow. Physiol. Rev. *48:* 688–707 (1968).
17 *Hedqvist, P. and Fredholm, B.B.:* Effects of adenosine on adrenergic neurotransmission; prejunctional inhibition and postjunctional enhancement. Arch. Pharmakol. *293:* 217–223 (1976).
18 *Herlihy, J.T.; Bockman, E.L.; Berne, R.M., and Rubio, R.:* Adenosine relaxation of isolated vascular smooth muscle. Am. J. Physiol. *230:* 1239–1243 (1976).
19 *Hughes, J. and Vane, J.R.:* An analysis of the responses of the isolated portal vein of the rabbit to electrical stimulation and to drugs. Br. J. Pharmacol. Chemother. *30:* 46–66 (1967).
20 *Irvin, J.L. and Irvin, E.M.:* The interaction of quinacrine with adenine nucleotides. J. biol. Chem. *210:* 45–56 (1954).
21 *Langer, S.Z.:* Presynaptic regulation of catecholamine release. J. Biochem. Pharmacol. *23:* 1793–1800 (1974).
22 *Needleman, P.; Minkes, M.S., and Douglas, J.R.:* Stimulation of prostaglandin biosynthesis by adenine nucleotides. Circulation Res. *34:* 455–460 (1974).
23 *Olson, L.; Ålund, M., and Norberg, K.-A.:* Fluorescence-microscopical demonstration of a population of gastro-intestinal nerve fibres with a selective affinity for quinacrine. Cell Tissue Res. *171:* 407–423 (1976).
24 *Paddle, B.M. and Burnstock, G.:* Release of ATP from perfused heart during coronary vasodilatation. Blood Vessels *12:* 1–10 (1974).
25 *Parkinson, P.I.:* The effect of graduated exercise on the concentration of adenine nucleotides in plasma. J. Physiol., Lond. *234:* 74P–76P (1973).
26 *Robinson, P.M.; McLean, J.R., and Burnstock, G.:* Ultrastructural identification of non-adrenergic inhibitory nerve fibres. J. Pharmac. exp. Ther. *179:* 149–160 (1971).
27 *Stjärne, L.:* Basic mechanisms and local feedback control of secretion of adrenergic and cholinergic neurotransmitters; in *Iversen, Iversen and Snyder* Handbook of psychopharmacology, vol. 6, pp. 180–233 (Plenum Publishing, New York 1975).
28 *Su, C.:* Neurogenic release of purine compounds in blood vessels. J. Pharmac. exp. Ther. *195:* 159–166 (1975).
29 *Su, C. and Lee, T.:* Regional variation of adrenergic and nonadrenergic nerves in blood vessels; in *Bevan et al.* Vascular neuroeffector mechanisms (1976).
30 *Triggle, D.J. and Triggle, C.R.:* Chemical pharmacology of the synapse (Academic Press, London 1976).
31 *Verhaeghe, R.H.; Vanhoutte, P.M., and Shepherd, J.T.:* Inhibition of sympathetic neurotransmission in canine blood vessels by adenosine and adenine nucleotides. Circulation Res. *40:* 208–215 (1977).
32 *Vizi, E.S. and Knoll, J.:* The inhibitory effect of adenosine and related nucleotides on the release of acetylcholine. Neuroscience *1:* 391–398 (1976).

G. Burnstock, MD, Department of Anatomy and Embryology, University College London, Gower Street, *London, WC1E 6BT* (England)

Mechanisms of Vasodilatation. Satellite Symp., 27th Int. Congr. Physiol. Sci., Wilrijk 1977, pp. 285–293 (Karger, Basel 1978)

Vasoactive Peptidergic Nerves in the Brain[1]

J.H. Hanko, L. Edvinsson and C. Owman

Department of Histology, University of Lund, and Department of Neurology, University Hospital of Lund, Lund

This review summarizes recent research on neuronal peptides and their role in neurogenic cerebrovascular mechanisms. Only peptides that have been localized in nerve tissue are considered and one peptide, vasoactive intestinal peptide (VIP), receives particular attention.

Since the early observations in 1931 on a peptide substance found in equine brain and gut by *von Euler and Gaddum* (10), the existence of vasodilatory peptides in nerve tissue has been implied. The further research on this peptide, named substance P (SP) was initially delayed for decades by the lack of purified material. During the past 10 years, SP has been chemically characterized by *Chang et al.* (4). A number of other neuronal peptides have since then been found and characterized; the development of immunohistochemical and radioimmunoassay techniques has greatly enhanced the research in this field. Some of the peptides have been proposed as conventional neurotransmitters; it has also been suggested that they participate in long-term regulation or modulation of both neuronal excitability and synaptic efficacy.

1. Substance P

SP is an undecapeptide, widely but unevenly distributed in the nervous system. In the central nervous system, the highest concentrations are found in the substantia nigra, hypothalamus, pineal gland and the dorsal gray matter of the spinal cord. Immunofluorescence studies on spinal segments from cat, rat and man reveal that it is contained predominantly in the processes of a subpopulation of small unmyelinated primary sensory neurons. The corresponding cell bodies containing SP have been localized to spinal sensory ganglia. Immunofluo-

[1] Supported by the Swedish Medical Research Council (Grant No. 04X-732) and the Swedish Tobacco Foundation.

rescence studies on brain have shown SP-positive cell bodies in the habenula, and stereotactical lesions in this area decrease the concentration of SP in the interpeduncular nucleus by 60%. Further stereotactical studies have suggested another tract of SP-containing cells running from the striatum to the substantia nigra (6). The presence of SP-positive cell bodies has been reported in the corpus striatum and the globus pallidum.

Microiontophoretically applied SP has been shown to excite nigral neurons, and local application *in vivo* in this area can induce rotating behaviour similar to that occurring on administration of dopamine. It has been speculated that the SP-induced turning behaviour results from stimulation of the ascending nigrostrial dopaminergic pathways. The SP-containing striato-nigral tract is accompanied by a GABA-mediated tract, and GABA analogues have been reported to antagonize the excitant actions of SP on rat cerebral cortical neurons, nigral neurons, and spinal motor neurons. The inhibitory effect of GABA on SP-induced neuronal excitation may not be specific. Injection of SP intracerebrally in mice reduces the response to pain stimuli. Animals made tolerant to morphine are also tolerant to SP, and this effect is blocked by the morphine antagonist, naloxone.

In the periphery, SP is present in the cell bodies and processes of Auerbach's and Meissner's plexuses in the gut (26). *Hökfelt et al.* (19) also describe SP in dermal nerves, and in this location the fibres are often found around vessels. SP can be extracted from almost all tissues of the body, including salivary glands, thyroid, trachea, pancreas, kidney, bladder and prostate (2). It is a stimulant of salivary secretion, inhibits insulin release from the pancreas, contracts all regions of the intestine, increases tracheobronchial muscle tone, stimulates adenylate cyclase activity in different areas of human and rat brain (7), increases the glycine concentration in the brain, and renal sodium excretion; it is also a potent vasodilator in adipose tissue and muscle (27). The stimulated adenylate cyclase activity, as well as the release of SP from isolated nerve endings and some of the effect on synaptic transmission in spinal segments, are reported to be calcium-dependent. *Fisher et al.* (12) have shown that a low dose of SP administered intravenously reduces the systemic arteriolar blood pressure in dogs; the fall in blood pressure lasts up to 20 min depending on the dose. Hitherto, this hypotensive peptide has not been found in association with cerebral vessels, and its significance in cerebrovascular regulation remains unsolved.

2. Vasoactive Intestinal Polypeptide

VIP was first isolated from porcine duodenum and upper small intestine in 1970 by *Said and Mutt* (29). It is a highly basic peptide, composed of 28 amino acids, and has a molecular weight of 3,827. Structurally, it is similar to glucagon, secretin and gastric inhibitory peptide (GIP). The peptide derives its name from the profound and long-lasting vasodilatory action on systematic administration (30, 31). It has several other effects, such as inhibition of histamine- and pentagastrin-stimulated gastric secretion, stimulation of exocrine pancreatic secretion and small intestine secretion, and stimulation of adenylate cyclase in the intestines as well as in the hepatic and fat cell membranes. VIP also stimulates myocardial contractility, glucogenolysis, lipolysis, insulin secretion and respiratory volume (29). Respiratory stimulation is reported to be almost abolished after chemoreceptor denervation. VIP appears to be effectively inactivated during the passage through the liver (20) and is reported to have a half-life of only just over 1 min (1).

With the radioimmunoassay technique, where VIP antibodies are produced in rabbits and purified VIP is labelled with ^{125}I, it is possible to demonstrate VIP in human plasma. Normal values are less than 200 pg/ml. A marked elevation of VIP in plasma has been demonstrated in patients with hepatic insufficiency, some patients with porto-caval shunts and in some patients with the WDHA syndrome (watery diarrhea, hypokalemia, achlorhydria).

VIP concentrations are several times higher in human cerebrospinal fluid than in plasma; there are significantly lower values in the cerebrospinal fluid of patients with cerebral atrophy when compared to that of normal persons.

Electrical stimulation of the vagal nerve and calcium infusion can release VIP from the gastrointestinal tract. With the immunofluorescence technique, the peptide has been localized in numerous nerve fibres in the gut wall (22). The submucosal and myenteric plexuses invariably contain a dense network of immunoreactive fibres and occasional strongly immunoreactive cell bodies. In the small and large intestines, nerve fibres are abundant in the mucosa and form a network around the crypts and extend up in the cores of the villi. The mucosal nerve fibres often appear to follow small blood vessels closely. Surgical vagotomy or chemical sympathectomy does not alter the number of VIP-containing nerves in the gut.

Examination of hypothalamic specimens from mice reveals the presence of a large collection of nerve cell bodies in the ventromedial part, probably in the arcuate nucleus. The cell bodies give off processes that can be followed for a short distance running lateral to the ventricle. Radioimmunoassay of extracts from mouse hypothalamus confirms the presence of significant quantities of VIP immunoreactivity. By gel-permeation chromatography, this material is found to be similar to highly purified porcine VIP. Immunoassayable VIP seems to be present also in other parts of the brain: frontal and occipital cortex of dog have the same concentration as the hypothalamus, whereas lower values are found in thalamus, cerebellum and brain stem.

Electron microscopic studies of the pial vascular innervation shows the existence of nerve terminals whose synaptic vesicles do not conform with those of either adrenergic or cholinergic nerves. These axons are characterized by the predominance of large vesicles, 85–200 mm in diameter, containing a moderately electron-dense core (25). Since VIP neurons have been demonstrated in gastrointestinal nerves with a perivascular localization, a search for similar neurons in the pial vessels was started.

Fibres containing VIP immunoreactivity are encountered in the adventitia and adventitia-medial border in most of the pial vessels (23). A moderate number of single fluorescent fibres or small fibre bundles run along the vertebral arteries, but none of these fibres seems to reach the muscle layer. A similar arrangement is found in the basilar artery. A well-developed plexus of fluorescent fibres with a varicosed appearance encloses the anterior cerebral artery, being located mainly at its adventitia-medial border. An almost equally dense plexus of VIP immunoreactive fibres is seen in the wall of the proximal portion of the middle cerebral artery and in the posterior communicating arteries. Occasional nerve fibres are observed in intracerebral arterioles.

VIP-fluorescent nerve cell bodies are found in the superior cervical ganglia in cats after local pretreatment with colchicine (17), but the VIP-reactive perivascular nerves in the brain are not affected by removal of these ganglia, an operation which regularly causes all adrenergic sympathetic nerves to disappear. It seems as if the VIP-immunoreactive nerves represent a new perivascular system in the brain, possibly of intrinsic origin.

Under *in vitro* conditions, VIP is without effect on the tone of resting cat middle cerebral arteries, but it produces a dose-related dilatation of vessels made to contract with either serotonin (23) or prostaglandin $F_{2\alpha}$ (8). Not infrequently, the dilatory response is preceded by a slight contraction. The dilatation is strong, increases gradually in strength during 15–20 min, and then decays. Doses of VIP as low as 3×10^{-10} *M* have a significant vasomotor effect. Full relaxation of tonically contracted vessels is usually achieved. Extracerebral vessels, such as maxillary arteries, also dilate when exposed to VIP. The dilatatory effect of VIP *in vitro* is not antagonized by agents which block adrenergic, cholinergic, histaminergic or opiate receptors. Only tetrodotoxin (in high concentrations) has hitherto been shown to block the VIP-induced vasodilatation.

Intravenous infusion of large amounts of VIP ($20\ \mu g \times kg^{-1} \times min^{-1}$) have been reported to render dogs hypotensive for 45 min (31). The blood flow in the mesenteric artery increases when VIP is infused intravenously (24). Administration of VIP at 3 μg/kg body weight via the femoral vein is reported to increase superior mesenteric arterial flow twofold, but this increase disappears after section of the splanchnic nerves (24).

Cerebral blood flow, measured with the ^{14}C-ethanol method, increases considerably when VIP is infused into the femoral vein (15). It is suggested that this new system of cerebrovascular nerves may play a significant physiological role in the cerebrovascular flow regulation, possibly by inducing changes of long duration in vascular smooth muscle tone.

3. Neurotensin

The discovery of a tridecapeptide, neurotensin (NT), in the bovine hypothalamus was made by *Carraway and Leeman* (3). They found that, in the rat, NT induced hypotension and relaxation of the duodenum. The principle was synthesized by the solid phase technique. Increasing evidence indicates that it has a wide spectrum of action: e.g. increase in vascular permeability (3), elevation of plasma glucose and glucagon in rats, reduction of plasma insulin, decrease in liver glucogen content, increase in liver glucogen phosphorylase A activity, and inhibition of pentagastrin-induced gastric acid secretion.

To judge from radioimmunoassay methods on brain tissues from rat and calf, NT immunoreactivity is high in the hypothalamus, in certain parts of the limbic system, the mamillary bodies and the corpus striatum. Marked variations in NT levels occur in different parts of the cerebral cortex. The highest values are found in the parahippocampal gyrus (40% of those in the medial hypothalamus). The lowest concentration of NT in the cerebral cortex is found in the frontal lobe. The lowest concentrations in the central nervous system occur in the brain stem, the spinal cord, the white matter of the cerebral cortex and the cerebellum (2–8% of those found in the anterior hypothalamus). Receptor binding for NT displays a regional distribution in calf brain which resembles in part the regional variations in the endogenous levels of the peptide. In the intact dog, the vascular effects of NT are first observed following intravenous infusion at a rate of 20 ng/kg·min (28). The fall in blood pressure is dose-dependent, and at infusion rates of 100 ng/kg·min, blood pressure decreases by about 10%. Tachyphylaxis to larger concentrations is observed when the dose interval is less than 1 h. Intravenous infusion of NT causes a dose-dependent vasoconstriction in denervated subcutaneous adipose tissue (28); the constriction is delayed in onset, occurs approximately 5 min after the start of the infusion and is preceded by a slight vasodilatation. The delayed action does not seem to be mediated by the central nervous system or the renin-angiotensin system. In the dog, the intestinal vasculature exhibits a biphasic reaction to NT, with a dose-dependent increase in blood flow followed by a slight vasoconstriction.

This vasoactive peptide has hitherto not been histochemically associated with cerebral vessels.

4. Angiotensins

Angiotensin II is an octapeptide derived from a circulating decapeptide precursor (angiotensin I) by the action of a converting enzyme. Peripheral actions of angiotensin II to regulate blood pressure and volume are well established. During the past decade, evidence has accumulated indicating that angiotensin II has specific effects in the central nervous system that are associated with pressure response, drinking behaviour, release of ADH and disruption of memory retention (11, 14). It is disputed whether these central effects are mediated via a peripherally synthesized peptide that passes the blood-brain barrier, or by centrally produced angiotensin II. Since the possible existence of a brain renin-angiotensin system was proposed, several publications have reported variable concentrations of the angiotensins in whole brain and in different brain segments; both biological assay and immunoassay methods have been used. Recently published immunohistochemical evidence for the existence of angiotensin II-containing nerve terminals in the brain and spinal cord in the rat have further stressed the possible role of this peptide in central blood pressure control. Since angiotensin II does not readily penetrate the blood-brain barrier, the pharmacological studies elucidating its role in central blood pressure control are difficult to evaluate. The existence of angiotenin II-containing nerves in the brain makes it possible to define more exactly the central functions of the peptide by interfering with the function of these neurons. Theoretically, the modulation of the vasomotor role of centrally acting angiotensin II could result not only in the well-known vasoconstrictor effect but also in vasodilatation.

Fuxe et al. (13) used antiserum obtained by coupling 5-isoleucin-angiotensin II to bovine serum albumin followed by injection of this complex together with complete Freund's adjuvant into rabbits. Nerve terminals containing specific angiotensin II immunofluorescence were found all over the brain and spinal cord. In most instances, the nerve terminals had a varicosed appearance, and only in the retrochiasmatic area were axons with specific immunofluorescence seen. No cell bodies were found.

The centrally mediated pressure response appears to be due primarily to activation of cells in the area postrema and subnucleus medialis with a concomitant increase in sympathetic tone.

5. Opiate Peptides

The existence of endogenous peptides with agonistic effects on opiate receptors has now been firmly established. *Hughes et al.* (18) have isolated and characterized two pentapeptides called methionine (met) and leucine (leu) enkephalin from porcine brain. Extracts of material comprising both pituitary and hypothalamic tissue contain a number of peptides with opiate activity called α-, β- and γ-endorphins. They are segmentes (amino acids 61–76, 61–91 and 61–77, respectively) of the so-called β-lipotropin that was isolated and characterized in 1964 from pituitary glands. Met-enkephalin has the same amino acid sequence as the amino acids 61–65 in β-endorphin. Biochemical assays of different brain areas have revealed similar regional distribution of opiate and enkephalin receptors. It is believed that the opiate peptides function as physiological receptor ligands that are displaceable by opiate compounds. The receptors are concentrated in discrete regions, some of which are associated with pain perception and emotions. Recently, *Elde et al.* (9) used antibodies to the enkephalins to show that they are located in nerve terminals in the same regions of the brain as the opiate receptors are found. A neurotransmitter role has been suggested for these peptides.

Although opiates and enkephalins relax isolated smooth muscle preparations, no physiological role in cerebrovascular regulation has been ascribed so far to these peptides. Met-enkephalin has a slight vasodilatatory effect on isolated cat middle cerebral arteries (16).

Recent studies on peripheral vessels (rabbit ear artery) indicate that the vasoconstriction induced by electrical activation of the perivascular nerves is efficiently inhibited by methionin-enkephalin and leucine-enkephalin (21). Inhibition of the ^{3}H-noradrenaline release and the unchanged response to exogenous noradrenaline suggest that the effect is presynaptic. The inability of naloxone to counteract the inhibitory effect suggests that the neuronal receptors in the central ear artery are not identical to other opiate receptors. It is possible that certain peptides through an action on the release of the noradrenergic transmitter, act as modulators of vascular smooth muscle tone.

6. Conclusion

Peptides localized to neuronal tissue often show common characteristics when observed in biological systems. Their actions usually have a slow onset, a long duration, and are multiphasic. Because of their multiple effects in different organ systems, it is often difficult to determine their physiological role. It has so far not been possible to fulfil for those substances all requirements that are traditionally demanded to accept their role as true transmitters. Indeed, it has not been convincingly shown that 'peptidergic neurons' constitute a special group of neurons that exclusively use peptides to influence their target cells. On the ultrastructural level, the large vesicles that are supposed to harbour neuronally active peptides have mostly been found together with smaller vesicles connected with traditional neurotransmitters.

The richness of the VIP nerve fibre plexus surrounding the cerebral vessels resembles the perivascular nerve plexuses of other transmitters with well-established importance in cerebrovascular regulation; this strongly suggests a role for this peptide in the control of cerebral blood flow. This impression is further strengthened by the potent *in vitro* activity of the peptide having significant effects in very low concentrations. Whether or not this peptide system is of intrinsic origin in the brain is still uncertain; such intrinsic systems have been proposed to be of importance in the post-contractional vasodilatation in striated muscle.

References

1 *Bryant, M.G.; Polak, J.M.; Modlin, I.; Bloom, S.R.; Alberquerque, R.H., and Pearse, A.G.E.:* Possible dual role for vasoactive intestinal peptide as gastrointestinal hormone and neurotransmitter substance. Lancet *i:* 991–993 (1976).

2 *Brodin, E. and Nilsson, G.:* Abstr. Nobel Symp., Stockholm 1976, vol. 37, p. 8.

3 *Carraway, R. and Leeman, S.E.:* The isolation of a new hypotensive peptide, neurotensin, from bovine hypothalami. J. biol. Chem. *248:* 6854–6861 (1973).

4 *Chang, M.; Leeman, S.E., and Niall, H.D.:* Amino acid sequence of substance P. Nature new Biol. *232:* 86–87 (1971).

5 *Dahlberg, K.; Brodin, E.; Sundler, F.; Nilsson, G., and Strandberg, K.:* Abstr. Nobel Symp., Stockholm 1976, vol. 37, p. 15.

6 *Davies, J. and Dray, A.:* Substance P in the substantia nigra. Brain Res. *107:* 623–630 (1976).

7 *Duffy, M.J.; Wong, J., and Powell, D.:* Stimulation of adenylate cyclase activity in different areas of human brain by substance P. Neuropharmacology *14:* 615–625 (1975).

8 *Edvinsson, L.; Hanko, J.H., and Owman, C.:* To be published.

9 *Elde, R.; Hökfelt, T.; Johansson, O., and Terenius, L.:* Immunohistochemical studies using antibodies to leucine-enkephalin: initial observations on the nervous system of the rat. Neuroscience *1:* 349–351 (1976).

10 *Euler, U.S. von and Gaddum, J.H.:* An unidentified depressor substance in certain tissue extracts. J. Physiol., Lond. *72:* 74–87 (1931).

11 *Ferrario, C.M.; Gildenberg, P.L., and Mc Cubbin, J.W.:* Cardiovascular effects of angiotension mediated by the central nervous system. Circulation Res. *30:* 257–262 (1972).

12 *Fisher, G.H.; Folkers, K.; Pernow, B., and Bowers, C.Y.:* Synthesis and some biological activities of the tyrosine-8-analog of substance P. J. med. Chem. *19:* 325–328 (1976).

13 *Fuxe, K.; Ganten, D.; Hökfelt, T., and Bolme, P.:* Immunohistochemical evidence for the existence of angiotensin II-containing nerve terminals in the brain and spinal cord in the rat. Neurosci. Lett. *2:* 229–234 (1976).

14 *Ganten, D.; Hutchinson, J.S., and Shelling, P.:* The intrinsic brain iso-renin-angiotensin system in the rat: its possible role in central mechanisms of blood pressure regulation. Clin. Sci. *48:* suppl. 2, pp. 265–268 (1975).

15 *Hanko, J.H.; Edvinsson, L., and Owman, C.:* To be published.

16 *Hanko, J.H.; Edvinsson, L., and Hardebo, J.E.:* To be published.

17 *Hanko, J.H.; Edvinsson, L.; Owman, C.; Håkanson, R., and Sundler, F.:* To be published.

18 *Hughes, J.; Smith, T.W.; Kosterlitz, H.W.; Fothergill, L.A.; Morgan, B.A., and Morris, H.R.:* Identification of two related pentapeptides from the brain with potent opiate agonist activity. Nature, Lond. *259:* 577–579 (1975).

19 *Hökfelt, T.; Kellerth, O.; Nilsson, G., and Pernow, B.:* Substance P: localization in the central nervous system and in some primary sensory neurons. Science *190:* 889–891 (1975).

20 *Kitamura, S.; Yoshida, T., and Said, S.I.:* Vasoactive intestinal polypeptide: inactivation in liver and potentiation in lung of anesthetized dogs. Proc. Soc. exp. Biol. Med. *148:* 25–29 (1975).

21 *Knoll, J.:* Neuronal peptide (enkephalin) receptors in the ear artery of the rabbit. Eur. J. Pharmacol. *39:* 403–407 (1976).

22 *Larsson, L.I.; Fahrenkrug, J.; Schaffalitzky de Mukadell, A.; Sundler, F., and*

Håkanson, R.: Localization of vasoactive intestinal polypeptide (VIP) to central and peripheral neurons. Proc. natn. Acad. Sci. USA *73:* 3197–3200 (1976).

23 *Larsson, L.I.; Edvinsson, L.; Fahrenkrug, J.; Håkanson, R.; Owman, C.; Schaffalitzky de Mukadell, A., and Sundler, F.:* Immunohistochemical localization of a vasodilatory polypeptide (VIP) in cerebrovascular nerves. Brain Res. *113:* 400–404 (1976).

24 *Matsuo, Y.:* Integration of hormonal and neural control of gastrointestinal secretion. Asian med. J. *19:* 612–629 (1976).

25 *Nielsen, K.C.; Owman, C., and Sporrong, B.:* Ultrastructure of the autonomic innervation apparatus in the main pial arteries of rats and cats. Brain Res. *27:* 25–30 (1971).

26 *Pearse, A.G.E. and Polak, J.M.:* Immunocytochemical localization of substance P in mammalian intestine. Histochemistry *41:* 375–385 (1975).

27 *Pernow, B. and Rosell, S.:* Effect of substance P on blood flow in canine adipose tissue and skeletal muscle. Acta physiol. scand. *93:* 139–148 (1975).

28 *Rosell, S.; Burcher, E.; Chang, D., and Folkers, K.:* Cardiovascular and metabolic actions of neurotensin and (Glu^4)-neurotensin. Acta physiol. scand. *98:* 484–491 (1976).

29 *Said, S.I. and Mutt, V.:* Polypeptide with broad biological activity: isolation from small intestine. Science *169:* 1217–1218 (1970).

30 *Said, S.I. and Mutt, V.:* Isolation from porcine-intestinal wall of a vasoactive octacosapeptide related to secretin and to glucagon. Eur. J. Biochem. *28:* 199–205 (1972).

31 *Said, S.I. and Mutt, V.:* Potent peripheral and splanchnic vasodilator peptide from normal gut. Nature, Lond. *225:* 863–865 (1970).

J.H. Hanko, MD, Department of Histology, University of Lund, *S–223 62 Lund* (Sweden)

General Comment and Conclusion

Mechanisms of Vasodilatation. Satellite Symp., 27th Int. Congr. Physiol. Sci., Wilrijk 1977, pp. 294–297 (Karger, Basel 1978)

Prospectives in Vasodilatation

A Few Answers but Many Questions

Paul M. Vanhoutte

Department of Internal Medicine, Universitaire Instelling Antwerpen, University of Antwerp, Wilrijk

In recent years, the once very lonely field of vascular smooth muscle physiology and pharmacology has become one of the major areas of interest in cardiovascular research. However, although problems related to vasodilatation have obvious physiological and clinical implications, most data available deal with how vascular smooth muscle contracts. This was the main reason for trying to assemble, in Wilrijk, a group of experts on the topic and to confront them with a knowledgeable audience, in the hope that the presentations, the discussions and the monograph resulting from it all would help the scientific community to assess mechanisms by which the degree of opening of the blood vessels is regulated in the intact organism, or possibly can be controlled for therapeutic purposes. As could be anticipated, this *First Symposium on Mechanisms of Vasodilatation* has clarified a number of interesting points, but at the same time has raised many new questions, or rephrased some very old ones.

Considering the vascular smooth muscle first, the knowledge of its structure, of the identity of the biochemical systems controlling the contractile proteins, and of the mechanical properties of the latter have reached a stage where we no longer feel that it is the 'parent pauvre' of the contractile cells (pages 12, 30, 48 and 79), although obviously not all the details on those topics have been elucidated. The importance of myogenic tone and of its modulation by physical factors has been stressed (pages 73 and 187), but do we really understand whether or why such physical influences are important and what sensor is involved? Central to the problem of the relaxation of vascular smooth muscle is the decrease in concentration of activator Ca^{++} in the cytosol. This can be achieved in part by lowering the influx of the extracellular ion, and the exciting possibility has been proposed that specific channels, opened by vasoconstrictor agonists, can be selectively antagonized (pages 137 and 144); this line of reasoning certainly has to be followed. The alternative way to lower the concentration of activator Ca^{++} is to accelerate its removal into cellular storage sites or towards the extracellular space. Concerning the former, a most stimulating observation because of its pathophysiological implications is that mitochondria may act as sinks for Ca^{++} in abnormal conditions (page 21). As far as the removal to the outside of the cell, interesting exchange systems have been proposed if not yet proven (page 37). One growing concern obviously is whether cyclic nucleotides play any role at all in the relaxation process; great caution is required when interpreting data obtained with so-called inhibitors of phosphodiesterase because of the multiplicity of their effects (pages 152 and 222). Although the electrophysiological behavior of vascular smooth muscle remains a controversial issue, there is now a fair amount of data suggesting that for certain naturally occurring substances, such as catecholamines (page 258) or vasodilator ions (pages 56 and 200), the trigger for relaxation may originate in hyperpolarization of the cell membrane. Although the vasodilator properties of many substances released by metabolically active tissues are established beyond doubt (pages 193–226), we do not know which one, or which exact combination we can nominate as causal factor of exercise hyperemia; neither do we know whether the continued increases in flow during prolonged activity is due to one of those factors, the role of which has been essentially examined during brief exercise. Another unanswered question is whether or not exercise and reactive hyperemia depend upon the same vasodilator trigger (page 1). It is striking in general that we still know very little about the mechanisms by which known and newly proposed endogenous vasodilators slow down the contractile process; this is certainly the case, for example, for acetylcholine (5, 6), adenine nucleotides (pages 214, 222, and 278), inhibitory peptides (page 285), or prostacycline and prostaglandins (pages 107 and 122). For the latter, we can still ask ourselves whether prostaglandin-like compounds are true modulators of the blood vessel function, are an ubiquitous curiosity, or are the last cry for help of the dying cell mistreated by the investigator or by a pathological process. In general, it seems that the mechanism of action of vasodilator therapeutic agents receives a lot more attention than that of endogenous mediators (pages 137–172), although it is surprising that we still have no completely satisfactory explanation of the vasodilator properties of the oldest group of them, the nitrates.

Turning to the control systems, the intricacy of the central nervous control of the sympathetic outflow (page 227) has certainly not been fully exposed, not to mention the centers regulating the outflow to the cholinergic, purinergic and peptidergic nerves (pages 266–293). The presence of peptidergic and purinergic nerves, and also possibly that of dopaminergic nerves (1), in the blood vessel wall is an exciting finding, the significance of which has to be elucidated. The main controller of vascular tone remains the sympathetic nervous system. There is good evidence that the sympathetic outflow can be modulated at the ganglionic level (page 241); the question here is whether this is part of a normal regulatory, or of a pathological, feedback loop. Being obviously biassed, I feel that by now it is clearly established that most of the endogenous substances which can cause vasodilatation by their direct effect on the smooth muscle cell also interrupt peripheral adrenergic transmission in the blood vessel wall. Again, one still, in many cases, has to prove that these inhibitory phenomena occur in the intact organism (page 248) (5). The finding that in certain blood vessels, released norepinephrine can activate β-adrenergic receptors mediating vasodilatation (page 258) is very exciting, but its physiological meaning, and its relationship with phenomena such as flushing remain to be established.

Different blood vessels react differently to vasoconstrictor substances (2–4, 6). The same holds for vasodilator drugs (page 129). It helps our understanding of the problem of vascular heterogeneity greatly to realize that age, or disease, may profoundly alter the sensitivity of the vascular neuroeffector (pages 173 and 181), even if we do not know the reason. Also the presence simultaneously of different neurohormones, which do not necessarily affect the smooth muscle in the same way (page 98), the hormonal impregnation (page 173), and the difference in the efficiency of local disposition mechanisms (page 89) are all factors which ultimately will explain many differences among the various parts of the vascular tree (6).

Many of the phenomena described in this monograph rest on observations obtained in isolated tissues, or in experimental models which differ considerably from the free-running cardiovascular system. Hence, the relevance of many of those to the intact organism, and to acute clinical problems, such as the improvement of the perfusion of ischemic tissues or the decrease of peripheral vascular resistances in hypertension and in spastic vascular disease, has to be established. As I have outlined briefly, while glancing over the proceedings of the *First Symposium on Mechanisms of Vasodilatation,* it is obvious that there are many questions left which need an answer. It is our hope that the reader will be prompted to tackle some of those, so that when we next meet to discuss vasodilatation the concluding remarks may not contain so many query marks.

References

1 *Bell, C. and Lang, W.J.:* Neural dopaminergic vasodilator control in the kidney. Nature new Biol. *246:* 27–29 (1973).

2 *Bohr, D.F.:* Individualities among vascular smooth muscles; in *Bajusz* Electrolytes and cardiovascular diseases, pp. 342–355 (Karger, Basel 1965).

3 *Shepherd, J.T. and Vanhoutte, P.M.:* Veins and their control, pp. 1–269 (Saunders, Philadelphia 1975).

4 *Somlyo, A.P. and Somlyo, A.V.:* Vascular smooth muscle. II. Pharmacology of normal and hypertensive vessels. Pharmacol. Rev. *22:* 251–353 (1970).

5 *Vanhoutte, P.M.:* Cholinergic inhibition of adrenergic transmission. Fed. Proc. Fed. Am. Socs exp. Biol. *36:* 2444–2449 (1977).

6 *Vanhoutte, P.M.:* Heterogeneity in vascular smooth muscle; in *Kaley and Altura* Microcirculation, vol. II (Academic Press, New York, in press 1978).

Dr. *P.M. Vanhoutte,* Department of Internal Medicine, Universitaire Instelling Antwerpen, Universiteitsplein 1, *B–2610 Wilrijk* (Belgium)

Author Index

Subject Index